130
Structure and Bonding

Series Editor: D. M. P. Mingos

Structure and Bonding
Series Editor: D. M. P. Mingos

Recently Published and Forthcoming Volumes

Contemporary Metal Boron Chemistry I
Volume Editors: Marder, T. B., Lin, Z.
Vol. 130, 2008

Recognition of Anions
Volume Editor: Vilar, R.
Vol. 129, 2008

Liquid Crystalline Functional Assemblies and Their Supramolecular Structures
Volume Editor: Kato, T.
Vol. 128, 2008

Organometallic and Coordination Chemistry of the Actinides
Volume Editor: Albrecht-Schmitt, T. E.
Vol. 127, 2008

Halogen Bonding
Fundamentals and Applications
Volume Editors: Metrangolo, P., Resnati, G.
Vol. 126, 2008

High Energy Density Materials
Volume Editor: Klapötke, T. H.
Vol. 125, 2007

Ferro- and Antiferroelectricity
Volume Editors: Dalal, N. S., Bussmann-Holder, A.
Vol. 124, 2007

Photofunctional Transition Metal Complexes
Volume Editor: V. W. W. Yam
Vol. 123, 2007

Single-Molecule Magnets and Related Phenomena
Volume Editor: Winpenny, R.
Vol. 122, 2006

Non-Covalent Multi-Porphyrin Assemblies
Synthesis and Properties
Volume Editor: Alessio, E.
Vol. 121, 2006

Recent Developments in Mercury Science
Volume Editor: Atwood, David A.
Vol. 120, 2006

Layered Double Hydroxides
Volume Editors: Duan, X., Evans, D. G.
Vol. 119, 2005

Semiconductor Nanocrystals and Silicate Nanoparticles
Volume Editors: Peng, X., Mingos, D. M. P.
Vol. 118, 2005

Magnetic Functions Beyond the Spin-Hamiltonian
Volume Editor: Mingos, D. M. P.
Vol. 117, 2005

Intermolecular Forces and Clusters II
Volume Editor: Wales, D. J.
Vol. 116, 2005

Intermolecular Forces and Clusters I
Volume Editor: Wales, D. J.
Vol. 115, 2005

Superconductivity in Complex Systems
Volume Editors: Müller, K. A., Bussmann-Holder, A.
Vol. 114, 2005

Principles and Applications of Density Functional Theory in Inorganic Chemistry II
Volume Editors:
Kaltsoyannis, N., McGrady, J. E.
Vol. 113, 2004

Contemporary Metal Boron Chemistry I

Borylenes, Boryls, Borane σ-Complexes, and Borohydrides

Volume Editors: Todd B. Marder · Zhenyang Lin

With contributions by
S. Aldridge · M. Besora · H. Braunschweig · D. L. Kays
C. Kollann · Z. Lin · A. Lledós · F. Seeler

Springer

The series *Structure and Bonding* publishes critical reviews on topics of research concerned with chemical structure and bonding. The scope of the series spans the entire Periodic Table. It focuses attention on new and developing areas of modern structural and theoretical chemistry such as nanostructures, molecular electronics, designed molecular solids, surfaces, metal clusters and supramolecular structures. Physical and spectroscopic techniques used to determine, examine and model structures fall within the purview of Structure and Bonding to the extent that the focus is on the scientific results obtained and not on specialist information concerning the techniques themselves. Issues associated with the development of bonding models and generalizations that illuminate the reactivity pathways and rates of chemical processes are also relevant.
As a rule, contributions are specially commissioned. The editors and publishers will, however, always be pleased to receive suggestions and supplementary information. Papers are accepted for *Structure and Bonding* in English.
In references *Structure and Bonding* is abbeviated *Struct Bond* and is cited as a journal.

Springer WWW home page: springer.com
Visit the Struct Bond content at springerlink.com

ISBN 978-3-540-78633-7 e-ISBN 978-3-540-78634-4
DOI 10.1007/978-3-540-78634-4

Structure and Bonding ISSN 0081-5993

Library of Congress Control Number: 2008924619

Cover design: WMXDesign GmbH, Heidelberg
Typesetting and Production: le-tex publishing services oHG, Leipzig

Printed on acid-free paper

9 8 7 6 5 4 3 2 1 0

springer.com

Structure and Bonding
Also Available Electronically

For all customers who have a standing order to Structure and Bonding, we offer the electronic version via SpringerLink free of charge. Please contact your librarian who can receive a password or free access to the full articles by registering at:

springerlink.com

If you do not have a subscription, you can still view the tables of contents of the volumes and the abstract of each article by going to the SpringerLink Homepage, clicking on "Browse by Online Libraries", then "Chemical Sciences", and finally choose Structure and Bonding.

You will find information about the

- Editorial Board
- Aims and Scope
- Instructions for Authors
- Sample Contribution

at springer.com using the search function.

Color figures are published in full color within the electronic version on SpringerLink.

Preface

The field of metal boron chemistry has been an especially active one since the 1960s. Much of the early work centered on the synthesis and structural characterization of polyhedral metallaborane and metallacarborane clusters, which will not be discussed in the current Volume. However, early work on metal boryl complexes by Nöth et al. also appeared during this period, although the first simple metal boryl complexes to be structurally characterized by single-crystal X-ray diffraction were not reported until 1990 (indeed, crystal structures of iridium hydrido boryl complexes were reported nearly simultaneously by Baker, Marder, et al. [1] and by Merola et al. [2]), and this new interest in the area arose specifically as a result of the report by Nöth et al. [3] on the rhodium-catalyzed hydroboration of alkenes using catecholborane. Although there had been previous reports of metal-catalyzed hydroboration employing polyhedral boranes or carboranes as the B–H source [4], the seminal paper by Nöth et al. in 1985 [3] marked the beginning of interest by the organic chemistry community in the application of this reaction in organic synthesis [5]. What ensued was a very rapid increase in the number of metal boryl complexes to be synthesized and structurally characterized, along with studies of their reactivity, in conjunction with an equally rapid growth in the development of catalytic hydroboration chemistry [5] and its asymmetric version [5e]. This was followed by a wide variety of other borylation reactions catalyzed by transition metals including diborations, silylborations, stannylborations, and thiaborations of unsaturated organic substrates [6], borylations of α,β-unsaturated carbonyl compounds [7], borylations of C–halide bonds [8], and most recently, borylations of C–H bonds in alkenes, arenes, heteroarenes, and even alkanes [9,10]. The C–halide and C–H borylation chemistry has been driven by the development and broad application of the Suzuki–Miyaura cross-coupling reaction [11], which is now one of the main C–C bond-forming reactions in the arsenal of the organic chemist. Thus, the synthetic and structural chemistry of metal boryl complexes has grown along with catalytic applications in which metal boryl complexes are key intermediates.

Likewise, the importance of metal borane σ-complexes as intermediates in many catalytic reactions has only recently been recognized [12], and this area is now a rapidly expanding one. This bonding mode is also now known for early, middle, and late transition metals, and is expected to be found more often (as was the case for dihydrogen complexes) as research continues.

Metal borylene complexes are another recent addition to the library of metal boron compounds, and these are now finding applications in organic synthesis, initially via their role as stoichiometric borylene transfer agents in the synthesis of borirenes from alkynes [13]. It is expected that additional applications of newly discovered metal borylene complexes will emerge as new routes to them are developed.

Borohydride complexes of metals, from across the periodic table, have been known for many years, and represent the other major class of mono-boron systems coordinated to metal centers. They can coordinate via one, two, or three hydrogen bridges to a single metal or can bridge between two metal atoms, and systems involving transition metals as well as lanthanides and actinides are well known [14]. Their structures, bonding, and dynamics have been of interest since their discovery. As the field is rather large, only mononuclear transition-metal complexes with BH_4^- ligands will be considered in the current review, although the principles discussed also apply to systems involving RBH_3^- and $R_2BH_2^-$ systems [15], for example.

Thus, this Volume brings together the main types of complexes involving single boron units coordinated to metal centers, including examples in which more than one of the boron ligands is involved. A word is required regarding an even more recently developed area of metal boron chemistry involving metal-to-boron dative bonds in complexes now known as "boratranes" [16]. These will be discussed in a future Volume.

Finally, we note that the chapter ordering in the current Volume places borylene complexes before boryl complexes. We have ordered the chapters according to the degree of metal–boron bonding, from highest to lowest, rather than following a historical perspective.

Durham and Hongkong, February 2008

Todd B. Marder
Zhenyang Lin

References

1. Baker RT, Ovenall DW, Calabrese JC, Westcott SA, Taylor NJ, Williams ID, Marder TB (1990) J Am Chem Soc 112:9399
2. Knorr JR, Merola JS (1990) Organometallics 9:3008
3. Männing D, Nöth H (1985) Angew Chem Int Ed Engl 24:878
4. (a) Hewes JD, Kreimendahl CW, Marder TB, Hawthorne MF (1984) J Am Chem Soc 106:5757; (b) Davan T, Corcoran EW, Sneddon LG (1982) Organometallics 2:1693; (c) Wilczynski R, Sneddon LG (1981) Inorg Chem 20:3955
5. For reviews see: (a) Burgess K, Ohlmeyer MJ (1991) Chem Rev 91:1179; (b) Burgess K, van der Donk WA (1994) In: King RB (ed) Encyclopedia of Inorganic Chemistry, vol 3. Wiley, Chichester, p 1420; (c) Fu GC, Evans DA, Muci AR (1997) In: Doyle MP (ed) Advances in Catalytic Processes. JAI, Greenwich, CT, p 95; (d) Beletskaya I, Pelter A (1997) Tetrahedron 53:4957; (e) Crudden CM, Edwards D (2003) Eur J Org Chem 4695

6. For reviews, see: (a) Marder TB, Norman NC (1998) Top Catal 5:63; (b) Ishiyama T, Miyaura N (1999) J Synth Org Chem Jpn 57:503; (c) Han LB, Tanaka M (1999) Chem Commun 395; (d) Ishiyama T, Miyaura N (2000) J Organomet Chem 611:392; (e) Onozawa S, Tanaka M (2002) J Synth Org Chem Jpn 60:826; (f) Dembitsky VM, Abu Ali H, Srebnik M (2003) Appl Organometal Chem 17:327; (g) Dembitsky VM, Abu Ali H, Srebnik M (2004) Adv Organomet Chem 51:193; (h) Ishiyama T, Miyaura N (2004) Chem Rec 3:271; (i) Beletskaya I, Moberg C (2006) Chem Rev 106:2320; (j) Ramírez J, Lillo V, Segarra AM, Fernández E (2007) C R Chimie 10:138; (k) Burks HE, Morken JP (2007) Chem Commun 4717
7. For a review, see: Marder TB (2008) In: Fairlamb IJS, Lynam J (eds) Specialist Periodical Report: Organometallic Chemistry. Royal Society of Chemistry, Cambridge 34:46
8. For a review, see: Ishiyama T, Miyaura N (2000) J Organomet Chem 611:392
9. (a) Iverson CN, Smith III MR (1999) J Am Chem Soc 121:7696; (b) Waltz KM, Hartwig JF (2000) J Am Chem Soc 122:11358; (c) Chen H, Schlecht S, Semple TC, Hartwig JF (2000) Science 287:1995; (d) Cho J-Y, Iverson CN, Smith III MR (2000) J Am Chem Soc 122:12868; (e) Tse MK, Cho J-Y, Smith III MR (2001) Org Lett 3:2831; (f) Ishiyama T, Ishida K, Takagi J, Miyaura N (2001) Chem Lett 1082; (g) Shimada S, Batsanov AS, Howard JAK, Marder TB (2001) Angew Chem Int Ed 40:2168; (h) Cho J-Y, Tse MK, Holmes D, Maleczka Jr RE, Smith III MR (2002) Science 295:305; (i) Ishiyama T, Tagaki J, Ishida K, Miyaura N, Anastasi NR, Hartwig JF (2002) J Am Chem Soc 124:390; (j) Kondo Y, Garcia-Cuadrado D, Hartwig JF, Boaen NK, Wagner NL, Hillmyer MA (2002) J Am Chem Soc 124:1164; (k) Takagi J, Sato K, Hartwig JF, Ishiyama T, Miyaura N (2002) Tetrahedron Lett 43:5649; (l) Ishiyama T, Takagi J, Hartwig JF, Miyaura N (2002) Angew Chem Int Ed 41:3056; (m) Wan X, Wang X, Luo Y, Takami S, Kubo M, Miyamoto A (2002) Organometallics 21:3703; (n) Webster CE, Fan Y, Hall MB, Kunz D, Hartwig JF (2003) J Am Chem Soc 125:858; (o) Lam WH, Lin Z (2003) Organometallics 22:473; (p) Tamura H, Yamazaki H, Sato H, Sakaki S (2003) J Am Chem Soc 125:16114; (q) Ishiyama T, Miyaura N (2003) J Organometal Chem 680:3; (r) Kurotobi K, Miyauchi M, Takakura K, Murafuji T, Sugihara Y (2003) Eur J Org Chem 3663; (s) Ishiyama T, Nobuta Y, Hartwig JF, Miyaura N (2003) Chem Commun 2924; (t) Ishiyama T, Takagi J, Yonekawa Y, Hartwig JF, Miyaura N (2003) Adv Synth Catal 345:1003; (u) Maleczka RE, Shi F, Holmes D, Smith MR (2003) J Am Chem Soc 125:7792; (v) Datta A, Köllhofer A, Plenio H (2004) Chem Commun 1508; (w) Lam WH, Lam KC, Lin Z, Shimada S, Perutz RN, Marder TB (2004) Dalton Trans 1556; (x) Martins K, Zapf A, Beller M (2004) J Mol Catal 207:21
10. (a) Coventry DN, Batsanov AS, Goeta AE, Howard JAKH, Marder TB, Perutz RN (2005) Chem Commun 2172; (b) Chotana GA, Rak MA, Smith MR (2005) J Am Chem Soc 127:10539; (c) Boller TM, Murphy JM, Hapke M, Ishiyama T, Miyaura N, Hartwig JF (2005) J Am Chem Soc 127:14263; (d) Bae CS, Hartwig JF, Chung HY, Harris NK, Switek KA, Hillmyer MA (2005) Angew Chem Int Ed 44:6410; (e) Hartwig JF, Cook KS, Hapke M, Incarvito CD, Fan YB, Webster CE, Hall MB (2005) J Am Chem Soc 127:2538; (f) Bae C, Hartwig JF, Harris NKB, Long RO, Anderson KS, Hillmyer MA (2005) J Am Chem Soc 127:767; (g) Ishiyama T (2005) J Synth Org Chem Jpn 63:440; (h) Hiroto S, Hisaki I, Shinokubo H, Osuka A (2005) Angew Chem Int Ed 44:6763; (i) Holmes D, Chotana GA, Maleczka RE, Smith MR (2006) Org Lett 8:1407; (j) Shi F, Smith MR, Maleczka RE (2006) Org Lett 8:1411; (k) Murphy JC, Lawrence JD, Kawamura K, Incarvito C, Hartwig JF (2006) J Am Chem Soc 128:13684; (l) Paul S, Chotana GA, Holmes D, Reichle RC, Maleczka RE, Smith MR (2006) J Am Chem Soc 128:15552; (m) Mkhalid IAI, Coventry DN, Albesa-Jové D, Batsanov AS, Howard JAK, Perutz RN, Marder TB (2006) Angew Chem Int Ed 45:489; (n) Murphy JM, Tzschucke CC, Hartwig JF (2007) Org Lett 9:757;

(o) Tzschucke CC, Murphy JM, Hartwig JF (2007) Org Lett 9:761; (p) Lo WF, Kaiser HM, Spannenberg A, Beller M, Tse MK (2007) Tetrahedron Lett 48:371; (q) Hata H, Yamaguchi S, Mori G, Nakazano S, Katoh T, Takatsu K, Hiroto S, Shinokubo H, Osuka A (2007) Chem Asian J 2:849

11. See, for example: (a) Suzuki A, Miyaura N (1995) Chem Rev 95:2457; (b) Suzuki A (1999) J Organomet Chem 576:147; (c) Phan NTS, Van Der Sluys M, Jones CW (2006) Adv Synth Catal 348:609
12. (a) Hartwig JF, Muhoro CN, He X, Eisenstein O, Bosque R, Maseras F (1996) J Am Chem Soc 118:10936; (b) Muhoro CN, He X, Hartwig JF (1999) J Am Chem Soc 121:5033; (c) See, for example, ref. [9w] and refererences therein; (d) Lin Z (2007) Coord Chem Rev 251:2280
13. Braunschweig H, Herbst T, Rais D, Seeler F (2005) Angew Chem Int Ed 44:7461
14. (a) Marks TJ, Kolb JR (1977) Chem Rev 77:263; (b) Xu Z, Lin Z (1996) Coord Chem Rev 156:139
15. See, for example: Westcott SA, Marder TB, Baker RT, Harlow RL, Calabrese JC, Lam KC, Lin Z (2004) Polyhedron 23:2665, and references therein
16. (a) Hill AF, Owen GR, White AJP, Williams DJ (1999) Angew Chem Int Ed 38:2759; (b) Crossley IR, Hill AF (2004) Organometallics 23:5656; (c) Pang KL, Quan SM, Parkin G (2006) Chem Commun 5105; (d) Blagg RJ, Charmant JPH, Connelly NG, Haddow MF, Orpen AG (2006) Chem Commun 2350; (e) Bontemps SB, Gornitzka H, Bouhadir G, Miqueu K, Bourissou D (2006) Angew Chem Int Ed 45:1611; (f) Crossley IR, Hill AF (2008) Dalton Trans 201

Contents

Struct Bond (2008) 130: 1–27
DOI 10.1007/430_2007_080

Published online: 21 February 2008

Transition Metal Borylene Complexes

Holger Braunschweig (✉) · Carsten Kollann · Fabian Seeler

Institut für Anorganische Chemie,
Bayerische Julius-Maximilians-Universität Würzburg,
Am Hubland, 97074 Würzburg, Germany
h.braunschweig@mail.uni-wuerzburg.de

Abstract Since the synthesis of the first structurally authentic examples of transition metal complexes with bridging borylene ligands in 1995, and transition metal complexes that feature terminally bound borylene ligands three years later, investigations into this field of chemistry have continued steadily. Being closely related to pivotal organometallic compounds, such as carbene, vinylidene, and especially carbonyl complexes, borylene complexes have attracted considerable interest both from experimental and theoretical points of view. Various methods for their preparation have been developed and ongoing studies on their reactivity have demonstrated an exciting chemistry. In particular, the nature of the metal–boron bond was subject to many structural and computational studies. This review summarizes the chemistry of transition metal borylene complexes, highlighting recent developments and emphasizing structural and electronic properties of these species.

Keywords Boron · Borylene complexes · Coordination chemistry · Transition metal complexes

1 Introduction

In the course of his ground-breaking work on boron subhalides, Timms reported in 1967 the preparation of fluoroborylene BF from elemental boron and BF3 at temperatures of about 2000 °C [1, 2]. In 1984, West generated the silylborylene B–$SiPh_3$ by irradiation of $(Ph_3Si)_3B$ in hydrocarbon matrices at −196 °C [3]. The borylene molecules turned out to be highly reactive species, which could only be obtained by applying drastic conditions. They could not be isolated as free molecules, but their formation was deduced from the structure of the trapping products. During the past decade, however, research in the chemistry of transition metal complexes of boron [4–7] made possible, for the first time, generating and stabilizing borylene moieties in the coordination sphere of various transition metals under ambient conditions [8–11]. Density functional theoretical studies had predicted the thermodynamic stability of borylene complexes with respect to homolytic dissociation of the metal–boron bond [12, 13]. The free borylene B–F is isoelectronic to N_2 and CO and thus, a comparison of the calculated valence orbital energies can be used to predict its ligating properties [14]. Generally, these ligands possess a lone pair in a σ HOMO with a *sp* lobe, convenient for σ donation into the empty dz^2 hybrid orbital on the metal fragment and two degenerate π^* LUMOs, accepting π-back-donation from the populated d_{xz} and d_{yz} orbitals.

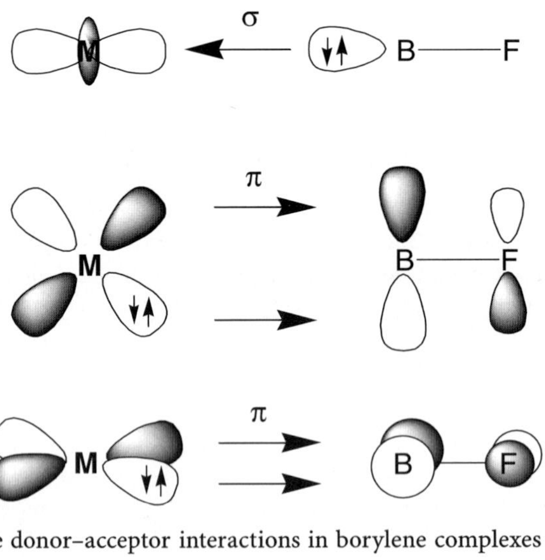

Scheme 1 Principle donor–acceptor interactions in borylene complexes L_nM–BF [12, 15]

In the case of the experimentally realized BNR_2-ligand, the symmetry is lowered and therefore the π^* LUMOs split, yielding a MO-diagram very similar to the well-known vinylidene-ligand. The calculated valence orbital energies of the σ HOMOs rise in the row $N_2 < CO < BF < BNH_2$, due to the decrease in electronegativity of the metal-binding atom, while the energies of the π^* LUMOs are very similar, resulting in an increase of the σ-donation ability and comparable π-backdonation.

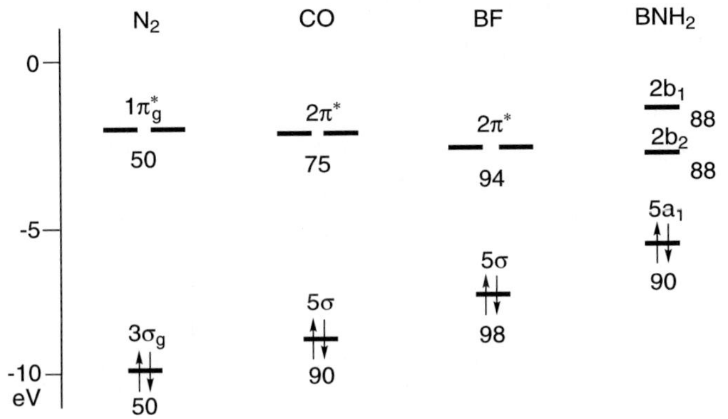

Fig. 1 Valence orbital energy of AE systems N_2, CO, BF, BNH_2. Below each level the percentage of atom A character is indicated [12, 15]

As a result, the thermodynamic stability of the M–BR bond with respect to homolytic dissociation is even higher than that of a M–CO bond, while the high polarity and the small HOMO–LUMO gap of borylenes B–R indicate a high reactivity towards nucleophiles and thus, a low kinetic stability. There-

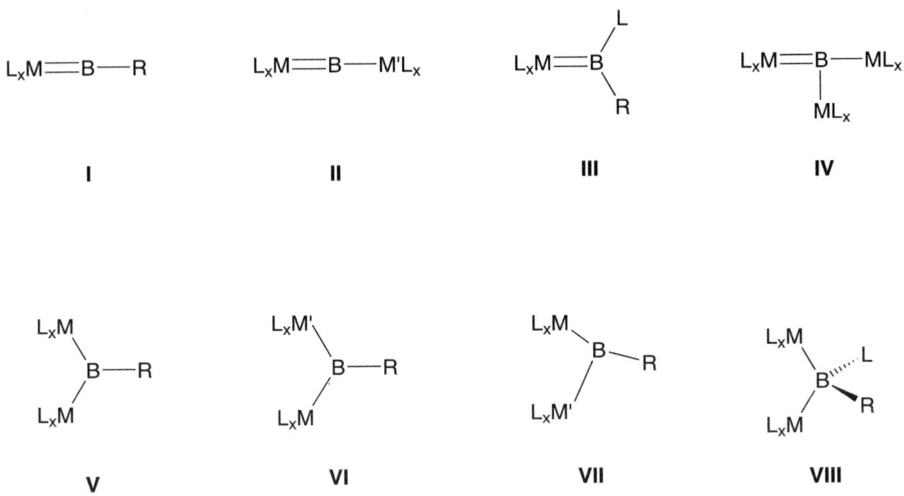

Scheme 2 Schematic representation of borylene complexes

fore, borylenes have to be stabilized by either electron-density releasing or bulky substituents [12, 15].

We proposed a classification scheme [16] for transition metal borylene complexes (Scheme 2) according to the coordination number of the boron atom and the number of metal–boron bonds, thus defining terminal borylene (**I**) complexes with boron in coordination number two as well as homodinuclear (**V**) and heterodinuclear bridging borylene complexes (**VI**) with boron in coordination number three. Lewis base-adducts of both terminal (**III**) and bridging (**VIII**) borylene complexes have also been realized. Furthermore, unusual coordination modes of boron-based ligands have been reported, including complexes with semi-bridging ligands (**VII**), and compounds displaying a "naked", i.e. non main group element-substituted boron atom in the coordination sphere of two (**II**) or three (**IV**) transition metals.

2 Terminal Borylene Complexes

2.1 Aminoborylene Complexes

Indeed, the electronically promising aminoborylene-ligand was used to prepare the structurally characterized terminal borylene complexes $[(OC)_5M{=}B{=}N(SiMe_3)_2]$ (**1**: M = Cr, **2**: M = W) by double salt elimination reactions of dianionic metal carbonylates $Na_2[M(CO)_5]$ with the dihaloboranes $X_2BN(SiMe_3)_2$ [17, 18] (Scheme 3). Both molecules are characterized by an almost linear M–B–N moiety (M = Cr, 177.4(4)°; M = W, 177.9(5)°) and short B–N (M = Cr, 135.3(6) pm; M = W, 133.8(8) pm) and M–B distances (M = Cr, 199.6(6) pm; M = W, 215.1(7) pm). The aminoborylene-ligand seems to have almost no *trans*-influence, which is reflected by the CO_{eq} and CO_{ax} geometries that are equivalent within experimental error. The bonding situation, derived from calculations on the model compound $[(OC)_5Cr{=}B{=}NH_2]$ and free BNH_2, was discussed by Bearends et al. [12]. Due to the similar π^* LUMO energies the authors assume a similar π-component as for CO. The equivalence of the B–N bond lengths in both calculated

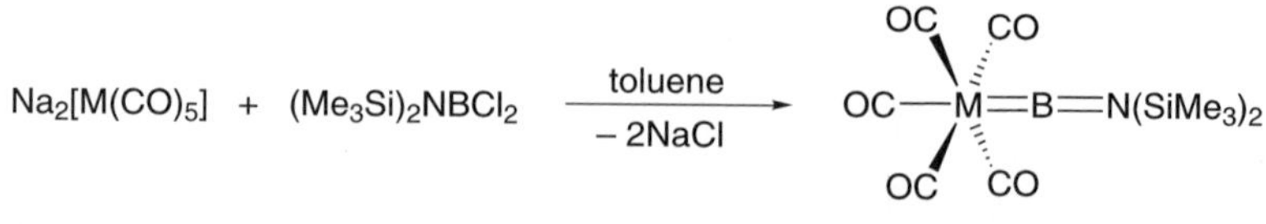

1, M = Cr
2, M = W

Scheme 3 Synthesis of terminal aminoborylene complexes

molecules is explained by the strong B–Cr σ-donation in the former, which compensates the lengthening effect of M–B π-backbonding.

The spectroscopic data of the analogue complex $[(OC)_4Fe{=}B{=}N(SiMe_3)_2]$ (**3**), also obtained via salt elimination, are described in the literature [17], but a structural characterization by X-ray diffraction is still missing. Various theoretical calculations were performed in order to predict the structure of this compound, leading to controversial results whether the axial [12] or the equatorial [19, 20] isomer is a minimum on the potential energy surface. Cowley and co-workers synthesized and structurally characterized the related complex $[(\eta^5\text{-}C_5Me_5)BFe(CO)_4]$ (**4**) [21]. The rather long Fe–B (201.0(3) pm) bond is similar to iron boryl complexes, and thus typical for a Fe–B single bond. The authors conclude that in **4**, σ-donation of the $(\eta^5\text{-}C_5Me_5)$B-fragment to the iron center is the only relevant interaction. Another model to describe the bonding situation in 4 is that of a *nido*-pentacarbahexaborane with an *exo*-iron(tetracarbonyl)substituent that is σ-bound to the apical boron atom [22].

The reactivity of the aminoborylene complexes **1** and **2** under thermal conditions leads to the formation of semi-bridged borylene or bis-borylene complexes (vide infra). The photochemical transfer of the borylene ligand is useful for the preparation of borirenes [23], turned out to be an alternative for the synthesis of **1**, and was also used for the synthesis of the first half-sandwich borylene complex $[(\eta^5\text{-}C_5H_5)(OC)_3V{=}B{=}N(SiMe_3)_2]$ (**5**) [24] (Scheme 4). In the crystal, the borylene ligand adopts the expected linear geometry (V–B–N 177.9(4)°) and the B–N bond (137.8(7) pm) is

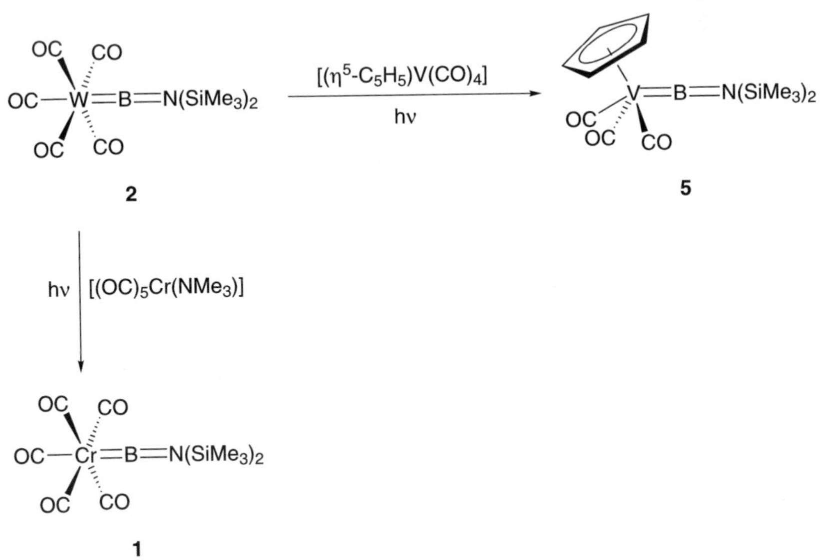

Scheme 4 Photochemical transfer of the aminoborylene ligand

comparably short as in **1** (135.3(6) pm) and **2** (133.8(8) pm). The V–B distance (195.9(6) pm) is very similar to the Cr–B distance in **1** (199.6(6) pm) and significantly smaller than the V–C double bond distance in carbene complexes of the type $[(\eta^5\text{-}C_5H_5)(OC)_3V{=}C(OR)R]$ [25]. The twist between the planar B–N–Si1–Si2 moiety and the plane defined by the cyclopentadienyl centroid, vanadium and boron amounts to 12° and was exactly reproduced by DFT calculations on **5**. The horizontal arrangement of the borylene moiety in **5** is 19 kJ mol^{-1} higher in energy than the observed twisted conformation. In the case of the model compound $[(\eta^5\text{-}C_5H_5)(OC)_3V{=}B{=}NH_2]$, the horizontal orientation is found to be the minimum structure, while the vertical arrangement is a transition state and about 9 kJ mol^{-1} higher in energy. Thus, the twist in **5** can be explained by the more bulky Me_3Si groups.

2.2
A Highly Electronically Unsaturated Borylene Complex

The stabilization of a borylene complex with bulky substituents lacking π-stabilizing properties was realized with the preparation of $[(OC)_5Cr{=}BSi(SiMe_3)_3]$ (**6**) from $Na_2[Cr(CO)_5]$ and $Cl_2BSi(SiMe_3)_3$ [26] (Scheme 5). The highly deshielded ^{11}B-NMR resonance ($\delta = 204.3$) and the thermal lability of **6** were both attributed to the absence of π-donating substituents at boron. The Cr–B bond (187.8(10) pm) is significantly shortened and the $Cr–C_{ax}$ (193.9(10) pm) distance elongated in comparison to **1** (Cr–B = 199.6(6) pm; $Cr–C_{ax}$ = 190.8(6) pm). Additionally, the $Cr–C_{ax}$ distance is longer than the average $Cr–C_{eq}$ value (189.4 pm). The B–Si and Si–Si distances and the undistorted B–Si–Si angles give no evidence for hyperconjugation and confirm the highly electronically unsaturated character of the boron atom. These structural data and the stronger umbrella effect, reflected by the average $B–Cr–C_{eq}$ (**6**: 85°; **1**; 88°) and $Cr–C_{eq}–O$ (**6**: 176°; **1**; 179°) angles, suggest a significantly enhanced Cr–B π-backbonding in **6** with respect to **1**. The structural properties of **6** were well confirmed by calculations on the model compounds $[(OC)_5Cr{=}BSiH_3]$ and $[(OC)_5Cr{=}BSi(SiH_3)_3]$. For the latter one, a Wiberg bond index of ca. 1 for the B–Cr bond and high Mulliken charges (B: 0.80; Cr: –1.73) were obtained, indicating a strong ionic character of the B–Cr bond. Earlier theoretical investigations on the bonding situation in

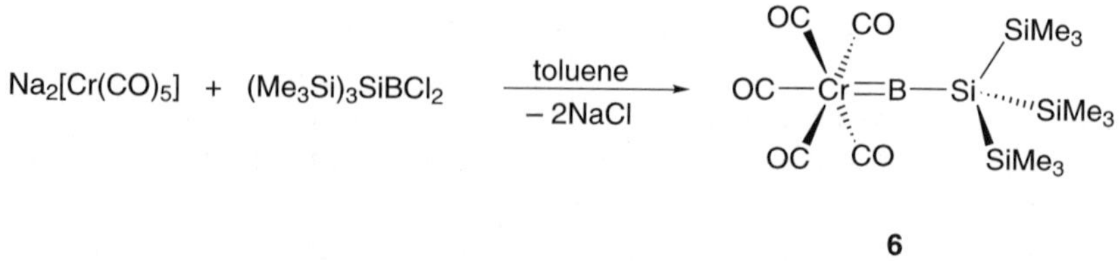

6

Scheme 5 Synthesis of a highly electronically unsaturated borylene complex transition metal complexes with group 13 diyl ligands also came to the result that a charge attraction between the negatively charged transition metal and the positively charged group 13 element is a main cause for the high dissociation energies of these bonds [14, 19, 27].

2.3 Cationic Borylene Complexes

Another successful route for the preparation of terminal borylene complexes, established by Aldridge and co-workers, is the abstraction of halides from asymmetric haloboryl complexes [28, 29]. The reaction of $[(\eta^5\text{-}C_5Me_5)(OC)_2Fe\{B(Br)Mes\}]$ (**7**) with $Na[BAr^f{}_4]$ in dichloromethane yields the cationic borylene complex $[(\eta^5\text{-}C_5Me_5)Fe(CO)_2(BMes)][BAr^f{}_4]$ (**8**) with precipitation of NaBr (Scheme 6). The ^{11}B-NMR resonance ($\delta = 145$) is low field shifted and the carbonyl stretching frequencies in the IR-spectrum (2055, 2013 cm^{-1}) are shifted to higher wavenumbers, compared to the boryl complex precursor **7** ($\delta = 113$) (2006, 1961 cm^{-1}) [30]. In the solid state, **8** is characterized by an almost linear Fe–B–C unit (178.3(6)°) and a very short Fe–B distance (179.2(8) pm), consistent with the presence of a Fe=B double bond. The perpendicular arrangement of the mesityl fragment with respect to the C_5Me_5–Fe–B plane allows Fe–B π-backdonation into one of the two vacant *p*-orbitals at boron, while the other orbital can be stabilized by π-interaction with the aromatic mesityl substituent. This binding model is confirmed by the significantly shortened B–C distance (149.1(10) pm) in comparison to that in **7** (156.9(3) pm). **8** reacts readily with several anionic and neutral nucleophiles [29].

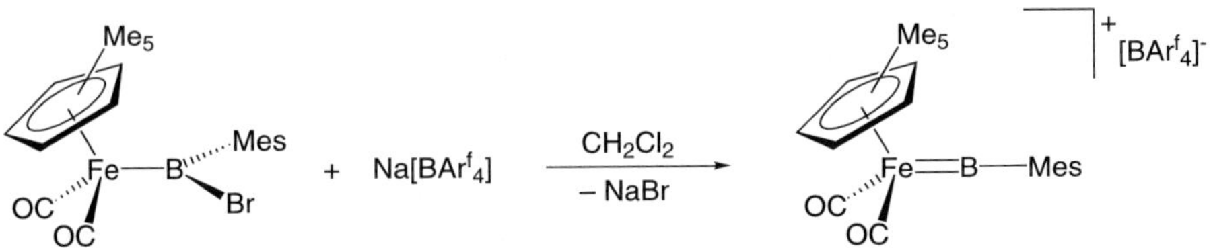

7 8

Scheme 6 Synthesis of a cationic borylene complex via halide abstraction

While the cationic aminoborylene complex $[(\eta^5\text{-}C_5H_5)Fe(CO)_2(BN\mathit{i}Pr_2)][BAr^f{}_4]$ (**9**), which shows interesting M–B bond metathesis chemistry [31], is a colorless oil, the analogue complex $[(\eta^5\text{-}C_5H_5)Fe(CO)_2(BNCy_2)][BAr^f{}_4]$ (**10**) was structurally characterized [32]. The Fe–B bond distance (185.9(6) pm) is longer than in **8** (179.2(8) pm), despite the more bulky C_5Me_5-substituent in the latter. This can be explained by the reduced π acidity of the boron atom in **10**, due to the π donating amino substituent. DFT-calculations on **10** show a significant Fe–B and B–N multiple-bond character and therefore **10** can be described as a B–N analogue to group 8 cationic vinylidene complexes of the type $[(\eta^5\text{-}C_5R_5)ML_2(=C=CR_2)]^+$.

2.4 Metalloborylene Complexes

All the aforementioned borylene complexes have in common that the substituent on boron is a main group element. The availability of the dichloroboryl complex $[(\eta^5\text{-}C_5Me_5)(OC)_2Fe(BCl_2)]$ (**11**) [33] as a precursor, allowed the synthesis of the first metalloborylene complexes $[(OC)_5Cr{=}B{-}Fe(CO)_2(\eta^5\text{-}C_5Me_5)]$ (**12**) and $[(OC)_4Fe{=}B{-}Fe(CO)_2(\eta^5\text{-}C_5Me_5)]$ (**13**), where boron is exclusively bound to transition metal centers [34, 35] (Scheme 7). The Cr–B–Fe (177.75(11)°) and Fe–B–Fe moieties (175.38(12)°) are near to linearity, as expected for the *sp*-hybridized boron atom. The Cr–B distance of **12** (197.5(2) pm) is intermediate between those of **1** (199.6(6) pm) and **6** (187.8(10) pm), indicating less π-bonding than in **6**. The Fe–B distance (186.17(19) pm) lies between those of the cationic borylene complex **8** (179.2(8) pm) [28], which displays a Fe=B double bond, and $[(\eta^5\text{-}C_5H_5)(OC)_2Fe(BCl_2)]$ (**14**: 194.2(3) pm) [36], which has the shortest reported Fe–B distance in a boryl complex. This indicates at least a modest π-component in the Fe–B bond of **12**, also confirmed by theoretical and structural studies, which suggest a noticeable π-component in **14**.

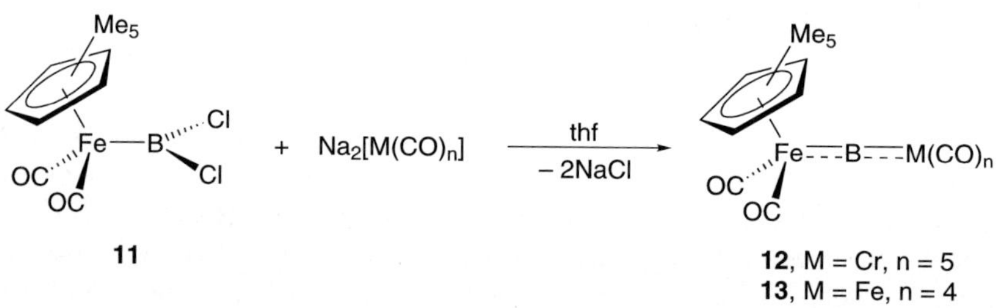

Scheme 7 Synthesis of metalloborylene complexes

2.5 Base-Stabilized Borylene Complexes

The first base-stabilized borylene complex, derived from the reaction of the dichloroboryl complex $[Os(BCl_2)Cl(CO)(PPh_3)_2]$ (**15**) with 8-aminoquinoline, was described by Roper et al. [37]. The reaction of the poorly soluble compound $[Os(=BNHC_9H_6N)Cl_2(CO)(PPh_3)_2]$ (**16**) with tetrabutylammonium iodide leads to the better soluble $[Os(=BNHC_9H_6N)ClI(CO)(PPh_3)_2]$ (**17**) by selective substitution of the chloride *trans* to the borylene ligand, giving evidence for its high *trans*-influence. **17** shows one broad signal in the ^{11}B-NMR spectrum at $\delta = 51.7$, which is only slightly high-field shifted with respect to that of **15** ($\delta = 52.5$). The ^{31}P-NMR data suggest the existence of two isomers, which can be explained by two different orientations of the base-stabilized borylene ligand. This assumption is further ascertained by a disorder between the *trans* positioned Cl and CO ligands in the crystal structure of **17**. The Os–B bond (205.5(8) pm) is shorter than in related six-coordinate osmium boryl complexes, but still longer than expected for an Os=B double bond. This can be explained by the contribution of several resonance structures with Os–B single bonds to the overall bonding and is also found in base-stabilized silylene complexes, wherein the metal–silicon bond distance is more typical for M–Si single bonds [38, 39]. The B–N1 distance (157.1(10) pm) is significantly longer than B–N2 (144.0(10) pm), but still shorter than the average of all B–N (pyridine type) bond lengths reported in the CCSD. Despite the presence of a stabilizing donor group, the boron center in **17** is highly electrophilic, which is reflected by its reactivity towards ethanol, yielding the corresponding boryl complex [Os{B(OEt)NHC$_9$H$_6$N}Cl(CO)(PPh$_3$)$_2$] (**18**).

The reaction of **15** with 2-aminopyridine yields a mixture of the amino-(chloro)boryl complex **19a** and the base-stabilized borylene complex **20a** [40]. **20a** reacts readily with ethanol, leading to the corresponding ethoxyborylene complex **20b**, which was crystallographically characterized. **19a** is stable in boiling ethanol but reacts, in the presence of Ag[SbF$_6$] in undried thf, to the hydroxyboryl complex **19b** (Scheme 8). The key structural parameters of **19b** and **20b** are very similar, which leads to the very important conclusion that there is only little difference in the nature of the Os–B bond in these types of base-stabilized borylene and boryl complexes.

The reaction of the T-shaped three coordinate platinum boryl complex *trans*-[(Cy$_3$P)$_2$Pt{B(Br)Fc}][BAr$^f{}_4$] (**21**) (Fc = ferrocenyl) with 4-methylpyridine leads to the formation of *trans*-[(Cy$_3$P)$_2$Pt(Br){B(Fc)(NC$_5$H$_4$Me-4)}] (**22**), the first example for a cationic, base-stabilized borylene complex [41] (Scheme 9). For both compounds, no signals could be observed in the ^{11}B-NMR spectrum, probably due to line-broadening as a consequence of unresolved coupling to the platinum and phosphorous centers. In the crystal, the plane of the borylene ligand in **22** adopts an almost orthogonal position (78.42(13)°) to the coordination plane of the platinum center. This is also found in **21** and analogous systems with carbene [42] or silylene [43] ligands and can be attributed to steric interactions with the bulky phosphine ligands. The Pt–Br distance is rather long (260.57(5) pm), which is consistent with the expected high *trans*-influence of the borylene ligand. The Pt–B bond distance (201.4(5) pm) is only slightly longer than in **21** (196.6(4) pm) and very similar to that of its precursor *trans*-[(Cy$_3$P)$_2$Pt(Br){B(Br)Fc}] (**23**) (199.63(34) pm) [44], comprising a four-coordinate Pt-center. This bond elongation is probably a consequence of the increased coordination number

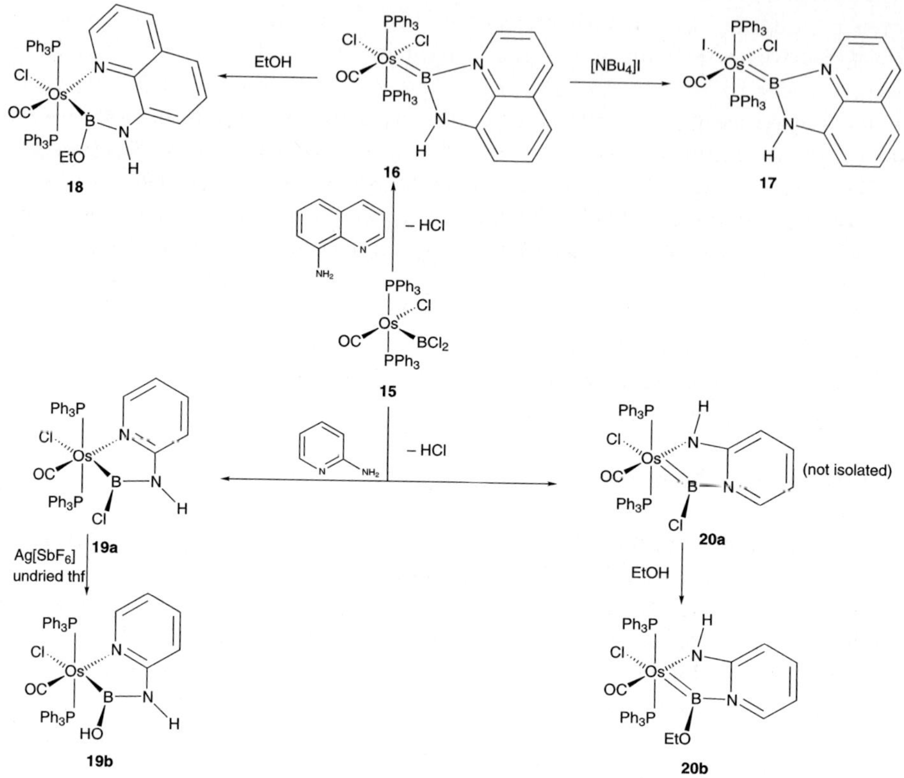

Scheme 8 Tethered osmium boryl and base-stabilized borylene complexes

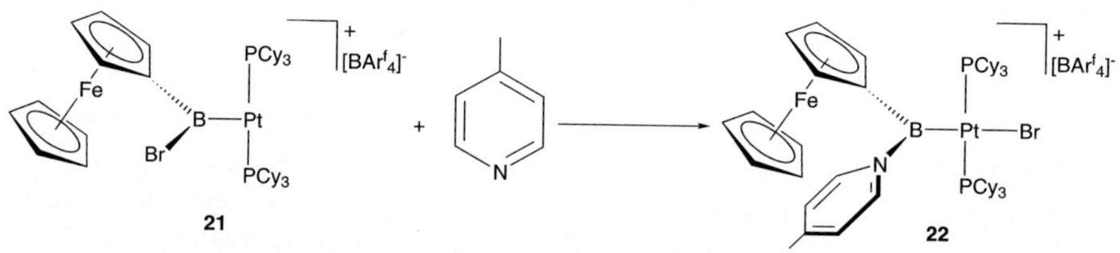

Scheme 9 Synthesis of a cationic base-stabilized borylene complex

at platinum and the enhanced steric bulk at the boron center. Thus, a classification of the bonding situation in the base-stabilised borylene and the corresponding boryl complexes solely on the basis of M–B distances is, as for the osmium complexes **19b** and **20b**, somewhat ambigious.

3
Bridging Borylene Complexes

The bridging borylene complexes (**V, VI**) are of some special relevance since they represent the first structurally authentic examples for the stabilization of the borylene ligand –BR in the coordination sphere of a transition metal. Species where the borylene unit adopts a bridging position between two metals represent the most numerous and well-studied class of these compounds.

3.1
Homodinuclear Bridging Borylene Complexes

3.1.1
Synthesis and Structure

Diboranes(4) and dihaloboranes are the traditional boron source for the synthesis of homodinuclear bridging borylene compounds. Reaction of $K[(\eta^5\text{-}C_5H_4Me)Mn(CO)_2H]$ with a variety of $B_2R_2Cl_2$ derivatives afforded the homodimetallic manganese compounds $[(\mu\text{-}BX)\{(\eta^5\text{-}C_5R_5)Mn(CO)_2\}_2]$ ($R = NR'_2$, *t*Bu) (**23a,b–26a,b**), which represent the first series of structurally authenticated borylene complexes (Scheme 10) [45–47].

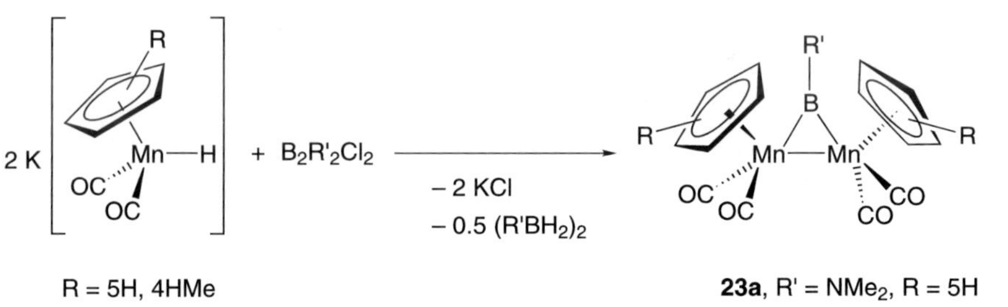

R = 5H, 4HMe

23a, R' = NMe$_2$, R = 5H
23b, R' = NMe$_2$, R = 4HMe
24, R' = N(CH$_2$)$_4$
25, R' = N(CH$_2$)$_5$
26a, R' = *t*Bu, R = 5H
26b, R' = *t*Bu, R = 4HMe

Scheme 10 Synthesis of bridging manganese borylene complexes

Due to the formation of two metal–boron bonds, the compounds **23a,b–26a,b** show characteristically deshielded ^{11}B-NMR signals in solution. In the case of aminoborylene complexes, resonances in the range of $\delta = 100–120$ are observed. For the *tert*-butylborylene complex **26a** an even more deshielded signal is found at $\delta = 170.0$. The observed CO stretching frequencies of **23a,b–26a,b** are comparable to those of the related μ-methylene complexes [48–50] and support their description as dimetallaboriranes. The results of the X-ray structure analysis of **23a** show that this molecule adopts approximate C_2 symmetry in the crystalline state. Both manganese atoms and boron form an isosceles triangle with B–Mn distances of 203 (1) pm and a Mn–Mn distance of 279.0(2) pm. The Me_2N group is only slightly rotated (8°) with respect to the Mn_2B plane and the B–N distance of 139 (1) pm is characteristic for a B=N double bond. The overall molecular structure data resemble those of the isoelectronic vinylidene manganese complexes [51], which can be considered as dimetallacyclopropane derivatives. It was found, that the aminoborylene complexes **23a,b–25** display an extremely low reactivity towards nucleophilic substitution at the boron center; they are practically inert against air and water. On the basis of quantum chemical calculations it was suggested, that this stability is either caused by steric protection or by reduction of the build-up of charge on the coordinated BR ligand. The latter should be achieved by restoring the balance between M–BR σ donation and π back donation, which means, in reality, by coordinating the borylene with suitable binuclear transition metal fragments, such as $[\{(\eta^5\text{-}C_5R_5)Mn(CO)_2\}_2]$ [12]. Taking into account the distinct sensitivity to air and water of the isostructural *tert*-butylborylene complex **26a** [63], in which the boron center is certainly more sterically protected than in **23a**, we propose a different explanation: the steric protection of the borylene moiety is mainly provided by the transition metal fragment, whereas the electronic stabilization in **23a,b** is caused by an efficient π donation from the nitrogen to the boron atom. The latter effect is assumed to be the main reason for the remarkable stability of **23a,b** in comparison to **26a,b**.

With a different approach, again based on the use of diboranes(4) and specifically of $B_2H_4 \cdot 2PMe_3$, Shimoi and Ogino succeeded in the synthesis of $[\{\mu\text{-BH}(PMe_3)\}(\mu\text{-CO})\{Co(CO)_3\}_2]$ (**27**) (Scheme 11) [52].

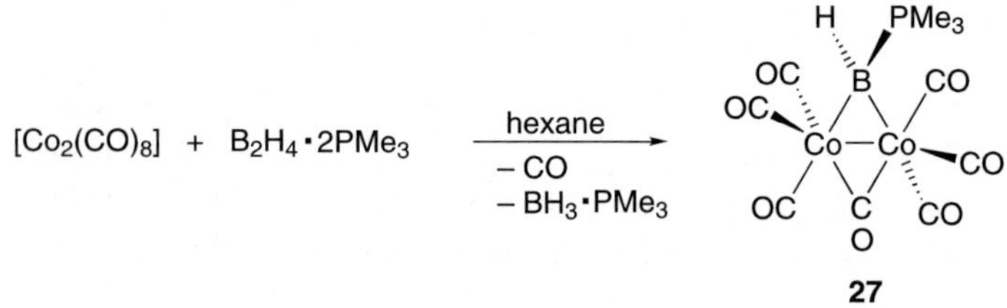

Scheme 11 Synthesis of the base-stabilized bridging borylene complex 27

Fragmentation of the starting diborane-bis(trimethylphosphine) adduct took place with the liberation of $BH_3 \cdot PMe_3$ and the generation of the parent borylene molecule BH. Stabilization of this extremely reactive species occurred upon coordination to the two cobalt centers and to the remaining PMe_3. To our knowledge, **27** is a unique example of a base-stabilized bridging borylene compound, that exhibits a tetrahedral coordinated boron atom. The higher coordination number of boron is indicated by an ^{11}B-NMR signal at $\delta = 17.5$, which is significantly high-field shifted with respect to that of **23a,b**. The solid state structure of **27** displays two $Co(CO)_3$ moieties symmetrically bridged by the carbonyl and borylene ligands. The Co–B bond lengths are 211.2(9) pm and 210.8(11) pm, respectively. These are shorter than that found in the base stabilized boryl complex $[Co(CO)_2(\eta^1\text{-dppm})(\mu\text{-dppm}\cdot BH_2)]$ (222.7(6) ppm) [53], but longer than those in the cobaltaborane cluster $[\{(\eta^5\text{-}C_5H_5)Co\}_3(\mu_3PPh)\ (\mu_3BPh)]$ (201.8(8)–206.5(8) pm) [54]. The interatomic distance of 248.6(2) pm indicates the existence of a single bond between the cobalt atoms.

The second pathway to homodimetallic bridging borylene complexes is the reaction of a dihaloborane with anionic transition metal compounds in a 1 : 2 ratio. According to Scheme 12, the complexes $[\{\mu\text{-BN}(SiMe_3)_2\}(\mu\text{-CO})\{(\eta^5\text{-}C_5H_4R)Fe(CO)\}_2]$ (**28a**, R = H, **28b**, R = Me) [55] and $[\{\mu\text{-BN}(SiMe_3)_2\}(\mu\text{-CO})\{(\eta^5\text{-}C_5H_5)Ru(CO)\}_2]$ (**29**) [56] were obtained from $(Me_3Si)_2NBCl_2$ and $Na[(\eta^5\text{-}C_5H_4R)M(CO)\}_2]$ (M = Fe, R = H, Me; M = Ru, R = H) with salt elimination and loss of one carbonyl group.

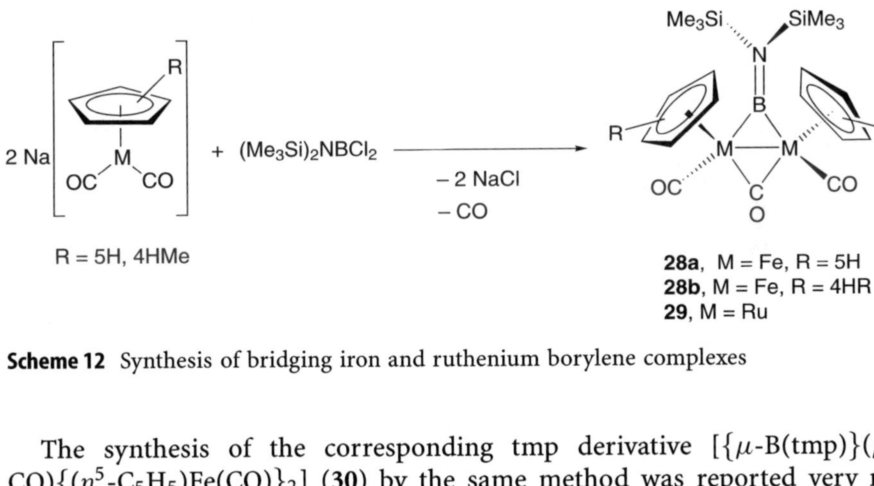

Scheme 12 Synthesis of bridging iron and ruthenium borylene complexes

The synthesis of the corresponding tmp derivative $[\{\mu\text{-B(tmp)}\}(\mu\text{-CO})\{(\eta^5\text{-}C_5H_5)Fe(CO)\}_2]$ (**30**) by the same method was reported very recently [57]. Due to the formation of two metal–boron-linkages the borylene complexes show strongly deshielded ^{11}B-NMR resonances ($\delta = 119$, **28a,b**; 105.9, **29**; 115.3, **30**). The results of an X-ray structure analysis of **28b** indicate a *trans* orientation of the MeC_5H_4 and CO groups with respect to the Fe_2B unit. These three atoms form an isosceles triangle with Fe–B distances of 200.7(3) pm and 200.2(3) pm, and a Fe–Fe distance of 254.8(1) pm. The Si–N–Si plane is twisted by 53.7(1)° with respect to the to the Fe–B–Fe plane due to steric interactions of the bulky Me_3Si groups with the MeC_5H_4 ligands. That leads to a weaker π-donation from the nitrogen to the boron atom, which is expressed in an extended B–N distance of 141.2(4) pm. As in the case of the abovementioned iron species, the vast majority of homodinuclear borylene compounds show a metal–metal bond and/or a second bridging ligand (especially CO) supporting the borylene bridge. The first complex with an unsupported borylene ligand [(μ-BMes){(η^5-C_5H_5)Fe$(CO)_2$}$_2$] (**31**) was reported by Aldridge in 2002 [58]. This compound was afforded by the reaction of the boryl complex [(η^5-C_5H_5)Fe$(CO)_2$BBrMes] with an excess of Na[(η^5-C_5H_5)Fe$(CO)_2$] (Scheme 13).

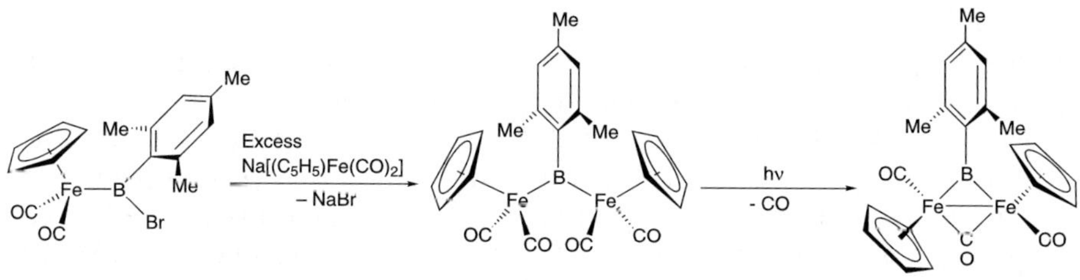

Scheme 13 Synthesis of a bridging borylene complex with an unsupported borylene ligand and photolytic conversion to the corresponding supported derivative

IR data for **31** indicate exclusively the presence of terminally bound carbonyl ligands. This finding is confirmed by the results of an X-ray structure determination. The Fe–Fe distance (380.2(10) pm) is too long for the presence of a covalent bond between the two metal centers and the Fe–B–Fe angle (130.8(5)°) is significantly wider than usually found in bridging borylene complexes featuring a MBM metallacycle, since the abovementioned complexes **23a**, **27** and **28b** show MBM angles in the range of 75–90°. The Fe–B bonds of **31** (209.0(10) pm, 209.1(10) pm) are considerably longer than those found in **28b** due to increased steric crowding at boron. The long Fe–B bonds, together with carbonyl stretching frequencies (2010 and 1949 cm^{-1}) which are very similar to those reported for [(η^5-C_5H_5)Fe$(CO)_2CH_3$] (2010 and 1958 cm^{-1}) [59] indicate the absence of a noteworthy π stabilization of the boron center through Fe–B backbonding. Photolysis of **31** results in the loss of one CO ligand and the formation of complex **32** [30] (Scheme 13). IR and crystallographic data confirm the bridging geometry with an Fe–Fe distance of 252.8(1) pm being similar to that found in **28b** (254.8(1) pm). **32** is formed exclusively as a single isomer with a *trans* orientation of cyclopentadienyl ligands. This geometry allows a sterically favourable coplanar orientation of both cyclopentadienyl ligands and the mesityl group.

Photochemically induced borylene transfer has already been discussed as a valuable synthetic method for the generation of terminal borylene complexes not accessible through salt elimination reactions (vide supra). Photolysis of [$(CO)_5W{=}BN(SiMe_3)_2$] (**2**) in the presence of [(η^5-C_5H_5)Re$(CO)_3$] yielded the

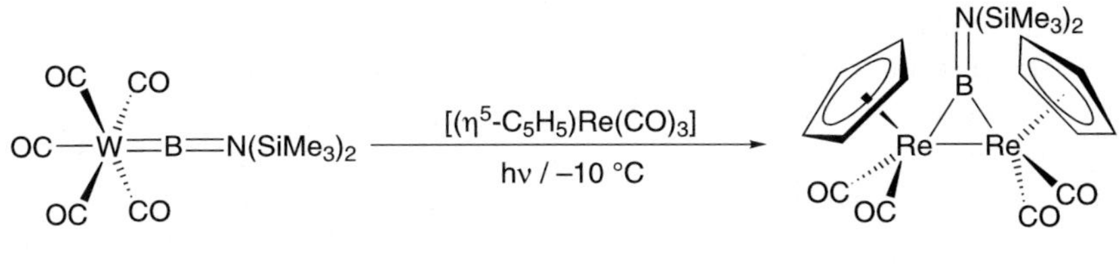

Scheme 14 Photolytic synthesis of the bridging borylene rhenium complex **33**

bridging borylene complex $[(\mu\text{-BN(SiMe}_3)_2)\{(\eta^5\text{-C}_5\text{R}_5)\text{Re(CO)}_2\}_2]$ (**33**) as the only boron containing species [18] (Scheme 14).

The structure of **33** derives from multinuclear NMR, IR and MS data since no X-ray structure determination could be carried out. All attempts to synthesize **33** via the same salt elimination route which leads to **28a,b** failed (Braunschweig H, Colling M, unpublished results). To our knowledge **33** represents the only example of a bridged borylene complex afforded by photolytic borylene transfer. Furthermore, transfer reactions of a borylene ligand from a bridged coordination mode to a transition metal fragment forming another bridging borylene complex or a terminal borylene complex have not been reported yet.

3.1.2 Reactivity

It has been found that the aminoborylene complexes **23a,b**–**25** display extremely low reactivity towards nucleophilic substitution at the boron center. Consequently, they show high stability against both air and moisture for long periods (vide supra). Treatment of **23b** with an excess of HCl, however, provided the corresponding chloroborylene complex $[(\mu\text{-BCl})\{(\eta^5\text{-C}_5\text{H}_4\text{Me})\text{Mn(CO)}_2\}_2]$ (**34**). The weakly π-stabilized complex **34** undergoes a number of substitution reactions at the boron center with protic reagents (Scheme 15) [46, 60]. Reaction of **34** with pyridine in the presence of $H[BF_4]$ results in the formation of the aminoborylene complex **41** via formal 1,4-hydroboration of pyridine [61, 62].

The Mn_2B metallacycle remained unchanged during all these reactions, whereas treatment of **23a,b** with powerful halogenating agents such as I_2 or SbF_3 led to decomposition (Braunschweig H, Colling M, unpublished results).

The *tert*-butylborylene complex **26a,b**, which still represents the only example of an alkylborylene complex, displays a significantly different reactivity compared to its amino-stabilized counterpart **23a,b**. **26a,b** decomposes quickly when it is exposed to air or moisture and a comparable substitution chemistry to that of **23a,b** (via **34**) cannot be realized. **23a,b** proved to be unreactive towards $[Pd(PCy_3)_2]$, whereas **26b** underwent

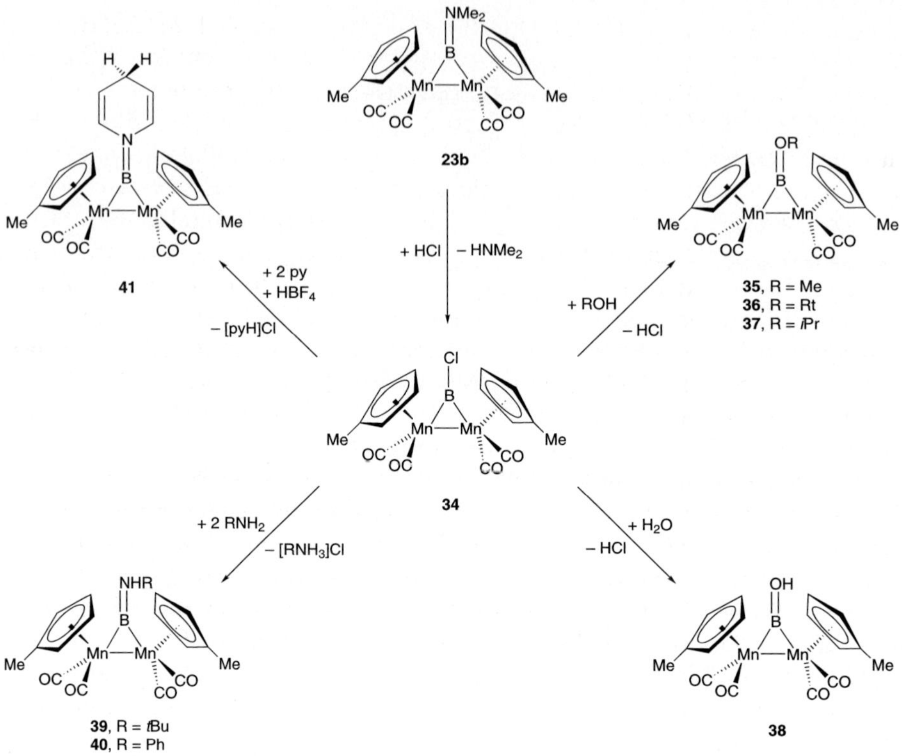

Scheme 15 Synthesis and reactivity of the chloroborylene complex **34**

a clean, albeit slow reaction with this highly unsaturated transition metal complex to form the unprecedented trimetallic species $[(\mu_3\text{-B}t\text{Bu})\{\{(\eta^5\text{-}C_5H_4Me)Mn(CO)_2\}\{Pd(PCy_3)_2\}\}]$ (**42**) [63] (Scheme 16).

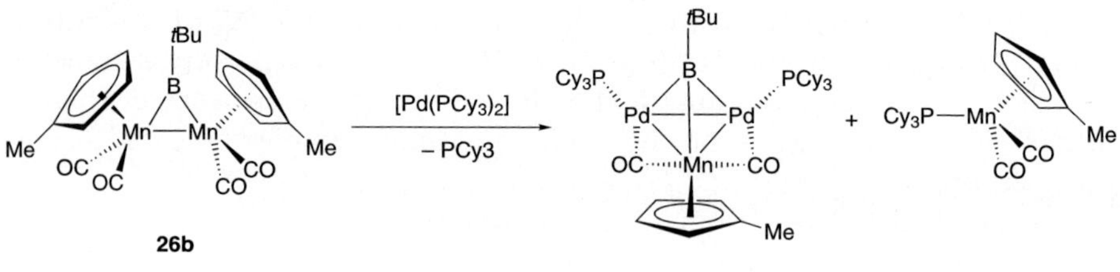

Scheme 16 Reaction of the bridging borylene complex **26b** with $[Pd(PCy_3)_2]$

The reaction was complete after three weeks, and no additional signals of further byproducts were detected by multinuclear NMR monitoring. Compound **42** can be viewed as the first example of a *hetero*-metallic μ_3-borylene complex. According to the results of a crystal structure determination **42** consists of a $MnPd_2$ isosceles triangle (bond lengths: Mn–Pd1 264.26(10) pm, Mn–Pd2 264.58(10) pm, Pd1–Pd2 284.23(7) pm) which is capped by the μ_3-borylene ligand. A formal electron count (each of the fragments {*t*BuB}, {$PdPCy_3$}, and {(η^5-C_5H_4Me)$Mn(CO)_2$} contributes two electrons to the cluster framework) is in agreement with a *hypercloso* metallaborane (a cluster that is two electrons short of a *closo* species). An alternative description bases on a more localized bonding picture. Natural bond orbital (NBO) calculations indicate a strong covalent Mn–B bond and weaker Pd–B interactions, while the Pd–Pd interactions are almost nonexistent. The latter result reflects the long distance between the two palladium atoms. The localized description is supported by the results of electron localization function (ELF) calculations, which revealed two fused basins representing two (3c, 2e) Mn–B–Pd bonds, whereas electron density over a triangular face of the {$BMnPd_2$} core indicative of a typical cluster, was not found.

The photochemistry of bridging borylene complexes differs significantly from that of their terminal counterparts. The aminoborylene complex **23a,b** proved to be entirely unreactive under photolytic conditions, which is in sharp contrast to the terminal aminoborylene complexes **1** and **2**. The chloro derivative **34**, however, yielded upon irradiation in the presence of a CO donor (i.e. $M(CO)_6$, M = Cr, Mo, W) the dimetalla-*nido*-tetraborane [B_2Cl_2{(η^5-C_5R_5)$Mn(CO)_2$}$_2$] (**43**) as shown in Scheme 17 [64]. The reaction provides a direct synthetic link between electron-precise borylene complexes and electron-deficient metallaboranes.

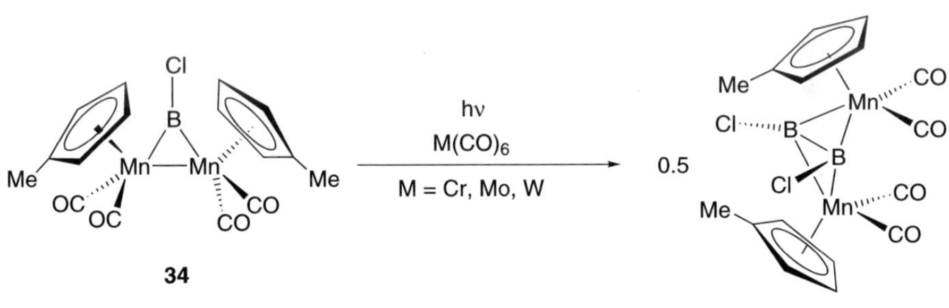

Scheme 17 Photolytic generation of [B_2Cl_2{(η^5-C_5R_5)$Mn(CO)_2$}$_2$] (**43**)

3.2
Heterodinuclear Bridging Borylene Complexes

All attempts at preparing heterodinuclear bridging borylene compounds from dihaloboranes, by sequential salt elimination steps with anionic complexes of different transition metals, proved unsuccessful. In most cases, the haloboryl complex simply does not react with a second equivalent of a metallate, regardless if it contains the same or a different transition metal. Instead, in a smaller number of reactions a substitution of the first transition metal fragment, than of the second halide atom is observed, due to the kinetic lability of the boron–metal bond. That leads to decomposition of both the boryl complex and the second metallate and usually occurs, when the second nucleophile is significantly stronger than the first one (Braunschweig H, Kollann C, Rais D, Seeler F, unpublished results). Following a different strategy, however, the mixed iron-palladium borylene compounds [(μ-BX)(μ-CO){(η^5-C_5Me_5)Fe(CO)}{(Cy_3P)Pd(Br)}] [X = Br, **44a**; X = Fc, **44b**] could recently be synthesized. To access these compounds, the oxidative addition reaction of B–Br bonds to low-valent transition metal complexes was employed. Specifically, upon reaction of the recently isolated bromoboryl complexes [(η^5-C_5Me_5)Fe$(CO)_2${B(Br)X}] (X = Br [44]; X = Fc [65]) with [Pd$(PCy_3)_2$], oxidative addition of a B–Br bond to the electron rich palladium center occurred, with liberation of one bulky PCy_3 ligand and formation of the desired heterodinuclear bridging borylene complexes (Scheme 18) [65]. The ^{11}B-NMR signals of **44a** and **44b**, δ = 136 and 118, respectively, reflect the bridging coordination mode of the bromo- and ferrocenylborylene ligands, displaying the expected low-field shift of the resonance that usually results from an increased number of metal–boron bonds.

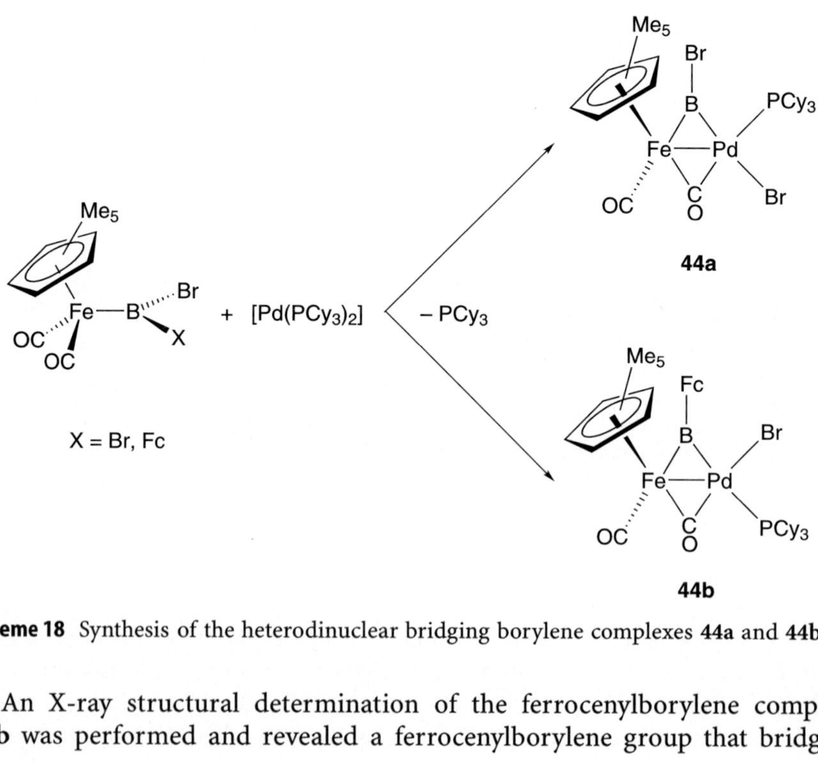

Scheme 18 Synthesis of the heterodinuclear bridging borylene complexes **44a** and **44b**

An X-ray structural determination of the ferrocenylborylene complex **44b** was performed and revealed a ferrocenylborylene group that bridges, together with a CO ligand, the metal centers of the $[(\eta^5\text{-}C_5Me_5)Fe(CO)]$ and $[(Cy_3P)Pd(Br)]$ complex fragments. The Fe–B distance is very short (190.3(3) pm); it is, in particular, shorter than that found in the bromoboryl precursor $[(\eta^5\text{-}C_5Me_5)Fe(CO)_2\{B(Br)Fc\}]$ (198.5(3) pm) [44]. The Fe–Pd bond distance (256.84(5) pm) is within the range of those reported for heterodinuclear compounds containing Fe–Pd linkages [66], whereas the Pd–B bond (209.0(3) pm) is longer than the ones reported for the two structurally characterized complexes (200.6(9) pm, 207.7(6) pm) that feature bonds between Pd and a three-coordinate boron atom [67,68]. The strong *trans* influence of the borylene ligand is reflected by the long bond between palladium and the phosphorus atom of the PCy_3 ligand (242.97(7) pm). The ferrocenylborylene and bromide ligands are in mutual *cis* disposition. This unexpected arrangement is likely caused by severe steric constraints imposed by the bulky tricyclohexylphosphine and ferrocenyl ligands. In contrast to that, the expected *trans* orientation of the bromide ligands at boron and at palladium in complex **44a** was proven by the results of ^{31}P-NMR experiments.

3.3 Metalloborylenes Stabilized by a Transition Metal Base

Molecular systems, such as the heterodinuclear bridging borylene complexes **44a** and **44b** just described, comprise two different transition-metal-ligand fragments linked by boron centers that feature a decreasing number of main group element substituents. The possibility of saturating the three classical valencies of boron by linking the boron center to three transition-metal-ligand fragments was recently explored and demonstrated by the full characterization of $[(\eta^5\text{-}C_5Me_5)(OC)Fe(\mu\text{-}CO)M(PCy_3)(\mu\text{-}Br)Pt(PCy_3)Br(\mu_3\text{-}B)]$ (M = Pd, **46**; M = Pt, **47**), featuring the unprecedented coordination of boron to three transition metals [69]. As a starting material for the buildup of the trimetallic systems, the heterodinuclear bridging bromoborylene complex $[(\mu\text{-}BBr)(\mu\text{-}CO)\{(\eta^5\text{-}C_5Me_5)Fe(CO)\}\{(Cy_3P)Pt(Br)\}]$ (**45**) was synthesized. **45** represents a platinum analogue of complex **44a**, with an increased stability in solution. That increased stability provides a clean reaction of the –BBr moiety with $[M(PCy_3)_2]$ (M = Pd, Pt), producing the new compounds $[(\eta^5\text{-}C_5Me_5)(OC)Fe(\mu\text{-}CO)M(PCy_3)(\mu\text{-}Br)Pt(PCy_3)Br(\mu_3\text{-}B)]$ (M = Pd, **46**; M = Pt^2, **47**) and loss of one equivalent of PCy_3 (Scheme 19).

In solution, **46** and **47** show broad singlets at $\delta = 144$ and 130, respectively, in the ^{11}B-NMR spectra. Both signals are low-field shifted with respect to that of the precursor **45** ($\delta = 108$), in agreement with the increased number of metal–boron bonds in the products. Accordingly, the ^{31}P-NMR spectra of the two compounds display two resonances, at $\delta = 31.9$ and 32.5 ($^1J(\text{Pt,P}) = 4559$ Hz) for **46**, and at $\delta = 31.1$ ($^1J(\text{Pt,P}) = 4703$ Hz) and 57.3 ($^1J(\text{Pt,P}) = 4626$ Hz) for **47**. The solid state structures of **46** and **47** display a boron atom coordinated by three transition metal atoms,

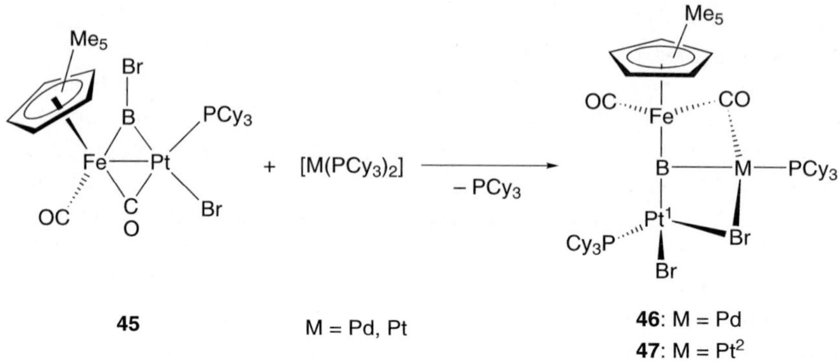

Scheme 19 Synthesis of the transition metal base-stabilized metallaborylene complexes **46** and **47**

with short Fe–B (**46**, 191.0(4) pm; **47**, 190.3(4) pm) and Pt^1–B distances (**46**, 192.3(4) pm; **47**, 193.8(4) pm). Instead of the expected trigonal planar geometry around a formally sp^2-hybridized boron center, the structures are characterized by a nearly linear arrangement of the iron, boron, and platinum atoms (Fe–B–Pt^1 = 168.7(2)°, **46**; 167.1(2)°, **47**), indicating formal *sp*-hybridization at boron. The iron and platinum disposition around boron is reminiscent of the structures of the metalloborylene complexes $[(\mu_2\text{-B})\{(\eta^5\text{-}C_5Me_5)Fe(CO)_2\}\{Cr(CO)_5\}]$ (**10**) (Fe–B–Cr = 177.75(11)°) and $[(\mu_2\text{-B})\{(\eta^5\text{-}C_5Me_5)Fe(CO)_2\}\{Fe(CO)_4\}]$ (**11**) (Fe–B–Fe = 175.38(12)°) (Sect. 2.4), accompanied, in **46** and **47**, by an additional boron–metal interaction. Indeed, a $[M(PCy_3)]$-complex fragment (M = Pd, **46**; M = Pt, **47**) approaches the central Fe–B–Pt core, resulting in long Pd–B (215.0(4) pm) and Pt^2–B (215.8(4) pm) linkages, supported by bridging carbonyl and bromide ligands to the outer transition metals. The final description of **46** and **47** as metalloborylenes of the type Fe–B=Pt, stabilized by interactions with a transition metal base of the type $[M(PCy_3)]$ (M = Pd, Pt), is supported by the results of theoretical calculations (DFT at B3LYP level) performed on the hypothetical metalloborylene complex $[(\mu_2\text{-B})\{(\eta^5\text{-}C_5H_5)Fe(CO)_2\}\{PdBr_2(PMe_3)\}]$, which served as a model system for compound **46** [69]. The dz^2-type HOMO of the $[Pd(PMe_3)]$ fragment, pointing along the Pd–P bond, can interact with six unoccupied orbitals of the hypothetical metalloborylene molecule $[(\eta^5\text{-}C_5H_5)(CO)_2Fe\text{–}B{=}PtBr_2(PMe_3)]$, partially localized on the Lewis acidic boron atom. Electron localization function (ELF) calculations on the hypothetical borylene complex $[(\eta^5\text{-}C_5H_5)(CO)_2Fe\text{–}B{=}PtBr_2(PMe_3)]$ show the presence of a ring attractor between the platinum and the boron atoms, which indicates a partial triple bond character of the B–Pt bond. Addition of a $[Pd(PMe_3)]$ moiety converts the ring attractor into a point attractor. The basin of the B–Fe bond is now stretched to include the palladium center, effectively representing a three-center two electron (3c, 2e) Fe–B–Pd bond.

3.4
Semi-Bridging Borylene Complexes

A number of theoretical studies have attempted to clarify the characteristic features of the transition metal–boron linkage in borylene complexes [11–14]. As a result, pronounced similarities between the borylene ligand BR and carbon monoxide CO have emerged, with respect to their ability to bind synergistically to transition metals [70, 71]. Such similarities, predicted at theoretical level, have gradually found experimental confirmation. The rich coordination chemistry of the carbonyl ligand, with terminal, symmetrically bridging, semi-bridging, and triply bridging coordination modes, finds now parallels in borylene chemistry. As already mentioned, symmetrically bridging borylene complexes represent the most numerous and well-studied class of borylene compounds. Appropriate synthetic routes to their terminal counterparts were subsequently developed, and cluster compounds of cobalt and ruthenium that feature triply bridging borylene ligands are known (Sect. 3.6). Compounds that feature semi-bridging borylene ligands represent the most recent accomplishment. The terminal borylene complexes $[(OC)_5M{=}B{=}N(SiMe_3)_2]$ (M = Cr, **1**; W, **2**) react cleanly with $[Pd(PCy_3)_2]$ at room temperature, leading to formation of the semi-bridging borylene compounds, $[\{\mu\text{-}BN(SiMe_3)_2\}(\mu\text{-}CO)\{M(CO)_4\}\{Pd(PCy_3)\}]$ (M = Cr, **48**; M = W, **49**), respectively (Scheme 20) [72].

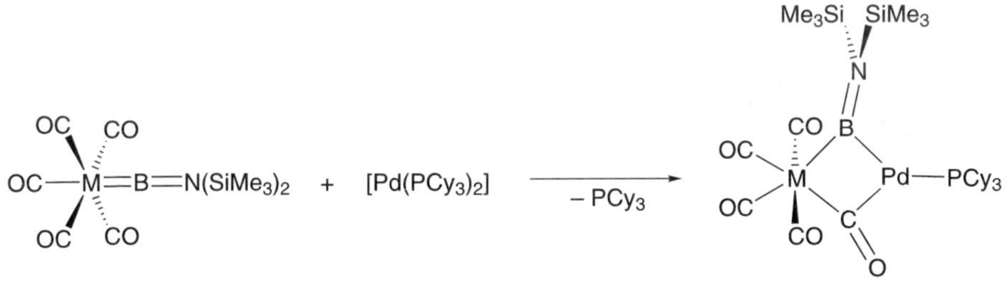

1: M = Cr
2: M = W

48: M = Cr
49: M = W

Scheme 20 Synthesis of the semi-bridging borylene complexes **48** and **49**

The ^{11}B-NMR data show downfield-shifted resonances at $\delta = 100$ ppm for **48** and $\delta = 97$ ppm for **49** (cf. $\delta = 92$ ppm for **1** and $\delta = 86$ ppm for **2**), suggesting the formation of additional boron–metal linkages. The exact connectivity of the atoms within **48** was determined by performing a single crystal X-ray diffraction analysis. The structural data revealed a virtually orthogonal orientation of the bis(trimethylsilyl)amino substituent at boron with respect to the plane containing the two transition metals and boron (angle between the planes: 85.8°). Such a disposition differs significantly from those observed in previously reported homodimetallic bridging borylene complexes, e.g. $[(\mu\text{-BNMe}_2)\{(\eta^5\text{-C}_5\text{H}_5)\text{Mn(CO)}_2\}_2]$ [45, 46] and $[\{\mu\text{-BN(SiMe}_3)_2\}(\mu\text{-CO})\{(\eta^5\text{-C}_5\text{H}_4\text{Me})\text{Fe(CO)}\}_2]$ [55], for which the structural parameters are in agreement with the presence of a formally *sp*^{2}-hybridized boron atom. Assuming the same hybridization of the boron atom in **48**, the orthogonal disposition of the amino substituent would rule out any effective N–B π-donation, and should result in considerable elongation of the N–B bond distance. However, despite the increased coordination number at boron, the B–N bond distance (137.7(2) pm) is only slightly longer than that observed in the terminal borylene complex **1** (135.3(6) pm), thus indicating significant double bond character of the B–N linkage. The bonding interactions that characterize **48** can be rationalized on the basis of the aforementioned analogy between borylene and carbon monoxide. More precisely, donation of electron density by the electronically saturated chromium or tungsten complexes **1** or **2** into the π^*-orbitals of the carbonyl and borylene units results in CO and $\text{B}{=}\text{N(SiMe}_3)_2$ forming a pair of non-compensating semi-bridging ligands. This is indicated by the relevant angular values (Pd–Cr–C(O): 54.2°, Cr–C–O: 165.5°; Pd–Cr–B: 49.8°, Cr–B–N: 152.3°), which are in accordance with the classification scheme for semi-bridging carbonyl ligands developed by Crabtree [73]. Compounds **48** and **49** can be viewed as the products of partial transfer of the borylene moiety $\text{BN(SiMe}_3)_2$ from the group 6 metals to palladium. A complete transfer under a variety of conditions has not been realized yet (Braunschweig H, Rais D, Uttinger K, unpublished results).

3.5 Bis(borylene) Complexes

Until very recently, the formation of the semi-bridging borylene compounds **48** and **49** represented the only instance of clean thermal, i.e., non-photolytic, reactivity of the terminal borylene complexes **1** and **2**. Interestingly, however, an unprecedented metal-to-metal borylene transfer occurred from $[\text{(OC)}_5\text{M}{=}\text{BN(SiMe}_3)_2]$ (M = Cr, **1**; M = W, **2**) to $[\text{RhCl(CO)}_2]_2$ at room temperature, yielding the tetranuclear bis(borylene) complex $[\text{Rh}_4\{\mu\text{-BN(SiMe}_3)_2\}_2\text{Cl}_4(\mu\text{-CO})(\text{CO})_4]$ (**50**) (Scheme 21) [74].

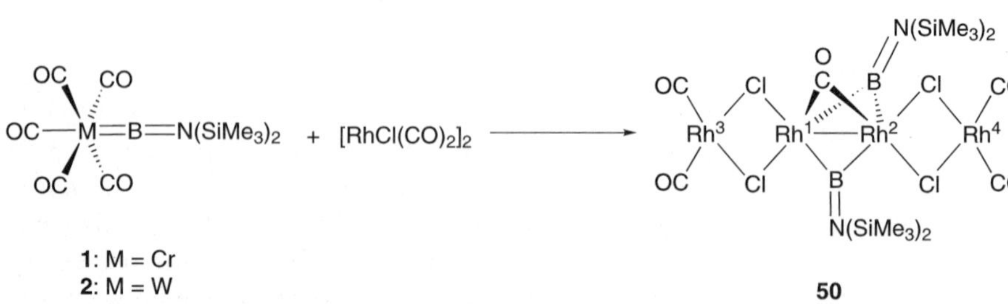

Scheme 21 Synthesis of the tetranuclear bis(borylene) complex **50**

As revealed by an X-ray diffraction study, the core of the molecule is composed by a chain of four rhodium atoms, with the internal metal centers bridged by two borylene moieties and a carbonyl ligand. The distance between the two central rhodium atoms is short (Rh^1–Rh^2 = 257.86(3) pm) and comparable to analogous distances found in dinuclear complexes in which the rhodium atoms are bridged by three carbonyl ligands [75]. The presence of a Rh–Rh bond may be assumed, as the formal electron count of each rhodium atom is odd. The distances between the central and the terminal rhodium atoms, Rh^1–Rh^3 (304.15(4) pm and Rh^2–Rh^4 (304.72(4) pm) are significantly longer than the Rh^1–Rh^2 distance. Surprisingly, the ^{11}B-NMR signal of compound **50** is found at $\delta = 74$, and is unusually high-field shifted with respect to that of the starting materials ($\delta = 92$, **1**; $\delta = 86$, **2**), contrarily to what could have been expected on account of the increased number of metal–boron bonds in the product (vide supra).

3.6 Triply Bridging Borylene Complexes

All aforementioned borylene complexes have in common that the bonding situation between the boron atom and the metal atom(s) is adequately described by electron precise 2c–2e bonds, although the description of the bonding situations of **42**, **46**, and **47** is worth of further investigations. An extension of the analogy between CO and BR to the triply bridging coordination mode requires a more delocalized bonding picture. The first reported compounds with a μ_3-BR moiety [$\{(\eta^5\text{-}C_5H_5)Co\}_3(\mu_3\text{-BPh})PPh)$] (**51**) [76] and [$\{(\eta^5\text{-}C_5Me_5)Ru\}_3(\mu_3\text{-BH})_3H_5$] (**52**) [77] have been classified as metal-rich metallaboranes, referring to the electron-deficient nature of these cluster compounds. New examples of ruthenium clusters that feature triply bridging borylene ligands were recently reported. Reaction of the cationic polyhydride triruthenium cluster [$\{(\mu\text{-H})_6(\eta^5\text{-}C_5Me_5)Ru\}_3$]X (X = BF_4, PF_6) with an equimolar amount of Na[BRH_3] (R = H, CN) resulted in the exclusive formation of the neutral species [$\{(\eta^5\text{-}C_5Me_5)Ru\}_3(\mu\text{-H})_3(\mu_3\text{-BR})$] (R = H, **53a**; R = CN, **53b**) (Scheme 22) [78].

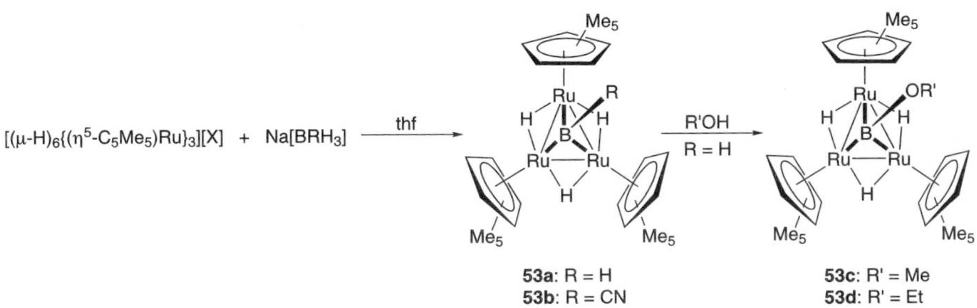

Scheme 22 Synthesis of ruthenium clusters featuring triply bridging borylene ligands

The ^{11}B-NMR resonances ($\delta = 131$, **53a**; 117 **53b**) are comparable to signals found for μ_2-borylene complexes (vide supra). Treatment of **53a** with methanol or ethanol afforded the corresponding triply bridging alkoxyborylene clusters $[\{(\eta^5\text{-}C_5Me_5)Ru\}_3(\mu\text{-}H)_3(\mu_3\text{-}BOR')]$ (R′ = Me, **53c**; R′ = Et, **53d**). The ^{11}B-NMR signals ($\delta = 78$, **53c**; 88 **53d**) are high field shifted with respect to the starting material due to π-donation from oxygen to boron, but the resonances are still at a much lower field than expected for compounds featuring a four-coordinated boron atom. The structure of **53d** was confirmed by the results of an X-ray diffraction analysis. The three ruthenium atoms form an equilateral triangle (with an average Ru–Ru distance of 267.7 pm), which is capped by an μ_3-borylene ligand. While the average B–Ru distance (215.4 pm) is in the usual range for B–Ru single bonds, the significantly short Ru–Ru distances reflect the unsaturated nature of the cluster.

4 Addendum

Since August 2006 the following relevant contributions have been published:

- Synthesis and Reactivity of Semi-bridging Borylene Complexes [79];
- Cationic Terminal Borylene Complexes: Interconversion of Amino and Alkoxy Borylenes by an Unprecedented Meerwein–Ponndorf Hydride Transfer [80];
- A single bonded cationic terminal borylene complex [81];
- BN-Analogues of Vinylidene Transition Metals Complexes: The Borylnitrene Isomer [82];
- Aminoborylene Complexes of Group 6 Elements and Iron: A Synthetic, Structural and Quantum Chemical Study [83];
- Synthesis and Structure of a Cationic Platinum Borylene Complex [84];
- Synthesis and Electronic Structure of a Ferroborirene [85];
- Stepwise Intermetal Borylene Transfer: Synthesis and Structure of Mono- and Dinuclear Cobalt Borylene Complexes [86];
- Synthesis and Characterization of Semi-Bridging Molybdenum Borylene Complexes [87];
- Synthesis and Electronic Structure of a Terminal Alkylborylene Complex [88];
- Synthesis and Structure of Bridged Haloborylene Complexes [89];
- Borylene Metathesis via [2+2]Cycloaddition [90];
- Cationic Terminal Aminoborylene Complexes: Controlled Stepwise Insertion into M=B and B=N Double Bonds [91];
- Insertion reactions of dicyclohexylcarbodiimide with amino-boranes, -boryls and -borylenes [92];

References

1. Timms PL (1967) J Am Chem Soc 89:1629
2. Timms PL (1973) Acc Chem Res 6:118
3. Pachaly B, West R (1984) Angew Chem Int Ed 23:454
4. Braunschweig H (1998) Angew Chem Int Ed 37:1786
5. Irvine GJ, Lesley MJG, Marder TB, Norman NC, Rice CR, Robins GR, Roper WR, Whittell GR, Wright LJ (1998) Chem Rev 98:2685
6. Braunschweig H, Colling M (2001) Coord Chem Rev 223:1
7. Aldridge S, Coombs DL (2004) Coord Chem Rev 248:535
8. Braunschweig H, Colling M (2000) J Organomet Chem 614/615:18
9. Braunschweig H, Colling M (2003) Eur J Inorg Chem 393
10. Braunschweig H, Rais D (2005) Heteroatom Chem 16:566
11. Braunschweig H (2004) Adv Organomet Chem 51:163
12. Ehlers AW, Baerends EJ, Bickelhaupt FM, Radius U (1998) Chem Eur J 4:210
13. Radius U, Bickelhaupt FM, Ehlers AW, Goldberg N, Hoffmann R (1998) Inorg Chem 37:1080
14. Macdonald CLB, Cowley AH (1999) J Am Chem Soc 121:12113
15. Boehme C, Uddin J, Frenking G (2000) Coord Chem Rev 197:249
16. Braunschweig H, Kollann C, Rais D (2006) Angew Chem Int Ed 45:5254
17. Braunschweig H, Kollann C, Englert U (1998) Angew Chem Int Ed 37:3179
18. Braunschweig H, Colling M, Kollann C, Stammler HG, Neumann B (2001) Angew Chem Int Ed 40:2298
19. Uddin J, Boehme C, Frenking G (2000) Organometallics 19:571
20. Chen Y, Frenking G (2001) J Chem Soc Dalton Trans 434
21. Cowley AH, Lomelí V, Voigt A (1998) J Am Chem Soc 120:6401
22. Greiwe P, Bethäuser A, Pritzkow H, Kühler T, Jutzi P, Siebert W (2000) Eur J Inorg Chem 9:1927
23. Braunschweig H, Herbst T, Rais D, Seeler F (2005) Angew Chem Int Ed 45:7461
24. Braunschweig H, Colling M, Hu C, Radacki K (2003) Angew Chem Int Ed 42:205
25. Erker G, Pfaff R, Krüger C, Werner S (1991) Organometallics 10:3559
26. Braunschweig H, Colling M, Kollann C, Merz K, Radacki K (2001) Angew Chem Int Ed 40:4198
27. Uddin J, Frenking G (2001) J Am Chem Soc 123:1683
28. Coombs DL, Aldridge S, Jones C, Willock DJ (2003) J Am Chem Soc 125:6356
29. Coombs DL, Aldridge S, Rossin A, Jones C, Willock DJ (2004) Organometallics 23:2911
30. Coombs DL, Aldridge S, Jones C (2002) J Chem Soc Dalton Trans 20:3851
31. Kays DL, Day JK, Ooi LL, Aldridge S (2005) Angew Chem Int Ed 44:7457
32. Aldridge S, Jones C, Gans-Eichler T, Stasch A, Kays DL, Coombs ND, Willock DJ (2006) Angew Chem Int Ed 45:6118
33. Braunschweig H, Radacki K, Seeler F, Whittell GR (2006) Organometallics 24:4605
34. Braunschweig H, Radacki K, Scheschkewitz D, Whittell GR (2005) Angew Chem Int Ed 44:1658
35. Braunschweig H, Whittell GR (2005) Chem Eur J 11:6128
36. Braunschweig H, Radacki K, Seeler F, Whittell GR (2004) Organometallics 23:4178
37. Irvine GF, Rickard CEF, Roper WR, Williamson A, Wright LJ (2000) Angew Chem Int Ed 39:948
38. Tilley TD (1991) In: Patai S, Rappoport Z (eds) The Silicon-Heteroatom Bond, chaps 9 and 10. Wiley, New York

39. Grumbine SK, Straus DA, Tilley TD, Rheingold AL (1995) Polyhedron 14:127
40. Rickard CEF, Roper WR, Williamson A, Wright LJ (2002) Organometallics 21:4862
41. Braunschweig H, Radacki K, Rais D, Scheschkewitz D (2005) Angew Chem Int Ed 44:5651
42. Schubert U (1984) Coord Chem Rev 55:261
43. Grumbine SK, Tilley TD (1993) J Am Chem Soc 115:7884
44. Braunschweig H, Radacki K, Rais D, Seeler F (2004) Organometallics 23:5545
45. Braunschweig H, Wagner T (1995) Angew Chem Int Ed 34:825
46. Braunschweig H, Ganter B (1997) J Organomet Chem 545:163
47. Braunschweig H, Koster M (1999) J Organomet Chem 588:231
48. Hermann WA, Reiter B, Biersack H (1975) J Organomet Chem 97:245
49. Creswick M, Bernal I, Hermann WA (1979) J Organomet Chem 172:C39
50. Hermann WA (1982) Adv Organomet Chem 20:159
51. Folting K, Huffman JC, Lewis LN, Caulton KG (1979) Inorg Chem 18:3483
52. Shimoi M, Ikubo S, Kawano Y (1998) J Am Chem Soc 120:4222
53. Elliot DJ, Levy CJ, Puddephatt RJ, Holah DG, Hughes AN, Magnuson VR, Moser IM (1990) Inorg Chem 29:5014
54. Feilong J, Fehlner TP, Rheingold AL (1988) Angew Chem Int Ed 27:424
55. Braunschweig H, Kollann C, Englert U (1998) Eur J Inorg Chem 465
56. Braunschweig H, Kollann C, Klinkhammer KW (1999) Eur J Inorg Chem 1523
57. Kays DL, Rossin A, Day JK, Ooi LL, Aldridge S (2006) Dalton 399
58. Aldrigde S, Coombs DL, Jones C (2002) Chem Commun 856
59. Pannell K, Wu CC, Long GJ (1980) J Organomet Chem 193:359
60. Braunschweig H, Müller M (1997) Chem Ber 130:1295
61. Braunschweig H, Colling M, Hu C (2003) Inorg Chem 42:941
62. Braunschweig H, Colling M, Hu C, Radacki K (2004) Inorg Chim Acta 357:1822
63. Braunschweig H, Burschka C, Burzler M, Metz S, Radacki K (2006) Angew Chem Int Ed 45:4352
64. Braunschweig H, Colling M, Hu C, Radacki K (2002) Angew Chem Int Ed 39:1415
65. Braunschweig H, Radacki K, Rais D, Seeler F, Uttinger K (2005) J Am Chem Soc 127:1386
66. Braunstein P, Durand J, Morise X, Tiripicchio A, Ugozzoli F (2000) Organometallics 19:444
67. Onozawa SY, Hatanaka Y, Sakakura T, Shimada S, Tanaka M (1996) Organometallics 15:5450
68. Onozawa SY, Tanaka M (2001) Organometallics 20:2956
69. Braunschweig H, Radacki K, Rais D, Seeler F (2006) Angew Chem Int Ed 45:1066
70. Barr RD, Marder TB, Orpen AG, Williams ID (1984) J Chem Soc Chem Commun 112
71. Werner H (1990) Angew Chem Int Ed 29:1077
72. Braunschweig H, Rais D, Uttinger K (2005) Angew Chem Int Ed 44:3763
73. Crabtree RH, Lavin M (1986) Inorg Chem 25:805
74. Braunschweig H, Forster M, Radacki K (2006) Angew Chem 45:2132
75. Heaton BT, Jacob C, Sampanthar JT (1998) J Chem Soc Dalton Trans 1403
76. Feilong J, Fehlner TP, Rheingold AL (1988) Angew Chem Int Ed 27:424
77. Lei X, Shang M, Fehlner TP (1998) Inorg Chem 37:3900
78. Okamura R, Tada K, Matsubara K, Oshima M, Suzuki H (2001) Organometallics 20:4772
79. Braunschweig H, Radacki K, Rais D, Uttinger K (2006) Organometallics 25:5159
80. Kays DL, Day JK, Aldridge S, Harrington RW, Clegg W (2006) Angew Chem Int Ed 45:3513

81. Vidovic D, Findlater M, Reeske G, Cowley AH (2006) Chem Commun 3786
82. Bettinger HF (2007) Inorg Chem 46:5188
83. Blank B, Braunschweig H, Colling-Hendelkens M, Kollann C, Radacki K, Rais D, Uttinger K, Whittell GR (2007) Chem Eur J 13:4770
84. Braunschweig H, Radacki K, Uttinger K (2007) Angew Chem Int Ed 46:3979
85. Braunschweig H, Frenking G, Radacki K, Seeler F, Fernández I (2007) Angew Chem Int Ed 46:5215
86. Braunschweig H, Forster M, Radacki K, Seeler F, Whittell GR (2007) Angew Chem Int Ed 46:5212
87. Braunschweig H, Radacki K, Uttinger K (2007) Eur J Inorg Chem 27:4350
88. Braunschweig H, Burzler M, Kupfer T, Radacki K, Seeler F (2007) Angew Chem Int Ed 46:7785
89. Bissinger P, Braunschweig H, Seeler F (2007) Organometallics 26:4700
90. Braunschweig H, Burzler M, Radacki K, Seeler F (2007) Angew Chem Int Ed 46:8071
91. Pierce GA, Aldridge S, Jones C, Gans-Eichler T, Stasch A, Coombs ND, Willock DJ (2007) Angew Chem Int Ed 46:2043
92. Pierce GA, Coombs ND, Willock DJ, Day JK, Stasch A, Aldridge S (2007) Dalton Trans 4405

Struct Bond (2008) 130: 29–122
DOI 10.1007/430_2007_079

Published online: 7 February 2008

Transition Metal Boryl Complexes

Deborah L. Kays[1] · Simon Aldridge[2] (✉)

[1]School of Chemistry, University of Nottingham, University Park, Nottingham NG7 2RD, UK

[2]Inorganic Chemistry, South Parks Road, Oxford OX1 3QR, UK
simon.aldridge@chem.ox.ac.uk

Abstract The article represents a review of the chemistry of transition metal complexes containing the boryl ($-BX_2$) ligand based on the chemical literature published up to and including the calendar year 2006. A brief compendium of papers published in the year 2007 is also included. Notwithstanding the ever-growing volume of literature detailing useful organic transformations in which these complexes have been implicated (e.g. C–H activation, hydroboration, diboration), the primary focus of this article is a review of issues of structure and bonding in boryl systems. Thus, the evidence for boryl ligands acting – to a greater or lesser extent – as sigma donors and/or pi acceptors is examined at length, as are issues relating to secondary interactions (e.g with hydride co-ligands) within the metal coordination sphere, and the consequent relationship to borane sigma complexes.

Keywords Bonding studies · Boron · Boryl ligand · Transition metal

Abbreviations

Ar^f	$C_6H_3(CF_3)_2$-3,5
9-BBN	9-borabicyclo[3.3.1]nonane
^{n}Bu	$CH_2CH_2CH_2CH_3$
^{t}Bu	$C(CH_3)_3$
3-^{t}Bucat	1,2-$O_2C_6H_3{}^t$Bu-3
4-^{t}Bucat	1,2-$O_2C_6H_3{}^t$Bu-4
3,5-^{t}Bu$_2$cat	1,2-$O_2C_6H_2{}^t$Bu$_2$-3,5
Bz	CH_2Ph
cat	Catecholato, 1,2-$O_2C_6H_4$
catCl$_4$	1,2-$O_2C_6Cl_4$

cod	1,4-cyclooctadiene
coe	Cyclooctene
Cp	Cyclopentadienyl, C_5H_5
Cp*	Pentamethylcyclopentadienyl, C_5Me_5
Cp′	Methylcyclopentadienyl, C_5H_4Me
Cy	Cyclohexyl
Cym	Cymantrenyl
depe	1,2-bis(diethylphosphino)ethane
DFT	Density Functional Theory
dme	1,2-dimethoxyethane
dmpe	1,2-bis(dimethylphosphino)ethane
dppb	1,4-bis(diphenylphosphino)butane
dppe	1,2-bis(diphenylphosphino)ethane
dppm	Bis(diphenylphosphino)methane
Et	CH_2CH_3
Fc	Ferrocenyl
Ind	Indenyl
IPr	1,3-bis(2,6-diisopropylphenyl)imidazole-2-ylidene
Me	CH_3
3-Mecat	1,2-$O_2C_6H_3$Me-3
4-Mecat	1,2-$O_2C_6H_3$Me-4
3,5-Me_2cat	1,2-$O_2C_6H_2Me_2$-3,5
Mes	Mesityl, 2,4,6-trimethylphenyl
4-Methiocat	1,2-$S_2C_6H_3$Me-4
NBO	Natural bond order
ped	$OCH_2CH(Ph)O$
4-pic	4-picoline
pin	Pinacolato, $OCMe_2CMe_2O$
Ph	Phenyl
^{i}Pr	$CH(CH_3)_2$
tart	Dimethyl tartrate
tbbpy	4,4′-ditertbutyl-2,2′-bipyridine
Tf	Triflate, CF_3SO_2
Thiocat	1,2-$S_2C_6H_4$
tmg	Trimethylene glycolato, $OCH_2CH_2CH_2O$
Tp*	Tris(3,5-dimethylpyrazolyl)hydroborate
Trip	2,4,6-triisopropylphenyl
triphos	$MeC(CH_2PPh_2)_3$
Xy	Xylyl

1 Introduction

Transition-metal complexes containing the boryl ($-BX_2$) ligand first appeared in the chemical literature in the 1960s in an extensive series of papers reported by Nöth and Schmid [1]. Subsequent developments (principally in the past 15 years) have been driven by the implication of boryl complexes in metal-catalyzed processes leading to the transfer of the BX_2 fragment to organic substrates, and aided by the more ready availability of crystallographic and quantum chemical techniques to clarify issues of structure and bonding. Thus, significant developments in the chemistry of the boryl ligand in the early 1990s stemmed from the involvement of group 9 metal complexes in metal-catalyzed hydroboration [2–5], and more recent work has centred on the involvement of boryl systems in the catalysis of diboration and C–H functionalization processes [6–10]. Such chemistry has been, at least in part, the motivation for studies aimed at a better understanding of the steric and electronic properties of the boryl ligand and how these influence the fundamental reaction processes characteristic of boryl complexes. Thus, a considerable body of experimental and computational data has been amassed attesting to the strong σ donor properties of the BX_2 ligand and to a *trans* influence equivalent to (or slightly greater than) that of the hydride ligand [11]. Ground-state structural and spectroscopic phenomena (e.g. bond lengths and coupling constants involving groups *trans* to the boryl ligand) have typically been evaluated in this regard. In a similar vein, structural parameters such as the M–B bond length (relative to the sum of the respective covalent radii), conformational preferences of the boryl ligand, and spectroscopic handles such as the stretching frequencies of ancillary carbonyl ligands, have been employed to quantify the potential for the boryl ligand to act as a π acceptor utilizing the formally vacant π symmetry orbital at boron (Fig. 1) [11]. In this regard a "competitive π-bonding" model akin to that proposed for metal carbene complexes has been advanced for BX_2 ligands [12]. Strong π donor boryl substituents X reduce the degree of M$\rightarrow$B π back-bonding in effect by elevating in energy the unoccupied boryl acceptor orbital (which is B–X π^* in character). Thus, even in the most favourable cases, π bonding is not thought to account for more than 20% of the overall M–B bonding density [13].

Investigation of these fundamental properties, along with a number of more specific issues, such as the potential for interaction of the Lewis acidic boryl ligand with other ligands in the metal coordination sphere (e.g. hydrides) [14], and the coordination geometry preference for five-coordinate boryl complexes of varying electron count [15], has been aided by the syntheses of families of closely related boryl complexes. Thus, a systematic appraisal of electronic and steric influences as a function of boryl ligand substituent (X) and the metal/ligand framework, and comparison with related ligand systems has been made possible for complexes of the types $(\eta^5\text{-}C_5R_5)ML_n(BX_2)$ (M = Mo, W, $n = 3$; M = Fe, Ru, $n = 2$), *cis*-$(R_3P)_2Pt(BX_2)_2$, *mer*-$(R_3P)_3Ir(Y)(Z)(BX_2)$, and for the valence isoelectronic series $(R_3P)_2Rh(Y)(Z)(BX_2)/(R_3P)_2(L)Os(Y)(BX_2)$. Such families of compounds have typically been synthesized by exploiting either oxidative addition (of B–B, B–H or B–halogen bonds) or salt elimination approaches in the M–B bond-forming step.

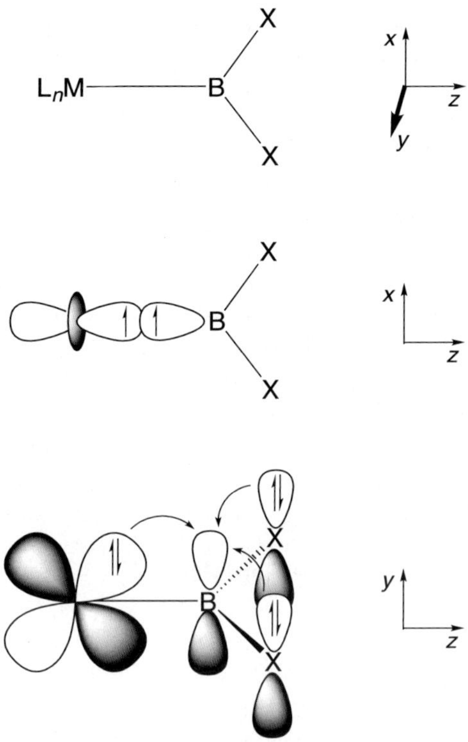

Fig. 1 Principal orbital interactions for transition metal boryl complexes – σ donor and π acceptor properties of the boryl ligand

In this article emphasis has been placed on a review of crystallographic, spectroscopic and quantum chemical studies which shed light on the fundamental structure and bonding characteristics of the boryl ligand. The chemistry of transition-metal boryl complexes was reviewed extensively in 1998 [11, 16–18], and a number of further review articles have been written in the interim [19–24]. Herein, we focus primarily on literature reports which have been published in the period 1998–2006, with due reference made to key systems reported in the pre-1998 period. The material includes three-

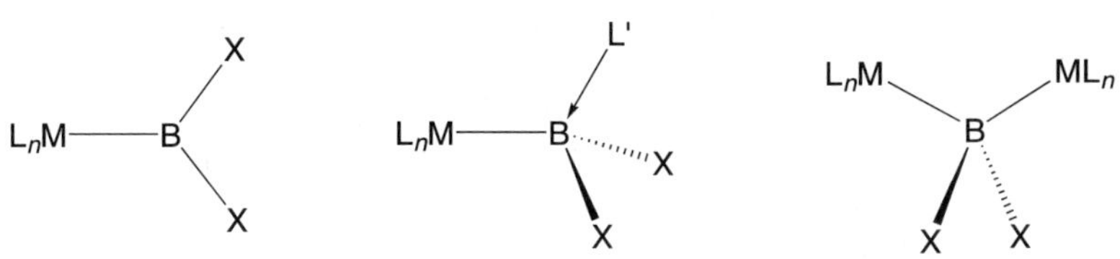

Fig. 2 Three-coordinate, base-stabilized four-coordinate and bridging modes of coordination of the boryl ligand

coordinate, base-stabilized four-coordinate and bridging (or semi-bridging) boryl ligand systems (Fig. 2) and has been organized by transition metal group.

2
Group 4 (Titanium, Zirconium and Hafnium)

To date no structurally authenticated boryl complexes of the group 4 metals have been reported in the literature, although the involvement of group 4 reagents in metal-catalyzed hydroboration has been the subject of considerable research, and a number of titanium σ-borane complexes have been characterized [18, 25–29]. Recent work has led to the synthesis of group 4 complexes containing related diaminogallyl ligands [30].

3
Group 5 (Vanadium, Niobium and Tantalum)

Although group 5 organometallic systems have been found to be of relevance in transition-metal catalyzed hydroboration reactions, structurally authenticated group 5 boryl complexes remain relatively few in number. Smith and co-workers, for example, have probed the mechanisms for the formation of niobium and tantalum mono- and bis(boryls) from propylene complex precursors, with concomitant formation of propyl boronate esters [31, 32]. Of particular interest from a structural viewpoint are the relative merits of alternative bonding descriptions for metal(V) boryl bis(hydrides) as borohydride complexes or as mono(hydride) σ-borane systems [31–34].

Early studies of niobium and tantalum complexes resulted in the syntheses of the complexes *endo*-$Cp_2NbH_2BX_2$ [BX_2 = Bcat, **5.1**; BX_2 = 9-BBN, **5.2**], and *endo*- and *exo*-Cp_2TaH_2Bcat (**5.3** and **5.4**, respectively), utilizing either anionic metal bis(hydride) or charge neutral tris(hydride) precursors [33, 34]. In the case of **5.1** and **5.2** a combination of crystallographic and ^{11}B NMR methods have been used to judge the relative merits of bonding descriptions featuring contributions from the three structural extremes **A–C** (Fig. 3) [33, 34]. Crystallographically, the Nb–B [2.292(5) and 2.40(1) Å], B–H [1.62(5), 1.69(4) and 1.38(7), 1.39(8) Å] and Nb–H distances [1.58(5), 1.62(4) and 1.80(7), 1.84(8) Å] and H–Nb–H angles [92(2) and 70(3)°] for the two compounds are consistent with Nb(V) boryl bis(hydride) and Nb(III) borohydride descriptions for **5.1** and **5.2**, respectively. These structural differences can be rationalized, at least in part, in terms of the differing degrees of interaction expected between the ancillary hydride ligands and the BX_2 fragments; this in turn is influenced by the relative Lewis acidities of the Bcat and 9-BBN moieties. Consistent with this, the ^{11}B chemical shift meas-

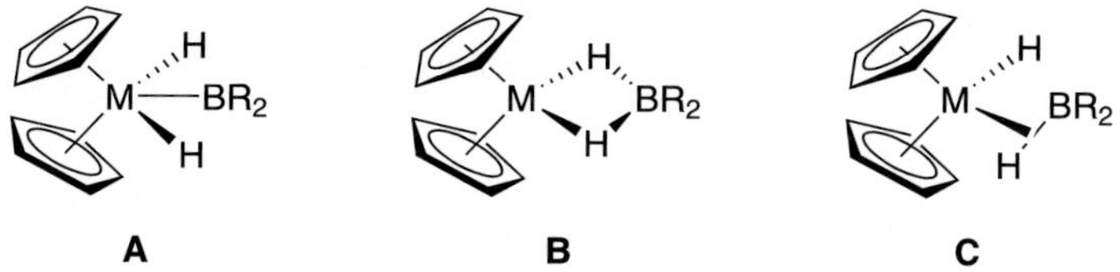

Fig. 3 Limiting models in the continuum of bonding in group 5 metallocene boryl complexes: d^0 boryl (**A**), d^2 borohydride (**B**) and d^2 σ borane (**C**)

ured for **5.1** in solution (δ_B 59) is as expected for a Bcat complex [11], while that for **5.2** (δ_B 57) is at much higher field than that reported for the Ir(III) boryl fac-$(Me_3P)_3Ir(H)_2$(9-BBN) (**9.27**, δ_B 106.2) [35], reflecting a significant contribution from bonding model **B** which features a four-coordinate boron ligand system. Interestingly, ^{1}H and ^{2}H NMR studies on **5.1** in solution are consistent with a rapid equilibrium between the two structures **A** and **B** [33].

A series of structural and mechanistic studies have also been performed on pentamethylcyclopentadienyl niobium and tantalum complexes [31, 32]. Thus, *endo*-$Cp^*_2MH_2BX_2$ (M = Nb, X_2 = cat, **5.5**; M = Nb, X_2 = 4-^{t}Bucat, **5.6**; M = Nb, X_2 = 3-^{t}Bucat, **5.7**; M = Ta, X_2 = cat, **5.8**; M = Ta, X_2 = 4-^{t}Bucat, **5.9**) were prepared for comparison of reactivity with the analogous Cp compounds **5.1**, **5.3** and **5.4**. Complex **5.7** has been described as a Nb(III) borohydride or as an HB(3-^{t}Bucat) adduct of Cp^*_2NbH, on the basis of a significantly longer Nb–B distance than that found in its Cp-substituted counterpart **5.1** [2.348(4) Å for **5.7**; 2.292(5) Å for **5.1**] [31]. The structures of **5.5**–**5.9** could not be as-

Table 1 Selected structural parameters and ^{11}B NMR chemical shifts for niobium and tantalum boryl complexes

Compound	*d*(M – B) (Å)	δ^{11}B	Refs.
endo-Cp_2NbH_2Bcat (**5.1**)	2.292(5)	59.2	[33]
endo-$Cp_2NbH_2BC_8H_{14}$ (**5.2**)	2.40(1)	57.0	[33]
endo-Cp_2TaH_2Bcat (**5.3**)	2.263(6)	70.0	[34]
exo-Cp_2TaH_2Bcat (**5.4**)	2.295(11)	64.7	[34]
endo-$Cp^*_2NbH_2$Bcat (**5.5**)	–	60.2	[31]
endo-$Cp^*_2NbH_2$[B(4-^{t}Bucat)] (**5.6**)	–	60.0	[31]
endo-$Cp^*_2NbH_2$[B(3-^{t}Bucat)] (**5.7**)	2.348(4)	60.9	[31]
endo-$Cp^*_2TaH_2$Bcat (**5.8**)	–	73.5	[31]
endo-$Cp^*_2TaH_2$[B(4-^{t}Bucat)] (**5.9**)	–	72.7	[31]
$Cp_2NbH(Bcat)_2$[a] (**5.10**)	2.29(1)	60, 65	[32]
$Cp_2NbH(\eta^2$-HBcat) (**5.11**)	–	59	[32]

[a] As the acetone solvate

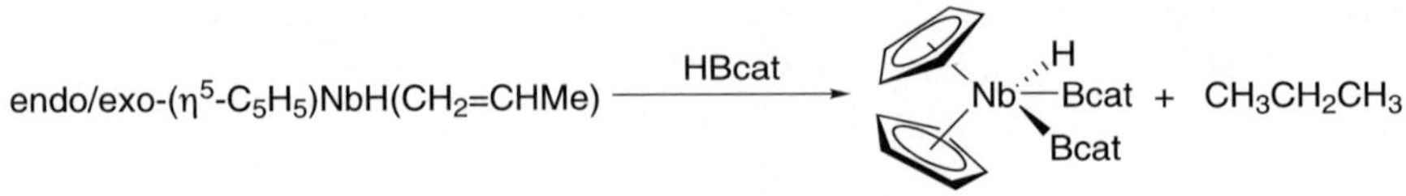

Scheme 1 Synthesis of diboryl complex $Cp_2NbH(Bcat)_2$ (**5.10**)

signed unambiguously on the basis of spectroscopic studies alone, although isotopic substitution revealed that tantalum complex **5.9** is probably best described as a d^0 boryl system. Reactivity studies have shown that **5.9** reacts very slowly compared to its niobium counterpart (**5.6**), a finding consistent with a description of **5.9** as a Ta(V) boryl which possesses a higher barrier to reductive elimination of HB(4-^{t}Bucat) than the corresponding niobium system.

Further study of the mechanisms involved in the reactions of alkene complexes $(\eta^5\text{-}C_5R_5)_2MH(CH_2{=}CHMe)$ (R = H, Me; M = Nb, Ta) have revealed different reaction pathways operating for the Cp and Cp* complexes [32]. Although the products for these reactions are analogous, the B–C bond forming pathways for Cp_2M and Cp_2^*M systems are different; B–H activation proceeds through an alkylidene intermediate for Cp_2Ta, but through alkyl intermediates for its permethylated analogue. Diboryl complex $Cp_2NbH(Bcat)_2$ (**5.10**) was formed (along with propane) in the reaction of *endo*- and *exo*-$Cp_2NbH(CH_2{=}CHMe)$ with HBcat (Scheme 1), suggesting that B–H activation in this system can lead to intermediates with the boryl ligand in the exo position. Interestingly, this is not found for the permethylated analogue, presumably due to the differing steric requirements of the Cp and Cp* ancillary ligands. The Nb(III) complex **5.11** $[Cp_2NbH(\eta^2\text{-}HBcat)]$ has been identified as an intermediate in this reaction.

4
Group 6 (Chromium, Molybdenum and Tungsten)

Although at present there are no structurally characterized boryl complexes of chromium [23, 24], there has been significant interest in systems featuring the heavier group 6 congeners, molybdenum and tungsten, due to their implication in the functionalization of hydrocarbons. Indeed, tungsten boryl complexes featuring catecholate substituents were among the first to be investigated in the stoichiometric borylation of alkanes, showing good selectivity for the terminal C–H bonds [36, 37]. A number of crystallographic and computational studies have been reported which shed light on issues of structure and bonding in these boryl systems, together with the mechanism of this highly unusual C–H functionalization reactivity [37–39].

Table 2 Selected spectroscopic and structural parameters for molybdenum and tungsten boryl complexes

Compound	d(M – B) (Å)	$\delta^{11}B$	ν(CO) (cm^{-1})	Refs.
$Cp^*Mo(CO)_3B(3,5\text{-}Me_2cat)$ (**6.1**)	–	55.4	2009, 1920	[37]
$Cp^*Mo(CO)_2(PXy_3)B(3,5\text{-}Me_2cat)$ (**6.2**)	–	58	1916, 1834	[37]
$CpMo(CO)_3B(NMe_2)B(NMe_2)Cl$ (**6.3**)	–	65.4, 38.0	1982, 1900, 1887	[41]
$CpMo(CO)_3B(NMe_2)B(NMe_2)Br$ (**6.4**)	–	65.6, 38.2	1985, 1901, 1887	[40]
$CpMo(CO)_3B(NMe_2)B(NMe_2)I$ (**6.5**)	–	66.0, 35.5	1992, 1916, 1895	[42, 43]
$Cp'Mo(CO)_3B(NMe_2)B(NMe_2)Br$ (**6.6**)	–	65.9, 38.2	1979, 1898, 1883	[41]
$CpMo(CO)_3B(NMe_2)B(NMe_2)OMe$ (**6.7**)	–	67.1, 33.7	1981, 1898, 1882	[41]
$CpMo(CO)_3B(NMe_2)B(NMe_2)OEt$ (**6.8**)	–	67.3, 33.4	[a]	[41]
$CpMo(CO)_2(PEt_3)B(NMe_2)B(NMe_2)Br$ (**6.9**)	–	68.8, 40.7	1932, 1885, 1853, 1812	[41]
$CpMo(CO)_3B(NMe_2)B(NMe_2)OC{\equiv}Mo(CO)_2Cp$ (**6.10**)	2.348(4)	65.2, 31.5	[a]	[42, 43]
$Cp^*Mo(CO)_3(BH_2{\cdot}PMe_3)$ (**6.11**)	2.497(5)	–24.6	1942, 1843	[46]
$Cp^*Mo(CO)_3[BH_2{\cdot}P(OMe)MeNCH_2CH_2NMe]$ (**6.12**)	2.472(4)	–29.6	1966, 1880	[47]
$Tp^*(CO)_2Mo\{\eta^2\text{-}B(Et)CH_2(C_6H_4Me\text{-}4)\}$ (**6.13**)	–	76	1908, 1830	[48, 49]
$Cp^*W(CO)_3Bcat$ (**6.14**)	–	52.4	2010, 1931, 1914	[37]
$Cp^*W(CO)_3[B(3,5\text{-}Me_2cat)]$ (**6.15**)	–	53	2002, 1920, 1900	[37]
$Cp^*W(CO)_3Bpin$ (**6.16**)	–	50	1999, 1916, 1900	[37]
$Cp^*W(CO)_2(PXy_3)B(3,5\text{-}Me_2cat)$ (**6.17**)	–	58	1914, 1832	[37]
$Cp^*W(CO)_2(PMe_3)B(3,5\text{-}Me_2cat)$ (**6.18**)	–	58	1906, 1823	[37]
$CpW(CO)_3B(NMe_2)B(NMe_2)Cl$ (**6.19**)	2.370(6)	69.5, 39.0	1988, 1932	[45]
$CpW(CO)_3B(NMe_2)B(NMe_2)Br$ (**6.20**)	–	62.7, 40.0	1989, 1909, 1892	[41]
$CpW(CO)_3B(NMe_2)B(NMe_2)I$ (**6.21**)	–	63.3, 37.4	1992, 1899, 1883	[42, 43]
$Cp'W(CO)_3B(NMe_2)B(NMe_2)Br$ (**6.22**)	–	63.2, 40.0	1987, 1907, 1889	[41]
$CpW(CO)_3B(NC_4H_8)B(NC_4H_8)Br$ (**6.23**)	–	60.2, 39.6	1985, 1900, 1884	[41]
$CpW(CO)_2(PMe_3)B(NMe_2)B(NMe_2)Cl$ (**6.24**)	2.327(3)	66.0, 41.8	1927, 1803	[54]

Table 2 (continued)

Compound	d(M – B) (Å)	δ^{11}B	ν(CO) (cm^{-1})	Refs.
$CpW(CO)_2(PPh_3)B(NMe_2)B(NMe_2)Cl$ (**6.25**)	–	64.8, 37.6	1929, 1805	[54]
$CpW(CO)_3B(NMe_2)B(NMe_2)OC{\equiv}W(CO)_2Cp$ (**6.26**)	-	62.7, 32.6	[a]	[42, 43]
$Cp^*W(CO)_3(BH_2 \cdot PMe_3)$ (**6.27**)	2.476(7)	–27.6	1943, 1848	[46]
$Tp^*(CO)_2W\{\eta^2\text{-}B(Et)CH_2(C_6H_4Me\text{-}4)\}$ (**6.28**)	2.07(1)	78	1896, 1817	[48, 49]
$Tp^*(CO)_2W\{\eta^2\text{-}B(Et)CH_2(Me)\}$ (**6.29**)	2.064(9)	77	1893, 1814	[48, 49]
$Tp^*(CO)_2W\{\eta^2\text{-}B(Ph)CH_2(C_6H_4Me\text{-}4)\}$ (**6.30**)	2.058(13)	–	[a]	[49]
$Tp^*(CO)_2W\{\eta^2\text{-}B(Ph)CH_2(Me)$ (**6.31**)	2.070(4)	–	[a]	[12]
$Cp_2WH(Bcat)$ (**6.32**)	2.190(7)	57.2	–	[52, 53]
$Cp_2WCl(Bcat)$ (**6.33**)		49	–	[52, 53]
$Cp_2W(4\text{-}^tBucat)_2$ (**6.34**)	2.19(1) 2.23(1)	59.3	–	[52, 53]

[a] Not given

4.1 Molybdenum Complexes

4.1.1 Catecholboryl and Related Complexes

The catecholboryl complexes $Cp^*Mo(CO)_2(L)[B(3,5\text{-}Me_2Cat)]$ [L = CO, **6.1**; L = PXy_3 = **6.2**] (Fig. 4) were synthesized as part of a systematic investigation into the borylation of alkanes [37]. The ^{11}B NMR chemical shifts for these systems (δ_B 55 and 58 for **6.1** and **6.2**, respectively) and carbonyl stretching frequencies for **6.1** (2009, 1920 cm^{-1}) are typical for boryl complexes bearing oxygen substituents, with the lower stretching frequencies for **6.2** (1916, 1834 cm^{-1}) reflecting the replacement of a CO ligand with a phosphine.

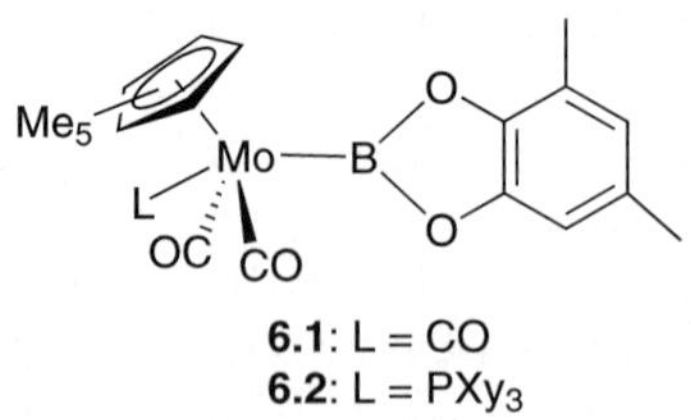

6.1: L = CO
6.2: L = PXy_3

Fig. 4 Molybdenum catecholboryl complexes $Cp^*Mo(CO)_2(L)[B(3,5\text{-}Me_2cat)$ [L = CO, **6.1**; L = PXy_3 = **6.2**]

4.1.2
Aminodiboran(4)yl Complexes

Molybdenum diborane(4)yl systems featuring amino substituents have been synthesized by the reaction $B_2(NMe_2)_2X_2$ (X = Cl, Br) with $K[(\eta^5\text{-}C_5H_4R)Mo(CO)_3]$; the complexes $(\eta^5\text{-}C_5H_4R)Mo(CO)_3B(NMe_2)B(NMe_2)X$ [R = H, X = Cl, **6.3**; R = H, X = Br, **6.4**; R = Me, X = Br, **6.6**] have been reported in which the B–B bond is retained from the diborane(4) starting material [40, 41]. The ^{11}B NMR chemical shifts for these complexes are typical for boryl systems containing amino substituents, with the boron atom directly bonded to the metal centre being downfield shifted in comparison to the β boron atom ($\delta_B = 65.4$, 38.0 for **6.3**; $\delta_B = 65.6$, 38.2 for **6.4**; $\delta_B = 65.9$, 38.2 for **6.6**). Compound **6.4** displays reactivity at the boron-bound halide and at the molybdenum centre which proceeds with retention of the metal–boron bond. Reaction of **6.4** with ROH yields $CpMo(CO)_3B(NMe_2)B(NMe_2)OR$ [R = Me, **6.7**; R = Et, **6.8**] and irradiation with PEt_3 gives rise to $CpMo(CO)_2(PEt_3)B(NMe_2)B(NMe_2)Br$ (**6.9**) (Scheme 2) [41].

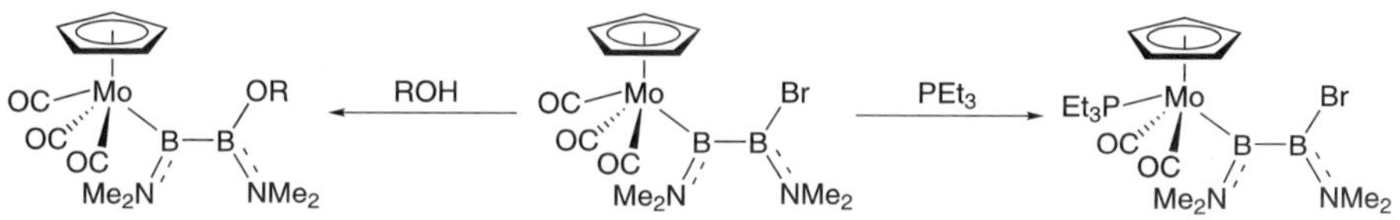

6.7: R = Me
6.8: R = Et
6.4
6.9

Scheme 2 Boron- and metal-centred substitution chemistry reported for the diborane(4)yl complex $CpMo(CO)_3B(NMe_2)B(NMe_2)Br$ (**6.4**)

The corresponding reactions with the iodo-diborane(4) reagent 1,2-$(Me_2N)_2I_2B_2$ are illustrative of one of the complications implicit in the salt elimination methodology, namely the existence of alternative sites of nucleophilicity within anionic organometallic reagents. Depending on conditions, the reactions of 1,2-$(Me_2N)_2I_2B_2$ with $K[(\eta^5\text{-}C_5H_5)M(CO)_3]$ (M = Mo, W) result in the formation of diboran(4)yl complexes (such as **6.5**), boryloxycarbyne complexes or mixed species (such as **6.10**) (Fig. 5) [42, 43]. Similar oxygen-centred reactivity has also been observed for nickel and triiron anionic nucleophiles [44]. That the molybdenum/tungsten boryl linkage represents the thermodynamically more stable configuration can be demonstrated by the conversion of the bis(boryloxycarbyne) systems to the mixed boryl/boryloxycarbyne species at elevated temperature. Complete conversion to the bis(boryl) dimer $CpM(CO)_3B(NMe_2)B(NMe_2)M(CO)_3Cp$ is thought to be prevented on steric grounds [42, 43]. The structure of $CpMo(CO)_3B(NMe_2)B(NMe_2)OC{\equiv}Mo(CO)_2Cp$ (**6.10**) has been determined crystallographically,

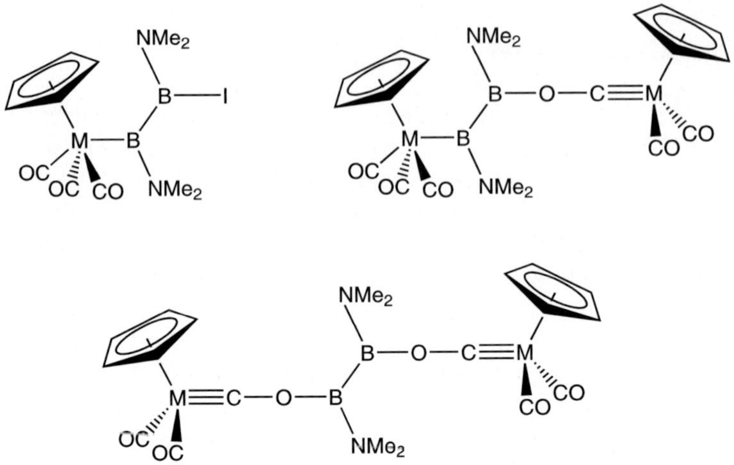

Fig. 5 Diboran(4)yl and boryloxycarbyne complexes derived from the reaction of $K[(\eta^5\text{-}C_5H_5)M(CO)_3]$ (M = Mo, W) with 1,2-$(Me_2N)_2I_2B_2$

thereby revealing a Mo–B bond length [2.348(4) Å] which is very long compared to the expected sum of the respective covalent radii (2.18 Å).[1]

4.1.3
Base-Stabilized Complexes

Crystallographically characterized phosphine- and phosphite-stabilized boryl complexes of molybdenum have been synthesized by the photochemical reaction of $Cp^*Mo(CO)_3Me$ with $BH_3{\cdot}PR_3$. The trimethylphosphine adduct $Cp^*Mo(CO)_3(BH_2{\cdot}PMe_3)$, (**6.11**, Fig. 6), was the first structurally authenticated molybdenum boryl complex to be reported, and features a four-coordinate boron centre which is characterized by a ^{11}B NMR resonance at δ_B –24.6 ppm [46]. As expected, the Mo–B distance of 2.497(5) Å is even longer than that reported for **6.10** [2.348(4) Å] which features a three-coordinate boryl ligand. This is consistent with a wholly σ bond in **6.11**, as expected for a boryl ligand lacking a vacant p orbital available for π back bonding. This observation is corroborated by the low carbonyl stretching frequencies (1942, 1843 cm^{-1}) measured for this compound (c.f. 2009, 1920 cm^{-1} for **6.1**). The structure of the related phosphite-stabilized boryl complex $Cp^*Mo(CO)_3[BH_2{\cdot}P(OMe)MeNCH_2CH_2NMe]$ (**6.12**, Fig. 6) has recently been reported [47]. The presence of the four coordinate boron centre, implied in solution by a ^{11}B resonance of –29.6 ppm, is confirmed in the solid state for **6.12** by X-ray crystallography. The Mo–B distance of 2.472(4) Å is slightly

[1] The figure of 2.18 Å has been determined from the value of 2.18 Å given for the related tungsten diboran(4)yl complex $CpW(CO)_3B(NMe_2)B(NMe_2)Cl$ in [45], together with the fact that the M–M distances in $Cp(OC)_3M-M(CO)_3Cp$ (M = Mo, W) are identical at 3.22 Å.

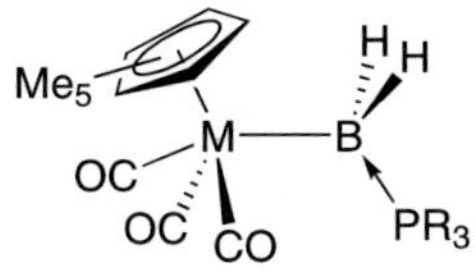

6.11: M = Mo, R_3 = Me_3
6.12: M = Mo, R_3 = OMe(MeNCH_2CH_2NMe)
6.27: M = W, R_3 = Me_3

Fig. 6 Base-stabilized boryl complexes of molybdenum and tungsten

shorter than that for **6.11**, but is still as expected for a complex containing a non-π acceptor boryl ligand. A range of fundamental modes of reactivity have been demonstrated for **6.12**, including B–H, B–Mo and B–P bond-cleavage chemistries. Reaction with MeI generates CpMo$(CO)_3$Me and $BH_2I{\cdot}P(OMe)MeNCH_2CH_2NMe$, thus providing further chemical evidence for marked $M^{\delta-}$–$B^{\delta+}$ polarization in compounds of this type.

4.1.4
Miscellaneous Boryl Complexes

The boryl system Tp*$(CO)_2$Mo$\{\eta^2$-B(Et)$CH_2(C_6H_4Me$-4)$\}$, (**6.13**, Fig. 7), has been synthesized by the reaction of the hydroborating agent "Et_2BH" with the Fischer carbyne complex Tp*$(CO)_2$Mo≡C(C_6H_4Me-4) [48, 49]; a number of related tungsten complexes have been synthesized which are discussed below (vide infra).

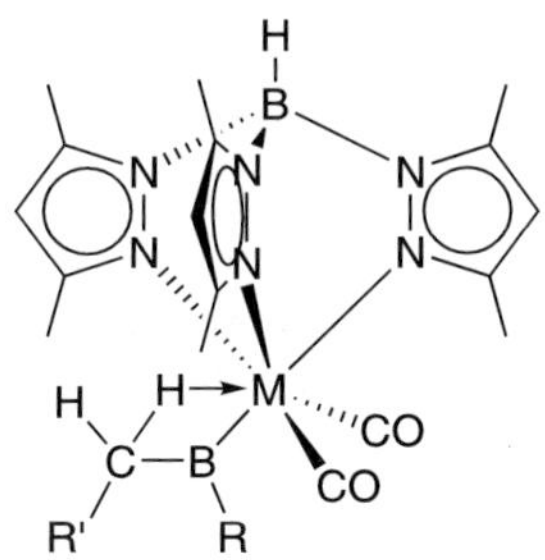

6.13: M = Mo, R = Et, R' = C_6H_4Me-4
6.28: M = W, R = Et, R' = C_6H_4Me-4
6.29: R = Et, R' = Me
6.30: R = Ph, R' = C_6H_4Me-4
6.31: R = Ph, R' = Me

Fig. 7 Molybdenum and tungsten boryl complexes of the type Tp*$(CO)_2$W$\{\eta^2$-B(R)CH_2R$'\}$

4.2 Tungsten Complexes

4.2.1 Catecholboryl and Related Complexes

As with molybdenum, a series of half-sandwich tungsten catecholboryl complexes have been investigated for their activity as hydrocarbon borylation reagents. The compounds $Cp^{*}W(CO)_2(L)BX_2$ (L = CO, X_2 = cat, **6.14**; L = CO, X_2 = 3-Mecat, **6.15**; L = CO, X_2 = pin, **6.16**; L = PXy_3, X_2 = 3-Mecat, **6.17**; L = PMe_3, X_2 = 3-Mecat, **6.18**, Fig. 8) are characterized by ^{11}B NMR chemical shifts (δ_B52–58) typical of boryl systems featuring aryloxy substituents [11, 37]. The photochemical functionalization of alkanes by tungsten boryl complex **6.15** was found to show remarkably high selectivity for the terminal position of *n*-pentane, generating the 1-pentylboronate ester in 85% yield. The importance of the blocking methyl groups in **6.15** (preventing attack at the sp^2 hybridized C–H bonds of the catechol and cyclopentadienyl substituents) was highlighted by the low yields obtained in the corresponding reaction with **6.14** (20% yield of the ester). Pinacolboryl complex **6.16** was also found to display high selectivity toward alkanes, producing the 1-pentylboronate ester in 72% yield. Experiments on **6.15**, **6.17** and **6.18** show that the borylation reaction proceeds via the same $Cp^{*}W(CO)_2[B(3\text{-Mecat})]$ intermediate. That phosphine-substituted boryls **6.17** and **6.18** produced lower yields of the boronate ester (59% and 32%, respectively) than **6.15** was rationalized by the parallel generation of $Cp^{*}W(CO)(PR_3)[B(3\text{-Mecat})]$ as an intermediate via CO dissociation [37].

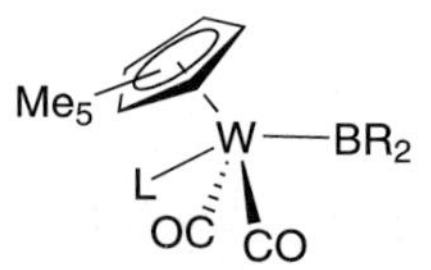

6.14: L = CO, R_2 = cat
6.15: L = CO, R_2 = 3,5-Me_2cat
6.16: L = CO, R_2 = pin
6.17: L = PXy_3, R_2 = 3,5-Me_2cat
6.18: L = PMe_3, R_2 = 3,5-Me_2cat

Fig. 8 Tungsten boryl complexes featuring oxygen substituents, **6.14–6.18**

The mechanism of stoichiometric CH borylation by complexes of this type has been the subject of several recent studies designed, in particular, to probe the underlying factors responsible for the unique ability of boryl systems to bring about alkane functionalization [37–39, 50, 51]. The *thermodynamics* of B–C bond formation have previously been noted to be favourable [50, 51], and the electrophilic properties of the boryl ligand are thought to offer favourable *kinetics* for generation of the B–C bond (due to the coupling of electrophilic and nucleophilic ligand fragments) [38]. Mechanistically, a radical mechanism is thought to be unlikely on the basis of the strong W–B bond. On the contrary, the initial formation of a 16-electron intermediate by loss of CO in the absence of a trapping agent, is thought to precede either sequential C–H oxidative addition/B–C reductive elimination or a σ bond metathesis step [37, 38]. Subsequent work has provided quantum chemical evidence for a σ bond metathesis mechanism, assisted by initial CH activation to form a tungsten alkyl species containing an ancillary σ borane ligand, via transfer of hydrogen from carbon to boron. The presence of the formally vacant boron *p* orbital is thought to assist in the CH bond-breaking process. The reaction then proceeds via rotation of the σ borane ligand such as to place the boron and alkyl carbon components in a mutually *cis* configuration prior to B–C reductive elimination [38]. An alternative mechanistic scheme, involving a two-step oxidative addition/reductive elimination process from the initially formed $CpW(CO)_2BX_2$ model intermediate has also been proposed [39]. This involves a further intermediate incorporating a formal tungsten(IV) centre and distinct boryl hydride and alkyl/aryl ligands to account for the formation of CH functionalization products in model benzene and methane systems [39].

Tungsten catecholboryl complexes based on a bent metallocene ligand framework have also been reported by Hartwig [52, 53]. Thus $Cp_2WX(Bcat)$ (X = H, **6.32**; X = Cl, **6.33**) and $Cp_2W(4\text{-}^tBucat)_2$ (**6.34**) have been synthesized and the structures of **6.32** and **6.34** in the solid state determined by X-ray diffraction. Boryl hydride complex **6.32** features no residual B···H interaction (as judged by a B···H separation of >2 Å) and features a sterically unfavoured boryl ligand environment with the catechol substituents pointing towards the Cp rings. This observation, together with a W–B bond length [2.191(6) Å] which is as expected on the basis of known W–C(sp^2) bond lengths (and the difference in the covalent radii of boron and carbon) led the authors to propose a weak W→B π back-bonding interaction. That such an interaction is relatively weak is attested to by the lack of observable line broadening in either the ^1H or ^{13}C NMR spectrum of the asymmetrically substituted 4-methylcatecholboryl derivative at −80 °C [53]. Both the alignments of the catecholboryl ligands and the W–B bond lengths [2.19(1), 2.23(1) Å] in **6.34** are very similar to those measured for **6.32**. Interestingly, the B–W–X angles for both compounds are 78(2)°, a value which is close to that calculated for an idealized d^2 Cp_2ML_2 system, and which therefore provides further evidence against any residual B···X interaction. That the angles for the two compounds are identical attests to the two-dimensional nature of the Bcat ligand, a feature which allows for relatively close "face on" approach, as observed in a number of platinum(II) bis(boryls) (vide infra) [53].

4.2.2 Aminoboryl and -Diboran(4)yl Complexes

The diborane(4)yl complexes (η^5-C_5H_4R)$W(CO)_3B(NR'_2)B(NR'_2)X$ [X = Br, R = H, NR'_2 = NMe_2, **6.20**; X = Br, R = Me, NR'_2 = NMe_2, **6.22**; X = Cl, R = H, NR'_2 = NC_4H_8, **6.23**] (Fig. 9) have been synthesized by the reaction of $B_2(NR'_2)_2X_2$ with the appropriate metal anion [41]. ^{11}B NMR resonances for both α and β boron atoms in **6.20, 6.22** and **6.23** are similar to those reported for the corresponding molybdenum complexes [41, 45], and similarly low carbonyl stretching frequencies have been taken as evidence for minimal W→B π back bonding. The related phosphine-substituted complexes $CpW(CO)_2(PR_3)B(NMe_2)B(NMe_2)Cl$ [R = Me, **6.24**; R = Ph, **6.25**] (Fig. 9) have been synthesized via the reaction of $B_2(NMe_2)_2Cl_2$ with $Li[CpW(CO)_2(PR_3)]$ [54]. The molecular structure of **6.24** shows that the phosphine and diborane(4)yl ligands occupy non-adjacent coordination sites in a four-legged piano stool geometry. Interestingly, the W–B distance [2.327(3) Å] is markedly shorter than that in $CpW(CO)_3B(NMe_2)B(NMe_2)Cl$ [2.370(8) Å], presumably resulting from increased M→B back-bonding from the more electron-rich phosphine-substituted tungsten centre [45]. The use of $B_2(NMe_2)_2I_2$ as the diborane(4) precursor leads to a range of diboran(4)yl and boryloxycarbyne products, as discussed above for the corresponding molybdenum systems [42, 43].

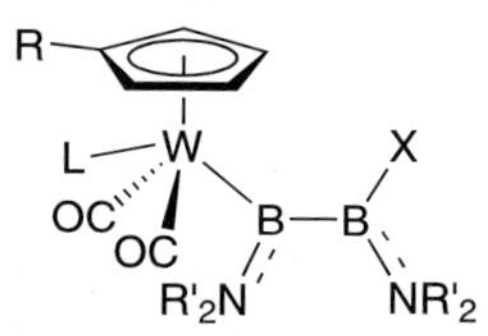

6.19: R = H, X = Cl, NR'_2 = NMe_2, L = CO
6.20: R = H, X = Br, NR'_2 = NMe_2, L = CO
6.21: R = H, X = I, NR'_2 = NMe_2, L = CO
6.22: R = Me, X = Br, NR'_2 = NMe_2, L = CO
6.23: R = H, X = Br, NR'_2 = NC_4H_8, L = CO
6.24: R = H, X = Cl, NR'_2 = NMe_2, L = PMe_3
6.25: R = H, X = Cl, NR'_2 = NMe_2, L = PPh_3

Fig. 9 Tungsten diborane(4)yl complexes **6.19–6.25**

4.2.3 Base-Stabilized Complexes

The base-stabilized compound $Cp^*W(CO)_3(BH_2{\cdot}PMe_3)$ (**6.27**, Fig. 6) contains a four-coordinate boron centre (δ_B −27.6) and features the longest W–B distance measured for a boryl complex [2.476(7) Å] [46]. The structural implication of a purely W–B σ bond is corroborated by the low carbonyl stretching frequencies (1943, 1848 cm^{-1}) for this compound. As with the related molybdenum complex **6.11**, the W–B bond in **6.27** has been described as being polarized in the sense $M^{\delta-}$–$B^{\delta+}$, this is reflected in the reactivities displayed by these Mo and W systems [46, 47].

4.2.3.1
Miscellaneous Boryl Complexes

The boryl systems $Tp^*(CO)_2W\{\eta^2\text{-}B(R)CH_2R'\}$ [R = Et, R′ = C_6H_4Me-4, **6.28**; R = Et, R′ = Me, **6.29**; R = Ph, R′ = C_6H_4Me-4, **6.30**; R = Ph, R′ = Me, **6.31**] (Fig. 7) have been synthesized by the reaction of the hydroborating agent "R_2BH" and Fischer carbyne complex $Tp^*(CO)_2W{\equiv}CR'$ [48, 49]. The W–B distances for **6.28–6.31** [2.064(9)–2.07(1) Å] are the shortest known for tungsten boryl complexes, reflecting not only the non π-donor alkyl substituents at boron, but also the presence of a supporting agostic interaction. ^{11}B NMR resonances for these systems and for related molybdenum complex **6.13** are in the order of 76–77 ppm, i.e. at somewhat higher field than other alkyl/aryl boryl complexes.

5
Group 7 (Manganese, Technetium and Rhenium)

Although boryl complexes of manganese and rhenium have been investigated in the activation of C–H bonds [55, 56], and related borylene systems containing Mn–B single bonds have been widely investigated [23, 24], structurally characterized examples of group 7 boryl systems are still comparatively rare, with only a handful of crystallographic studies having been reported. Of particular interest are half-sandwich manganese complexes containing both hydride and boryl ligands, for which limiting descriptions as manganese(III) boryl hydrides or as manganese(I) σ-borane complexes are conceivable (Fig. 10).

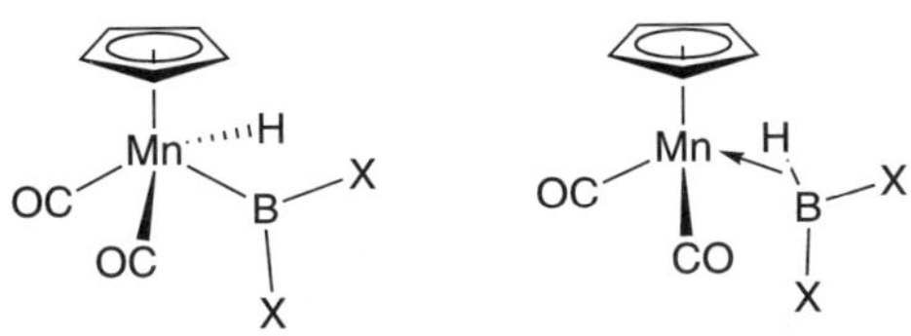

Fig. 10 Limiting descriptions of a half-sandwich manganese complex containing both hydride and boryl ligands as a boryl hydride or σ-borane complex

Table 3 Selected spectroscopic and structural parameters for manganese and rhenium boryl complexes

Compound	d(M – B) (Å)	δ^{11}B	ν(CO) (cm^{-1})	Refs.
$(OC)_5MnBcat$ (**7.1**)	2.108(6)	49	2112, 2009	[55, 56]
$(OC)_5MnBO_2C_6Cl_4$ (**7.2**)	–	50.6	2120, 2036	[58]
$(OC)_5MnBCl_2$ (**7.3**)	–	94.2	2115, 2052, 2012	[60]
$(OC)_5MnBBr_2$ (**7.4**)	–	92.9	2116, 2051, 2013	[60]
$\{(OC)_5MnBBr\}_2O$ (**7.5**)	2.093(2)	–	–	[60]
$Cp'Mn(CO)_2H[B\{Si(SiMe_3)_3\}Cl]$ (**7.6**)	2.138(16)	105.2	1978, 1913	[61]
$Cp'Mn(CO)_2H[B\{Ge(SiMe_3)_3\}Cl]$ (**7.7**)	–	105.82	1975, 1914	[61]
$Cp'Mn(CO)_2H(Bcat)$ (**7.8**)	2.083(2)	46	1995, 1937	[59]
$Cp'Mn(CO)_2H(Bpin)$ (**7.9**)	2.149(2)	45	1983, 1921	[59]
$Cp'Mn(CO)_2H(BCy_2)$ (**7.10**)	2.187(2)	104	1967 1901	[59]
$Cp'Mn(CO)_2H(BMe_2)$ (**7.11**)	–	101	1975, 1910	[59]
$(OC)_5MnB(Mes)Br$ (**7.12**)	–	119.4	2101, 2046, 2015, 1981, 1952	[62]
$(OC)_5MnB(C_6F_5)_2$ (**7.13**)	–	130.7	2137, 2050, 2015, 1950	[63]
$Cp'Mn(CO)_2(H)B(C_6F_5)_2$ (**7.14**)	–	84.2	1986, 1922	[63]
$(OC)_4Mn(PMe_2Ph)(BH_2{\cdot}PMe_3)$ (**7.15**)	2.314(2)	−29.4	2009, 1926, 1906, 1894	[65]
$(OC)_4Mn(PEt_3)(BH_2{\cdot}PMe_3)$ (**7.16**)	–	−29.6	2004, 1910, 1889	[65]
$(OC)_5ReBcat$ (**7.17**)	–	44	2129, 2051, 2016	[57]
$Cp^*Re(CO)_2H(Bpin)$ **7.19**	–	46	1981, 1924	[59]

5.1 Manganese Complexes

5.1.1 Catecholboryl and Related Complexes

Boryl complexes featuring the $[Mn(CO)_5]$ moiety were first reported as early as 1995, the manganese catecholboryl system $(OC)_5MnBcat$ (**7.1**, Fig. 11) having been synthesized by the reaction of ClBcat with $Na[Mn(CO)_5]$, and investigated in the functionalization of arenes and alkenes [55–57]. The Mn–B distance of 2.108(6) Å measured for **7.1** is not particularly short, when compared to the sum of the respective covalent radii (2.01 Å [11]). Moreover, the Mn–C distance for the carbonyl ligand *trans* to Bcat [1.841(6) Å] falls within the

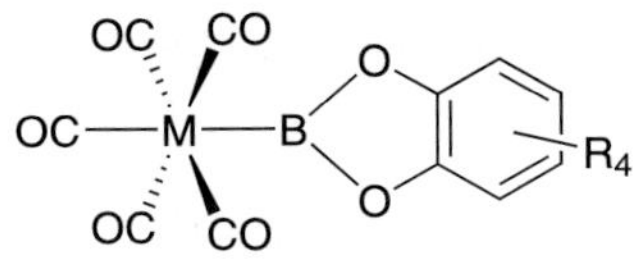

7.1: M = Mn, R = H
7.2: M = Mn, R = Cl
7.17: M = Re, R = H

Fig. 11 Group 7 boryl complexes featuring catecholate substituents

range of the other four Mn–C distances [1.832(8)–1.870(7) Å]. This situation contrasts with that found in related hydride and alkyl complexes containing the $[Mn(CO)_5]$ fragment, in which the *trans* Mn–C distance is typically *shorter* than the other four [55]. Such an observation is consistent with the boryl ligand having greater σ donor and/or π acceptor strength than hydride and alkyl ligands. The analogous halogenated complex, $(OC)_5MnBO_2C_6Cl_4$ (**7.2**), has been synthesized by a similar route, although no structural data are available [58].

Catecholboryl and related systems containing the half sandwich (η^5-C_5H_4Me)$Mn(CO)_2H$ fragment have also been reported [59]. Thus, (η^5-C_5H_4Me)$Mn(CO)_2H$(Bcat) (**7.8**) and (η^5-C_5H_4Me)$Mn(CO)_2H$(Bpin) (**7.9**) have been synthesized by two routes, namely photolytic substitution of CO by HBcat (or HBpin) and salt elimination from K[(η^5-C_5H_4Me)$Mn(CO)_2H$] and the corresponding chloroborane. Moreover, the borane can be displaced by competing two-electron donors such as diphenylacetylene or triphenylstannane, with kinetic data for the former reaction yielding an activation enthalpy for borane displacement of ca. 24 kcal mol^{-1}. This data, together with structural and spectroscopic parameters for **7.8, 7.9** and the related dicyclohexylboryl system (η^5-C_5H_4Me)$Mn(CO)_2H(BCy_2)$ (**7.10**) are consistent with a weakened but significant residual B···H interaction. Thus, the B–H distances [1.24(2)–1.31(2) Å] and H–Mn–B angles [33.2(7)–38.2(6) Å] are consistent with elongation of the B–H bond compared to the parent borane, but with the B–H ligand occupying a single coordination site within a three-legged piano-stool geometry. Consistent with this the B–H coupling constants (ca. 95 Hz) imply a direct one-bond interaction. Within the series of structurally characterized complexes **7.8–7.10**, both the IR-measured carbonyl stretching frequencies and borane ligand orientations reflect the fact that dialkylborane B–H bonds are better σ donors than those of dialkoxy/diaryloxyboranes. Thus, the stretching frequencies measured for **7.10** (1967, 1901 cm^{-1}) are significantly lower than those for **7.8** (1995, 1937 cm^{-1}), and the HBCat ligand in **7.8** adopts an alignment closer to classical η^2 binding than does the $HBCy_2$ ligand in **7.10**. Presumably the stronger σ donor B–H bond in $HBCy_2$ leads to the predominant interaction being through the hydridic H atom and therefore to an alignment closer to "end on" [59].

5.1.1.1 Haloboryl Complexes

The dihaloboryl complexes $(OC)_5MnBX_2$ (X = Cl, **7.3**; X = Br, **7.4**, Scheme 3), have been reported very recently [60], although the analogous iodide complex, $(OC)_5MnBI_2$, could not be isolated. ^{11}B NMR chemical shifts for these systems (δ_B 94.2 and 92.9 for **7.3** and **7.4**, respectively) reflect the presence of relatively poor π donor halide substituents on the boron centre, and these complexes give rise to carbonyl stretching frequencies similar to those for catecholate systems **7.1** and **7.2** (e.g. 2112, 2009 for **7.1**; 2115, 2052 and 2012 cm^{-1} for **7.3**; 2116, 2051 and 2013 cm^{-1} for **7.4**), indicating some degree of Mn–B π back bonding [55, 57, 58]. Although complexes **7.3** and **7.4** were not structurally authenticated, the crystal structure of the bis(boryl)oxide species $[(OC)_5MnBBr]_2O$ (**7.5**, Scheme 3) has been reported; **7.5** is presumably formed by partial hydrolysis of **7.2**. The Mn–B distance for **7.5** [2.093(2) Å] is similar to that for catecholboryl complex **7.1** [2.108(6) Å] [55, 57].

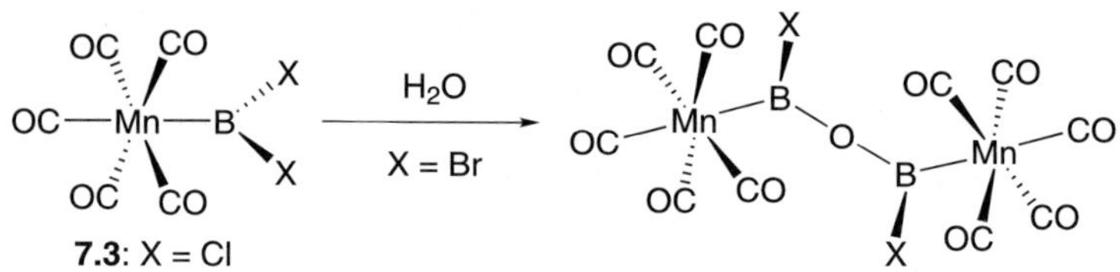

7.3: X = Cl
7.4: X = Br

7.5

Scheme 3 Haloboryl complexes 7.3–7.5

The silyl and germylboryl complexes $Cp'Mn(CO)_2H[B(ER_n)Cl]$ [ER_n = $Si(SiMe_3)_3$, **7.6**; ER_n = $Ge(SiMe_3)_3$, **7.7**, Fig. 12] containing pendant halide substituents have also been reported by Braunschweig and co-workers [61]. Here too, the lack of π-donating substituents is reflected in ^{11}B NMR chemical shifts (δ_B = 105.2 and 105.8 for **7.6** and **7.7**, respectively) which are shifted significantly downfield with respect to the catecholate system **7.8** (46 ppm). The fact that the boron centres in **7.6** and **7.7** are not as deshielded as the corresponding iron systems [δ_B 141.2 for $CpFe(CO)_2B\{Si(SiMe_3)_3\}Cl$ (**8.44**); δ_B 139.0 for $CpFe(CO)_2B\{Ge(SiMe_3)_3\}Cl$ (**8.45**)] can be ascribed to the presence of a Mn–H–B bridge present in the manganese complexes. Complex **7.6** displays a long Mn–B distance [2.138(16) Å], similar to those reported for **7.8**–**7.10**, and the relatively low carbonyl stretching frequencies (1978, 1913 cm^{-1}) are similar to those measured for $(\eta^5\text{-}C_5H_4Me)Mn(CO)_2H(BCy_2)$ (**7.10**; 1967, 1901 cm^{-1}) [59]. A detailed analysis of the bonding situation in **7.6** and further direct comparison with **7.8**–**7.10** is precluded by the difficulty in locating

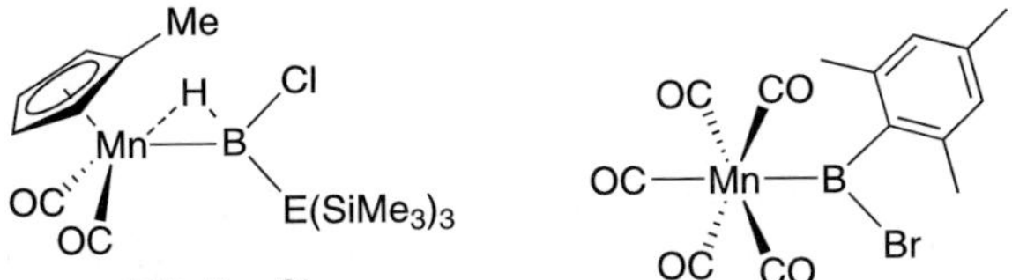

7.6: E = Si
7.7: E = Ge

7.12

Fig. 12 Manganese haloboryl complexes 7.6, 7.7 and 7.12

the hydride ligand from the diffraction data. Aryl(halo)boryl complexes of manganese have also been investigated, although these have proved to be more labile than their $CpFe(CO)_2$ counterparts, and no structural data are yet availablc. Mesityl(bromo)boryl complex $(OC)_5MnB(Mes)Br$ (**7.12**, Fig. 12) has been found to contain a very deshielded boron centre ($\delta_B = 119.4$), reflecting the presence of poorly π-donating boryl substituents [62].

5.1.2 Dialkyl- and Diarylboryl Complexes

The dialkylboryl systems (η^5-C_5H_4Me)$Mn(CO)_2H(BCy_2)$ (**7.10**) and (η^5-C_5H_4Me)$Mn(CO)_2H(BMe_2)$ (**7.11**) have been reported by Hartwig and Schlecht, and the former complex characterized in the solid state by X-ray diffraction (vide supra) [59]. The bis(pentafluorophenyl)boryl ligand has been employed in boryl complexes featuring both $Mn(CO)_5$ and $Cp'Mn(CO)_2H$ moieties. $(OC)_5MnB(C_6F_5)_2$ (**7.13**, Fig. 13) displays a low field ^{11}B NMR chemical shift (δ_B 130.7), characteristic of boryl complexes featuring weakly π donating substituents [63]. Carbonyl stretching frequencies for **7.13** (2137, 2050, 2015, 1950 cm^{-1}) are blue shifted in comparison to those for catecholate systems **7.1** and **7.2**, consistent with an enhanced degree of Mn–B π back bonding [55, 57], although no structural data are yet available for this complex. $Cp'Mn(CO)_2(H)B(C_6F_5)_2$ (**7.14**) is thermally labile and has been characterized spectroscopically. The ^{11}B NMR resonance for **7.14** (δ_B 84.2) is significantly upfield with respect to that of **7.13**, reflecting, at least in part the presence of a bridging hydride in **7.14**.

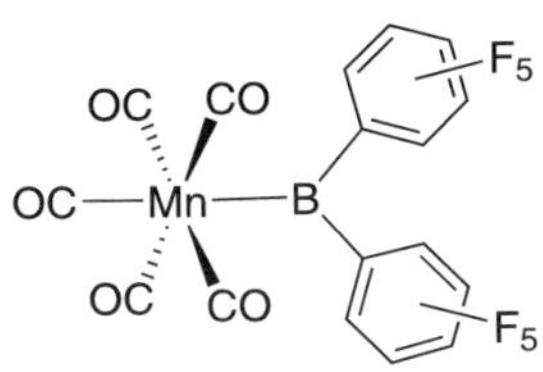

7.13

Fig. 13 Bis(pentafluorophenyl)boryl manganese complex 7.13

5.1.3 Base-Stabilized Complexes

As part of a wide-ranging study of the reactivity of base-stabilized boranes towards unsaturated metal fragments, Shimoi and co-workers have isolated several manganese σ borane systems. The complexes $CpMn(CO)_2(BH_3{\cdot}L)$ ($L = NMe_3$ **7.20**; $L = PMe_3$, **7.21**) were synthesized by the reaction of $CpMn(CO)_3$ with the appropriate borane under photolytic conditions [64]. An "end on" mode of coordination of the B–H bonds in each case is revealed by wide Mn–H–B angles [142(3) and 129(3)° for **7.20** and **7.21**, respectively] and relatively long Mn···B contacts [2.682(3) and 2.573(2) Å]. These distances can be compared to the corresponding Mn–B distance of 2.083(2) Å measured for the corresponding σ complex of the three-coordinate borane HBcat (i.e. **7.8**), and which features an essentially "side on" η^2 BH ligand [59]. In the case of the four-coordinate ligands $BH_3{\cdot}L$ these structural features are thought to be indicative of a borane complex featuring strong BH σ donation, but negligible back-bonding into the BH σ^* orbital [64].

In related work, the base-stabilized manganese boryl complexes $(OC)_4Mn(PR_3)(BH_2{\cdot}PMe_3)$ [$PR_3 = PMe_2Ph$, **7.15**; $PR_3 = PEt_3$, **7.16**] have been prepared by the reaction of $(OC)_4Mn(PR_3)$ with $BH_3{\cdot}PMe_3$ (Scheme 4) [65]. High-field ^{11}B NMR chemical shifts (δ_B –29.4 and –29.6 for **7.15** and **7.16**, respectively) are characteristic of the four-coordinate boron centres present in these systems. The Mn–B distance for **7.15** [2.314(2) Å] is significantly longer than that measured for three-coordinate boryl systems, Mn–B π interaction being precluded by the coordination of PMe_3. The solid-state structure of **7.15** shows that two of the *cis* carbonyl ligands are tilted significantly towards the boryl group, and as with other base-stabilized boryl systems the charge polarization in the complex is thought to resemble the contact ion pair $[(OC)_4Mn(PMe_2Ph)]^-[BH_2{\cdot}PMe_3]^+$. Interestingly, **7.15** can be protonated to give the cationic σ borane complex $[(OC)_4Mn(PR_3)(BH_3{\cdot}PMe_3)]^+[BAr^f{}_4]^-$ the DFT calculated structure of which features a very long Mn···B contact (2.780 Å) and an "end on" coordinated $BH_3{\cdot}PMe_3$ ligand reminiscent of that found in **7.21**.

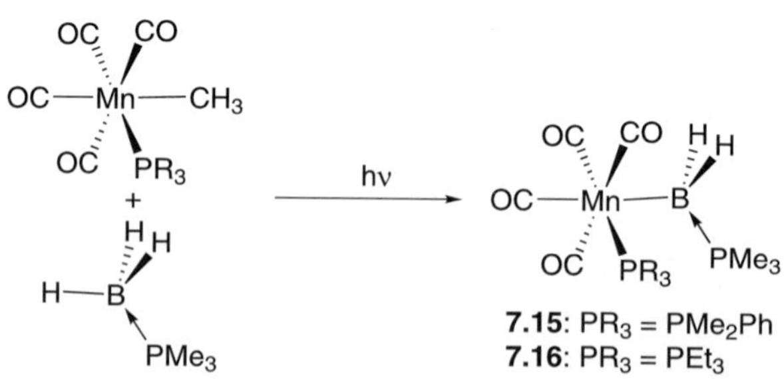

Scheme 4 Synthesis of base-stabilized manganese boryl complexes **7.15** and **7.16**

5.2
Rhenium Complexes

Although the catecholboryl complex $(OC)_5ReBcat$ (**7.17**, Fig. 11) has been implicated in the functionalization of hydrocarbons, rhenium boryl complexes are still relatively rare [55, 57, 59]. The ^{11}B NMR chemical shift and carbonyl stretching values for **7.17** [δ_B 44; $\nu(CO) = 2129, 2051, 2016\,cm^{-1}$] are similar to that of manganese analogue **7.1**. The half-sandwich complexes $Cp^*Re(CO)_2XBpin$ (X = H, **7.18**; X = Bpin, **7.19**) have also been investigated. The former is thought to contain a σ borane ligand, by comparison of spectroscopic properties with the related manganese complex **7.9** [59]. Bis(boryl) complex **7.19** has been isolated (as the *trans* isomer) from the reaction of $Cp^*Re(CO)_3$ with B_2pin_2 and has been shown to be a viable intermediate in the $Cp^*Re(CO)_3$ catalyzed photolytic conversion of pentane to 1-pentylBpin [56].

6
Group 8 (Iron, Ruthenium and Osmium)

There has been much interest in the synthesis of boryl complexes of the group 8 metals, with over 60 structurally authenticated examples having been reported to date. In particular, two families of compounds offer a systematic basis on which to study the fundamental properties of the boryl ligand (Fig. 14). A range of three-legged piano-stool complexes of the type $(\eta^5$-$C_5R_5)Fe(CO)_2BX_2$ has been reported, offering wide variation in the nature of X. A number of spectroscopic and structural features of these systems (viz. carbonyl stretching frequencies, Fe–B bond lengths and the torsion angle between the centroid-Fe–B and BX_2 planes) have been used as probes of electronic structure, focussing in particular on the extent to which the BX_2 ligand can act as a π acid [22]. Likewise, a series of complexes of the type $(R_3P)_2(L)Os(Y)(BX_2)$ has also been reported in which there is considerable variation in the nature of the BX_2 ligand [12].

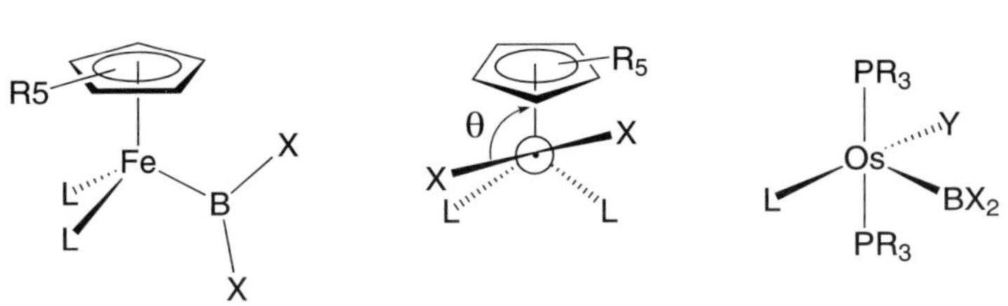

Fig. 14 Widely investigated half-sandwich iron and five-coordinate osmium boryl systems (showing definition of the torsion angle θ for the former)

6.1
Iron Complexes

The strongly nucleophilic properties of the cyclopentadienyliron anions $[(\eta^5\text{-}C_5R_5)Fe(CO)_2]^-$ ($R_5 = H_5$, Me_5, H_4Me) has allowed access to boryl complexes bearing many different substituents via salt elimination chemistry, thereby allowing systematic study of the effects of variation of substituent on the bonding in these systems. In addition to this, the catecholboryl complexes $(\eta^5\text{-}C_5R_5)Fe(CO)_2Bcat$ were among the first systems to be implicated in the stoichiometric borylation of carbon–hydrogen bonds [55, 57].

6.1.1
Catecholboryl and Other Oxygen-Containing Complexes

The structure of the catecholboryl compound $CpFe(CO)_2Bcat$ (**8.1**) was reported as early as 1993 by Hartwig [66], and further examined in 1999, along with that of $Cp^*Fe(CO)_2Bcat$ (**8.2**, Scheme 5) [57]. The larger torsion angle θ (Fig. 14) for **8.3** compared to **8.1** [7.9° and 26.7° for **8.1** and **8.3**, respectively] has been ascribed to the greater steric bulk of the Cp* ligand which forces the catecholate substituent out of the Cp centroid-Fe–B plane. The attendant lengthening of the Fe–B bond for **8.3** [1.959(6) Å and 1.980(2) Å for **8.1** and **8.3**, respectively] is consistent not only with this increase in steric congestion, but also with a decrease in the Fe→B π interaction resulting from overlap of the HOMO of the $[(\eta^5\text{-}C_5R_5)Fe(CO)_2]^+$ fragment with the formally vacant p orbital at boron, which is maximized for a torsion angle of 0°(Fig. 15) [67]. DFT studies on these and related systems have shown that the orientational preference of the BX_2 ligand is typically very small [13]. Although the barrier to rotation about the Fe–B bond [calculated to be typically >5 kJ mol^{-1} for $CpFe(CO)_2BO_2C_2H_2$] is not strictly a measure of π bond strength, but rather a measure of the *change* in the π-bonding contribution as a function of torsion angle (Fig. 15), DFT calculations imply that the π component to the Fe–B bond in catecholboryl systems is typically of the order of 10% [13]. Replacement of both competing π acid

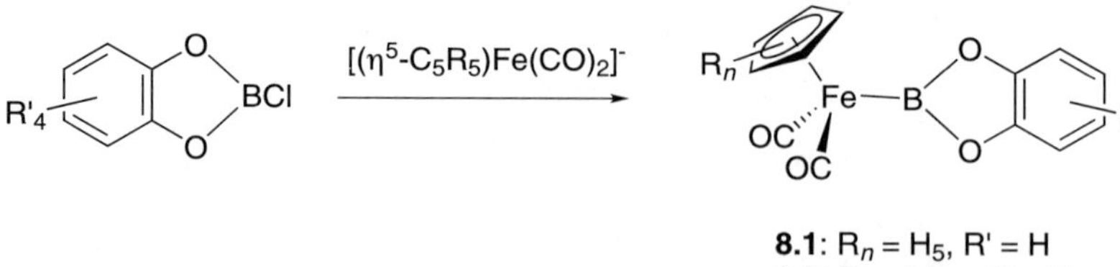

8.1: $R_n = H_5$, R' = H
8.3: $R_n = Me_5$, R' = H
8.8: $R_n = H_5$, R' = Cl
8.9: $R_n = H_4Me$, R' = Cl

Scheme 5 Synthesis of catecholboryl systems **8.1**, **8.3**, **8.8** and **8.9**

Table 4 Selected spectroscopic and structural parameters for iron boryl complexes

Compound	d(Fe – B) (Å)	Torsion angle (°)	δ^{11}B	ν(CO) (cm^{-1})	Refs.
$CpFe(CO)_2Bcat$ (**8.1**)	1.959(6)	7.9	51.8	2024, 1971	[57, 66]
$CpFe(CO)_2B(3,5\text{-}^tBu_2cat)$ (**8.2**)	–	–	51	2022, 1969	[37]
$Cp^*Fe(CO)_2Bcat$ (**8.3**)	1.980(2)	26.7	54.3	1996, 1940	[57]
$Cp^*Fe(CO)_2B(3,5\text{-}^tBu_2cat)$ (**8.4**)	–	–	53.7	2003, 1949	[37]
$Cp^*Fe(CO)_2B(3,5\text{-}Me_2cat)$ (**8.5**)	–	–	54	1996, 1939	[37]
$CpFe(CO)(PMe_3)Bcat$ (**8.6**)	–	–	57.5	1927	[57]
$CpFe(PMe_3)_2Bcat$ (**8.7**)	–	–	60	–	[57]
$CpFe(CO)_2BOC_6Cl_4$ (**8.8**)	–	–	55.2	2008, 1949	[58]
$Cp'Fe(CO)_2BOC_6Cl_4$ (**8.9**)	1.967(6)	26.6	53.4	2004, 1945	[58]
$CpFe(CO)_2BO_2C_6H_2O_2BFe(CO)_2Cp$ (**8.10**)	1.971(2)	82.2	48	2006, 1954	[68, 69]
$Cp'Fe(CO)_2BO_2C_6H_2O_2BFe(CO)_2Cp'$ (**8.11**)	1.973(2)	87.7	51	2001, 1943	[69]
$Cp^*Fe(CO)_2BO_2C_6H_2O_2BFe(CO)_2Cp^*$ (**8.12**)	–	–	53	2001, 1943	[69]
$CpFe(CO)_2BO_2C_5H_8O_2BFe(CO)_2Cp$ (**8.13**)	2.030(5)	43.5	45.3	1998, 1932	[69]
$Cp^*Fe(CO)_2Btmg$ (**8.14**)	2.024(4)	92.9	47.0	1971, 1910	[70]
$Fe(CO)_4(Bcat)_2$ (**8.15**)	–	–	45	2117, 2050, 2037, 2000	[73]
$Fe(CO)_4\{B(4\text{-}^tBucat)\}_2$ (**8.16**)	2.208(7)	–	45	2113, 2044, 2029, 2001	[73]
$Fe(CO)_4\{B(3,5\text{-}^tBu_2cat)\}_2$ (**8.17**)	–	–	45	2111, 2047, 2031, 1998	[73]
$Li[Fe(CO)_4(Bcat)]$ (**8.18**)	–	–	55	2016, 1907, 1821	[73]
$Li[Fe(CO)_4\{B(4\text{-}^tBucat)\}]$ (**8.19**)	–	–	54.8	2020, 1931, 1878, 1819	[73]

Table 4 (continued)

Compound	d(Fe – B) (Å)	Torsion angle (°)	δ^{11}B	ν(CO) (cm^{-1})	Refs.
Li[$Fe(CO)_4${B(3,5-$^t Bu_2$cat)}] (**8.20**)	–	–	54.3	2021, 1945, 1873, 1823	[73]
CpFe$(CO)_2$B(NMe_2)Cl (**8.21**)	–	–	56.4	2006, 1946	[79]
Cp′Fe$(CO)_2$B(NMe_2)Cl (**8.22**)	–	–	56.9	2002, 1944	[79]
Cp*Fe$(CO)_2$B(NMe_2)Cl (**8.23**)	2.027(5)	87.4	59.1	1988, 1928	[79]
CpFe$(CO)_2$B(NMe_2)Br (**8.24**)	–	–	54.5	b	[80]
Cp′Fe$(CO)_2$B(NMe_2)Br (**8.25**)	–	–	57.2	b	[80]
Cp*Fe$(CO)_2$B(NMe_2)Br (**8.26**)	–	–	57.2	1991, 1933	[80]
CpFe$(CO)_2$B(N$^i Pr_2$)Cl (**8.27**)	2.054(4)	83.7	55.4	2001, 1941	[75, 81]
CpFe$(CO)_2$B(NCy_2)Cl (**8.28**)	2.053(3)	84.1	56	2000, 1939	[83]
CpFe$(CO)_2$B(NBz_2)Cl (**8.29**)	2.036(3)	86.6	59	1996, 1933	[268]
CpFe$(CO)_2$B(NPh_2)Cl (**8.30**)	2.022(3)	82.2	60	2011, 1950	[268]
[CpFe$(CO)_2${B(N$^i Pr_2$)$(OPPh_3)$}]$^+$[BAr^f_4]$^-$ (**8.31**)	2.057(4)	72.5	48.9	2004, 1949	[75]
CpFe$(CO)_2$B(NMe_2)B(NMe_2)Cl (**8.32**)	2.090(3)		69.5[a]	1988, 1932 39.0	[45]
CpFe$(CO)_2$B(NC_4H_8)B(NC_4H_8)Cl (**8.33**)	–	–	66.5[a]	1980, 1922 38.4	[86]
CpFe$(CO)_2$B(NC_5H_{10})B(NC_5H_{10})Cl (**8.34**)	–	–	68.0[a]	1981, 1927 37.4	[86]
CpFe$(CO)_2BCl_2$ (**8.35**)	1.942(3)	77.7	90.7	2022, 1963, 2026, 1974	[58, 60, 74]
CpFe$(CO)_2$$(BCl_2$·4pic) (**8.36**)	2.1326(14)	177.3	18.8	1976, 1916	[74]
CpFe$(CO)_2BF_2$ (**8.37**)	–	–	47.1	2022, 1965	[60, 75]
CpFe$(CO)_2BBr_2$ (**8.38**)	–	–	85.9	2045, 2000	[60]
Cp*Fe$(CO)_2BF_2$ (**8.39**)	–	–	48.3	2002, 1946	[60, 75]

Table 4 (continued)

Compound	d(Fe – B) (Å)	Torsion angle (°)	$\delta^{11}B$	ν(CO) (cm^{-1})	Refs.
$Cp^*Fe(CO)_2BCl_2$ (**8.40**)	–	–	94.9	2006, 1955	[60]
$Cp^*Fe(CO)_2BBr_2$ (**8.41**)	–	–	91.0	2022, 1975	[60]
$Cp^*Fe(CO)_2(BCl_2{\cdot}4pic)$ (**8.42**)	2.129(3)	–	21.3	1958, 1899	[60]
$Cp^*Fe(CO)_2(BBr_2{\cdot}4pic)$ (**8.43**)	2.106(7)	–	14.3	1963, 1904	[60]
$CpFe(CO)_2B(Mes)Br$ (**8.44**)	1.964(5)	10.2	111.4	2016, 1962	[62, 88]
$Cp'Fe(CO)_2B(Mes)Br$ (**8.45**)	1.962(4)	3.8	111.3	2009, 1961	[62, 88]
$Cp^*Fe(CO)_2B(Mes)Br$ (**8.46**)	1.972(2)	2.3	113.2	2006, 1961	[62]
$Cp^*Fe(CO)_2B(Ph)Cl$ (**8.47**)	2.005(10)	77.6	111	1995, 1929	[90]
$CpFe(CO)_2B(2,6\text{-}Trip_2C_6H_3)Br$ (**8.48**)	–	–	114.9	2013, 1951	[89]
$CpFe(CO)_2B(Mes)SPh$ (**8.49**)	–	–	103.5	2000, 1935	[89]
$CpFe(CO)_2B(Mes)OC_6H_4^tBu\text{-}4$ (**8.50**)	2.040(2)	4.8	80.3	1997, 1933	[62, 89]
$CpFe(CO)_2B(Mes)O^tBu$ (**8.51**)	2.056(2)	2.8	72.9	1987, 1918	[89]
$Cp^*Fe(CO)_2B(Mes)F$ (**8.52**)	2.017(3)	37.6	90.4	1989, 1931	[82]
$Cp^*Fe(CO)_2B(Mes)Cl$ (**8.53**)	1.985(2)	179.5	112.1	1996, 1937	[82]
$Cp^*Fe(CO)_2B(Mes)I$ (**8.54**)	–	–	110.7	2005, 1955	[82]
$CpFe(CO)_2B(OMes)Cl$ (**8.55**)	1.977(4)	67.1	61.5	2002, 1940	[75]
$CpFe(CO)_2B(OMes)Br$ (**8.56**)	–	–	63.7	1971, 1925	[75]
$Cp^*Fe(CO)_2B(OMes)Cl$ (**8.57**)	1.977(4)	87.6	59.3	2015, 1961	[75]
$CpFe(CO)_2B(OMes)OC_6H_4^tBu\text{-}4$ (**8.58**)	–	–	47.4	2006, 1944	[75]
$CpFe(CO)_2B(OMes)SPh$ (**8.59**)	2.034(4)	57.0	69.1	1996, 1930	[75]
$CpFe(CO)_2B\{Si(SiMe_3)_3\}Cl$ (**8.60**)	1.964(8)		141.2	2009, 1956	[61]
$CpFe(CO)_2B\{Ge(SiMe_3)_3\}Cl$ (**8.61**)	1.985(11)		139.0	2009, 1956	[61]
$CpFe(CO)_2B\{Si(SiMe_3)_3\}F$ (**8.62**)	1.983(9)	83.5	113.2	2001, 1946	[61]
$CpFe(CO)_2B(Fc)Br$ (**8.63**)	1.997(2)	15.2	99.1	2015, 1955	[92]
$Cp^*Fe(CO)_2B(Fc)Br$ (**8.64**)	1.972(3), 1.985(3)	88.8, 89.0	103.0	1995, 1934	[92]

Table 4 (continued)

Compound	d(Fe – B) (Å)	Torsion angle (°)	δ^{11}B	ν(CO) (cm^{-1})	Refs.
$CpFe(CO)_2B(Cym)Br$ (**8.65**)	–	–	102.2	2045, 2024, 1999, 1940	[95]
$CpFe(CO)_2B(Cym)Br$ (**8.66**)	1.983(4)	88.7	106.1	2023, 1976, 1923	[95]
$CpFe(CO)_2B(C_6F_5)_2$ (**8.67**)	1.964(4), 1.965(5)	27.9, 28.2	121.5	2014, 1968	[63, 96]
$Cp^*Fe(CO)_2B(C_6F_5)_2$ (**8.68**)	1.968(5)	43.1	121.3	2004, 1954	[63]
$CpFe(CO)_2B(C_6H_5)_2$ (**8.69**)	2.034(3)	75	121	2021, 1951	[66]
$Cp^*Fe(CO)_2$(9-BBN) (**8.70**)	–	–	125.6	1981, 1927	[37]
$\{Cp'Fe(CO)_2\}_2(B_3N_3H_3Cl)$ (**8.71**)	2.041(1), 2.045(1)	90.6, 94.5	53[a]	1996, 1937	[87]
$Cp^*Fe(CO)_2(BH_2{\cdot}PMe_3)$ (**8.72**)	–	–	–25.1	1932, 1869	[98]
$(\eta^5$-$C_5Me_4Et)Fe(CO)_2(BH_2{\cdot}PMe_3)$ (**8.73**)	2.195(14)		–	–	[98]

[a] For the metal-bound boron atom

[b] Not given

Fig. 15 Potential π interactions between $[(C_5R_5)ML_2]$ and $[BX_2]$ fragments: **a** involving the boron-based π orbital and the a'' symmetry HOMO of the $[(C_5R_5)ML_2]$ fragment; **b** involving the boron-based π orbital and the HOMO-2 of the $[(C_5R_5)ML_2]$ fragment (of a' symmetry)

ancillary carbonyl ligands by model phosphines is calculated to increase the degree of Fe$\rightarrow$B backbonding to ca. 15%. However, although phosphine-substituted boryl complexes of the type CpFeL(L′)Bcat [L = CO, L′ = PMe_3, **8.6**; L = L′ = PMe_3, **8.7**] have been reported no structural data are as yet available [57].

Given the interest in iron boryl complexes bearing catecholate and related substituents, a number of related modified ligand systems have been reported. A series of iron boryl complexes containing catechol-type substituents and featuring methyl or tertiary butyl groups at the 3 and 5 positions have been synthesized as part of a thorough study into the mechanism of stoichiometric alkane borylation. Although half-sandwich iron complexes were found to be significantly less effective at CH borylation than their ruthenium or tungsten counterparts, these studies did reveal that steric blocking of CH bonds at the sp^2 carbon centres of both cyclopentadienyl and catechol substituents was required for alkane CH functionalization to occur [37]. The tetrachloro-catecholboryl complexes $(\eta^5\text{-}C_5H_4R)Fe(CO)_2BOC_6Cl_4$ (R = H = **8.8**; R = Me = **8.9**) have also been reported (Scheme 5) [58]. The Fe–B bond length measured for **8.9** [1.967(6) Å] is similar to that of **8.1**, suggesting a similarly weak π-bonding interaction, although perchlorination of the catechol substituent is found to lead to somewhat lower carbonyl stretching frequencies for **8.8** compared to **8.1** (2004, 1945 vs. 2024, 1971 cm^{-1}). The boryl ligand in the related methylcyclopentadienyl complex **8.9** is orientated at an angle of 26.6(5)° to the Cp centroid-Fe–B plane, the conformation adopted in this complex being found to be influenced by a weak C–H···O hydrogen bond between the methyl group on the Cp′ ligand and an oxygen atom of the boryl ligand [58]. That such a relatively weak interaction is able to determine the torsion angle about the Fe–B bond further testifies to the weak orientational influence of Fe$\rightarrow$B π bonding.

Bifunctional boryl systems based on a polyhydroxbenzene framework have also been studied; the structurally characterized symmetrically bridged complexes $(\eta^5\text{-}C_5R_4R')Fe(CO)_2BO_2C_6H_2O_2BFe(CO)_2(\eta^5\text{-}C_5R_4R')$ (R = R′ = H

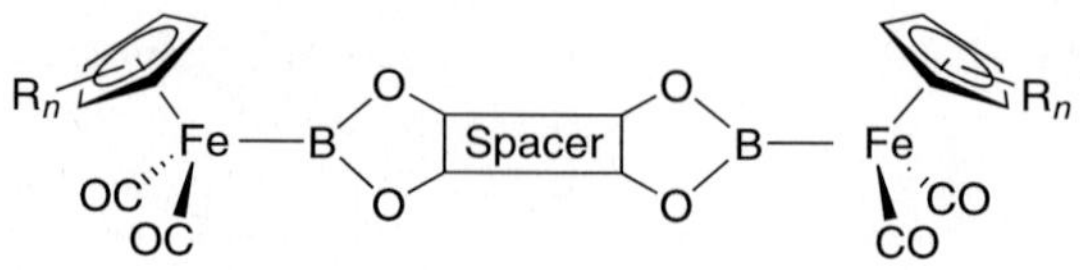

8.10: Spacer = C_6H_2, $R_n = H_5$
8.11: Spacer = C_6H_2, $R_n = H_4Me$
8.12: Spacer = C_6H_2, $R_n = Me_5$
8.13: Spacer = C_5H_8, $R_n = H_5$

Fig. 16 Bifunctional boryl systems based on a tetraalkoxyl ligand framework

8.10; R = H, R′ = Me **8.11**; R = R′ = Me **8.12**) can thus be compared with their monofunctional analogues (Fig. 16) [68, 69]. The bridged systems **8.10** and **8.11** are structurally very similar to each other, with essentially identical (short) Fe–B lengths [1.971(2) and 1.973(2) Å for **8.10** and **8.11**, respectively]; the orientation of the boryl ligands in these complexes is, however, very different from that found in compound **8.1**. The torsion angles between the Cp centroid-Fe–B and O–B–O planes for **8.10** and **8.11** are 82.2(1) and 87.7(7)°, respectively, implying that any π back bonding in these compounds involves not the HOMO of the $[(\eta^5\text{-}C_5H_4R)Fe(CO)_2]^+$ fragment, but the lower lying HOMO-2 (Fig. 15). π Back bonding of this type has been shown by Hoffmann and co-workers to be less efficient [67], a fact reflected in the lower carbonyl stretching frequencies measured for **8.10** and **8.11** compared to **8.1** (2006, 1954 for **8.10**, respectively c.f. 2024, 1971 cm^{-1} for **8.1**). The structural data for these bridged complexes (taken in comparison with those for the superficially very similar ligand systems **8.1**, **8.3** and **8.9**) provide further evidence for the weak orientational preference of the boryl ligand, with factors such as crystal packing forces being thought to offer sufficient energetic incentive to tip the balance between different ligand alignments [13].

A related dinuclear boryl complex containing a saturated bridging framework, $CpFe(CO)_2BO_2C_5H_8O_2BFe(CO)_2Cp$ (**8.18**), has also been reported (Fig. 16), which features a significantly longer Fe–B bond length than **8.1**, **8.10** and **8.11** [2.030(5) Å] [68, 69]. This lengthening is thought to be due to increased O→B π donation for **8.13** compared to aryloxyboryl systems, resulting in a less π acidic boron centre, which reduces the extent of π back donation from iron. This relative lack of π back bonding is supported by the low carbonyl stretching frequencies for **8.13** (1998, 1932 cm^{-1}). A non-bridged analogue of **8.13**, $Cp^*Fe(CO)_2Btmg$, **8.14**, has also been reported, for which the Fe–B length and carbonyl stretching frequencies [2.024(4) Å and 1971, 1910 cm^{-1}, respectively] provide further evidence of the weaker π acceptor properties of alkoxoboryl ligands compared with their aryloxo analogues [70]. Similar conclusions concerning the electronic properties of alkoxy and aryloxy boryl have been reached for related osmium systems (vide infra) [71, 72]. The bonding in aryloxy- and alkoxyboryl complexes **8.10** and **8.13** has been further probed by ^{57}Fe Mössbauer spectroscopy, a tool previously employed to investigate σ and π ligand properties in complexes of the type $CpFe(CO)_2X$. The low isomer shifts (0.02 mm s^{-1} and 0.00 mm s^{-1}, respectively) taken together with the relatively low carbonyl stretching frequencies provide further evidence for a ligand with strong σ donor and weak π acceptor properties [68, 69].

A number of catecholboryl complexes have been reported for non-Cp iron systems [73]. In particular complexes of the type *cis*-$Fe(CO)_4(BX_2)_2$ have been synthesized featuring catecholate and related boryl substituents via two routes: (i) a double salt elimination methodology utilizing $[Fe(CO)_4]^{2-}$ and chloroboranes as the iron and boron containing precursors; and (ii) oxidative addition of diborane(4) reagents such as B_2cat_2 to the photolytically generated $[Fe(CO)_4]$ fragment. The structure of $Fe(CO)_4\{B(4\text{-}^tBucat)\}_2$ (**8.16**) in the solid state has been elucidated by X-ray diffraction and reveals a distorted octahedral coordination geometry. The relatively acute B–Fe–B angle [82.8(4)°] mirrors that found in a range of related *cis*, bis(catecholboryl) systems and is thought to be due to the relatively flat, two-dimensional nature of such ligands. In addition the Fe–B distances [2.028(7) Å] are markedly longer than those found in half-sandwich iron Bcat systems (Table 4), presumably due to the presence of a less electron-rich metal centre featuring a wholly π-acid carbonyl ancillary ligand set [73]. In addition, the observation that the Fe–C distances *trans* to the boryl ligand (mean 1.82 Å) are *longer* than the *cis* Fe–C distances (mean 1.80 Å), contrasts with the observation of shorter *trans* Fe–C distances for related complexes containing hydride ligands. As such the stronger *trans* influence of the Bcat ligand over hydride is clearly demonstrated.

6.1.2 Haloboryl Complexes

(i) Dihaloboryl Complexes

The first dihaloboryl complex of iron, $CpFe(CO)_2BCl_2$ (**8.35**, Fig. 17) was reported in 2002, and although no structural characterization was obtained, further reaction with dilithiocatechol was exploited to demonstrate a second route to $CpFe(CO)_2Bcat$ (**8.1**), in a rare example of reaction at the boryl ligand centre with retention of the Fe–B bond [58]. The structure of **8.35** was subsequently reported, along with that of the Lewis acid/base adduct $CpFe(CO)_2BCl_2{\cdot}$(4-pic) (**8.36**, Fig. 17) in 2004 [60, 74]. Compound **8.35** displays a very short Fe–B distance [1.942(3) Å c.f. 2.045 Å for the sum of the respective covalent radii] [11], which with carbonyl stretching frequency values of 2022 and 1963 cm^{-1} indicates the presence of modest π back bonding. Furthermore, the torsion angle of 77.7° indicates that this π interaction involves the HOMO-2 orbital of the metal fragment [67]. Complexes **8.35** and **8.36** allow for the structural comparison of otherwise identical three-coordinate and base-stabilized four-coordinate boryl complexes; a significantly longer Fe–B distance [2.1326(14) Å] and red-shifted carbonyl stretching frequencies (1976, 1916 cm^{-1}) are measured for **8.36.** Furthermore, the extent of Fe–B bond lengthening (ca. 10%) was found to be greater than that from rehybridization at boron alone, leading to the conclusion that there must be significant π contribution to the Fe–B bond in **8.35.**

Recently, further dihaloboryl complexes (η^5-C_5R_5)Fe(CO)$_2$BX$_2$ (R = H, X = F, **8.37**; X = Br, **8.38**; R = Me, X = F, **8.39**; X = Cl, **8.40**; X = Br, **8.41**) have been reported, allowing for comparison of structure/bonding as a function of halide substituent (Fig. 17) [60, 75]. The corresponding iodo complexes (η^5-C_5R_5)Fe(CO)$_2$BI$_2$ (R = H, **8.74**; R = Me, **8.75**) have also been synthesized, but were reported to be too labile for complete characterization [60]. The carbonyl stretching frequencies for bromoboryl complexes **8.38** (2045, 2000 cm^{-1}) and **8.41** (2022, 1975 cm^{-1}) are at a much higher wavenumber than their corresponding difluoro- and dichloroboryl counterparts (2022, 1965 cm^{-1} for **8.37**; 2002, 1946 cm^{-1} for **8.39**; 2026, 1974 cm^{-1} for **8.35**; 2006, 1955 cm^{-1} for **8.40**), reflecting the decrease in π-donating ability of the halides in the order F > Cl > Br. The Lewis acid-base adducts Cp*Fe(CO)$_2$BX$_2$·(4-pic) (X = Cl, **8.42**; X = Br, **8.43**) (Fig. 17), display similar Fe–B distances to CpFe(CO)$_2$BCl$_2$ · (4-pic) (**8.36**) [2.129(3), 2.106(7) and 2.1326(14) Å for **8.42**, **8.43** and **8.36**, respectively], consistent with a uniform lack of Fe→B π backbonding. The dichloroboryl ligand has also been structurally characterized adopting a bridging mode of coordination between two transition-metal fragments. Reaction of **8.40** with Pd(PCy$_3$)$_2$ produces the dinuclear system Cp*Fe(μ-CO)$_2$(μ-BCl$_2$)Pd(PCy$_3$), which is discussed in Section 10.2 below [76]. In addition, complex **8.40** proves to be a valuable precursor for the synthesis of interesting systems featuring boron as a bridging atom between two metal centres [77, 78].

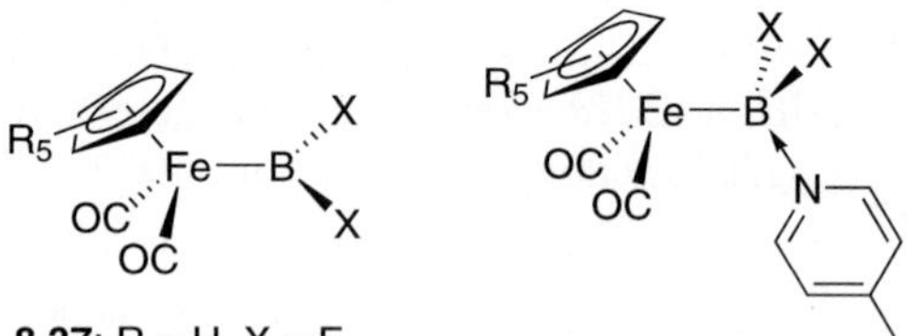

8.37: R = H, X = F
8.35: R = H, X = Cl
8.38: R = H, X = Br
8.39: R = Me, X = F
8.40: R = Me, X = Cl
8.41: R = Me, X = Br

8.36: R = H, X = Cl
8.42: R = Me, X = Cl
8.43: R = Me, X = Br

Fig. 17 Base-free and 4-picoline stabilized dihaloboryl systems **8.35–8.43**

(ii) Amino(halo)boryl Complexes

Amino(halo)boryl systems have been reported in the literature featuring a range of different amino substituents. The first such compounds were synthesized by Braunschweig and co-workers in 1998 (Fig. 18), and the crystal structure of $Cp^*Fe(CO)_2B(NMe_2)Cl$ (**8.23**) determined [79, 80]. The relatively long Fe–B length [2.027(5) Å] is consistent with little Fe–B π back bonding, the π acceptor properties of the boryl ligand being limited by competing B–N π bonding. These inferences are corroborated by carbonyl stretching frequencies [1988, 1928 and 1991, 1933 cm^{-1} for **8.23** and **8.35**] which are significantly lower than, for example, the corresponding dihaloboryl systems [i.e. 2006, 1955 and 2022, 1975 cm^{-1} for dichloro- and dibromoboryl complexes **8.40** and **8.41**, respectively]. Relatively minor substituent effects are observed as a function of the amino group. Thus, the diarylaminoboryl system $CpFe(CO)_2B(NPh_2)Cl$ displays a slightly shorter Fe–B bond [2.022(3) Å] and slightly higher carbonyl stretching frequencies [2011, 1950 cm^{-1}] than related dialkylaminoboryls [e.g. 2.054(4) Å and 2001, 1941 cm^{-1} for $CpFe(CO)_2B(N^iPr_2)Cl$] [81, 268].

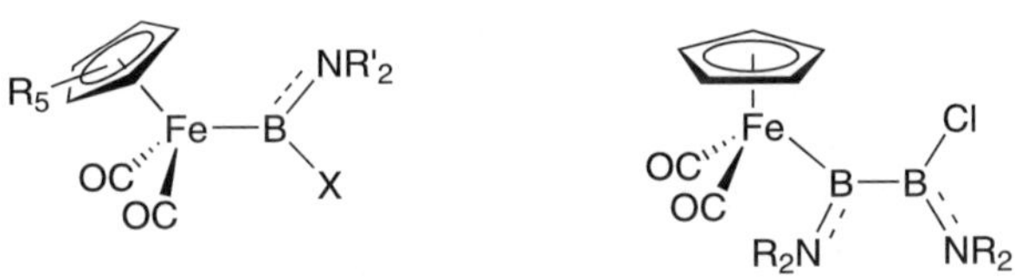

8.21: $R_5 = H_5$, R' = Me, X = Cl
8.22: $R_5 = H_4Me$, R' = Me, X = Cl
8.23: $R_5 = Me$, R' = Me, X = Cl
8.24: $R_5 = H_5$, R' = Me, X = Br
8.25: $R_5 = H_4Me$, R' = Me, X = Br
8.26: $R_5 = Me_5$, R' = Me, X = Br
8.27: $R_5 = H_5$, R' = ^{i}Pr, X = Cl
8.28: $R_5 = H_5$, R' = Cy, X = Cl
8.29: $R_5 = H_5$, R' = Bz, X = Cl
8.30: $R_5 = H_5$, R' = Ph, X = Cl

8.32: $R_2 = Me_2$
8.33: $R_2 = C_4H_8$
8.34: $R_2 = C_5H_{10}$

Fig. 18 Iron amino(halo)boryl and aminodiboran(4)yl systems

The stabilization derived from π interaction between the nitrogen and boron atoms in amino(halo)boryl complexes has made them amenable to the generation of cationic borylene systems via halide abstraction methodology [81, 82, 268]. The first such use of amino-containing compounds for this purpose was documented in 2004, the reaction of **8.26** with $Na[BAr^f_4]$ being shown to generate $[Cp^*Fe(CO)_2B(NMe_2)]^+[BAr^f_4]^-$. Because of the lack of steric shielding around the low-coordinate boron centre this dimethylaminoborylene complex could only be characterized spectroscopically at −20 °C and by its reaction with [PPN]Cl to generate the known amino(chloro)boryl complex **8.23** [82]. Subsequently amino(halo)boryl precursors featuring more sterically bulky *N*-substituents have been targeted, including $CpFe(CO)_2B(NR_2)Cl$ (R = ^{i}Pr, **8.27**; R = Cy, **8.28**; R = Bz, **8.29**; R = Ph, **8.30**) [81, 83, 268]. The increased steric bulk of the amino substituents in **8.27** and **8.28** has allowed the isolation of the corresponding cationic terminal borylene complexes $[CpFe(CO)_2B(NR_2)]^+[BAr^f{}_4]^-$ (R = ^{i}Pr, **8.76**; R = Cy, **8.77**) and $[Cp'Fe(CO)_2B(N^iPr_2)]^+[BAr^f{}_4]^-$ (**8.78**) as room temperature stable materials generated by halide abstraction. **8.77** and **8.78** represent the first structurally characterized cationic aminoborylene complexes, with FeB bond lengths [e.g. 1.859(6) Å for **8.77**] intermediate between the double bond found in $[Cp^*Fe(CO)_2BMes]^+[BAr^f{}_4]^-$ [1.792(8) Å] and the single bond found in $[CpFe(CO)_2BCp^*]^+[AlCl_4]^-$ [1.977(3) Å] [82–84] (Aldridge S, Pierce GA, unpublished results). Further reaction chemistry of **8.76** and **8.77** has been investigated, including examples which lead to the formation of boryl or base-stabilized borylene complexes [75, 81, 83, 85]. Reaction of **8.76** with Ph_3PO, for example, leads to the formation of the adduct $[CpFe(CO)_2\{B(N^iPr_2)(OPPh_3)]^+[BAr^f{}_4]^-$ (**8.31**) which has been described as an amino(oxy)boryl complex featuring a pendant cationic phosphorus centre [75, 81] on the basis of Fe–B and P–O distances [2.057(4) and 1.540(2) Å, respectively] consistent with formal Fe–B and P–O single bonds (Scheme 6).

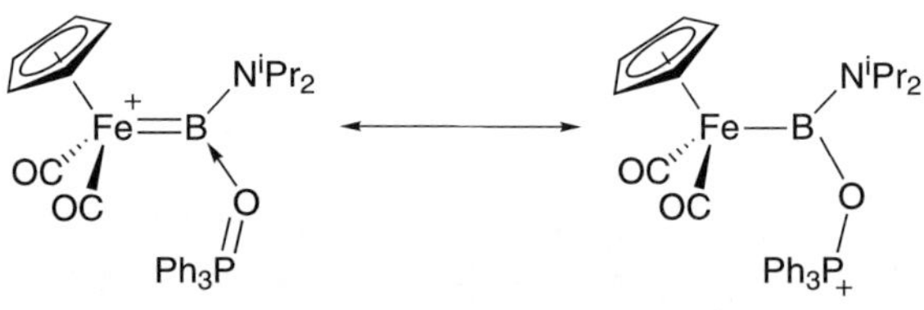

Scheme 6 Formal resonance contributions to the structure of $[CpFe(CO)_2\{B(N^iPr_2)(OPPh_3)]^+[BAr^f_4]^-$ (**8.31**)

Aminodiboran(4)yl complexes of iron have also been reported, exploiting synthetic approaches used for analogous group 6 systems. Thus, the 1,2-bis(dimethylamino), 1,2-dipyrrolidino and 1,2-dipiperidino-substituted complexes $CpFe(CO)_2B(NR_2)B(NR_2)Cl$ (R_2 = Me_2, **8.32**; R_2 = C_4H_8, **8.33**; R_2 = C_5H_{10}, **8.34**) have been synthesized from the corresponding diaminodichlorodiboranes(4) (Fig. 18), and the crystal structure of **8.32** obtained [45, 86]. The very long Fe–B bond length measured for **8.32** [2.090(3) Å] and the low carbonyl stretching frequencies for all three compounds (1988, 1932; 1980, 1922 and 1981, 1927 cm^{-1} for **8.32, 8.33** and **8.34**, respectively) are consistent with a bonding model incorporating little or no Fe–B π backbonding [45, 86].

Of related interest is the η^1-borazine complex $\{Cp'Fe(CO)_2\}_2(B_3N_3H_3Cl)$ (**8.71**) which has been synthesized by the reaction of trichloroborazine, $(ClBNH)_3$, with two equivalents of $Na[Cp'Fe(CO)_2]$ (Scheme 7) [87]. The relatively long Fe–B distances for **8.71** [2.041(1) and 2.045(1) Å] and low carbonyl stretching frequencies (1996, 1937 cm^{-1}) are consistent with little Fe–B π backbonding in this complex.

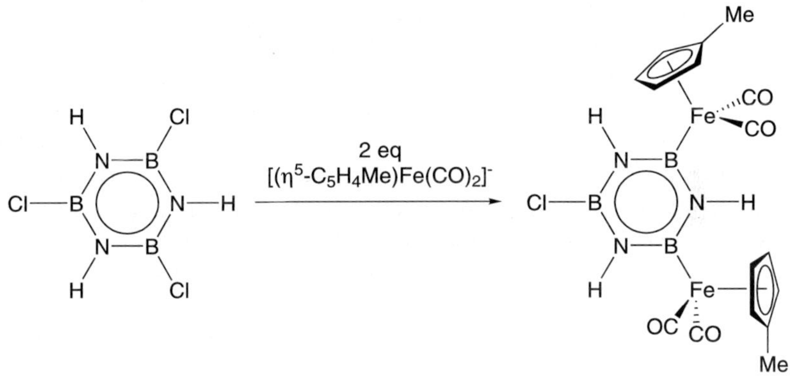

Scheme 7 Synthesis of the η^1 borazine complex **8.71**

(iii) Aryl(halo)boryl Complexes

First reported in 2002, aryl(halo)boryl complexes of iron represent versatile precursors for a number of related ligand systems via boron-centred reactions which proceed with retention of the Fe–B bond [62, 88–90]. Complexes bearing differing aryl substituents have been synthesized via the general route shown in Scheme 8, allowing for varying degrees of steric bulk at the boron centre.

The mesityl(bromo)boryl complexes $(\eta^5\text{-}C_5R_5)Fe(CO)_2B(Mes)Br$ ($R_5 = H_5$, **8.44**; $R_5 = H_4Me$, **8.45**; $R_5 = Me_5$, **8.46**) all have relatively short Fe–B distances [1.964(5) 1.962(4) and 1.972(2) Å, respectively] and this together with torsion angles close to 0° and relatively high carbonyl stretching frequencies (2016, 1962; 2009, 1961 and 2006, 1961 cm^{-1}, for **8.44**, **8.45** and **8.46** respectively) are consistent with modest Fe–B π back bonding [62, 88]. In each case, the mesityl substituent and the boryl ligand plane are orthogonal to each other [$Fe - B - C_{ipso} - C_{ortho}$ = 91.9(3), 88.3(2) and 89.5(2)°, **8.44**, **8.45** and **8.46** respectively], signifying little π interaction between these two moieties; this presumably is a consequence of the steric bulk of both the metal fragment and mesityl substituent. Further comparison can be made with the remaining members of the series of mesityl(halo)boryl complexes $Cp^*Fe(CO)_2B(Mes)X$

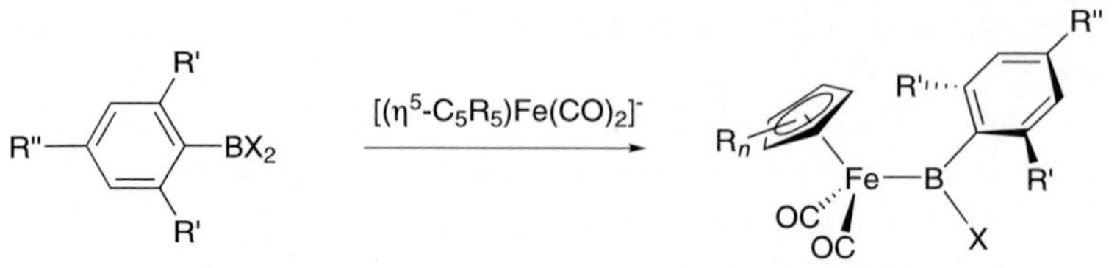

8.44: $R_n = H_5$, R' = R'' = Me, X = Br
8.45: $R_n = H_4Me$, R' = R'' = Me, X = Br
8.46: $R_n = Me_5$, R' = R'' = Me, X = Br
8.47: $R_n = Me_5$, R' = R'' = H, X = Cl
8.48: $R_n = H_5$, R' = 2,4,6-$^iPr_3C_6H_2$, R'' = H, X = Br

Scheme 8 Synthesis of aryl(halo)boryl complexes **8.44–8.48**

(X = F, **8.52**; X = Cl, **8.53**; X = I, **8.54**, Table 5), which have been synthesized by the reaction of the cationic borylene $[Cp^*Fe(CO)_2(BMes)]^+[BAr^f{}_4]^-$ with an appropriate source of the respective halide ion (Scheme 9) [82, 91]. The Fe–B lengths increase on going from bromoboryl **8.46** to chloro- and fluoroboryl analogues **8.53** and **8.52**, which can be rationalized in terms of the reduction of both σ donor and π acceptor properties of the boryl ligands in the order – B(Mes)Br > – B(Mes)Cl > – B(Mes)F. The longest Fe–B distance is exhibited by fluoroboryl **8.52**; the essentially σ only Fe–B bond present allows for rotation about this linkage, thereby enabling the molecule to reduce the steric interaction between the bulky metal and boryl substituents. Indeed, such is the rotational freedom in **8.52** that the conformation of the molecule in the solid state is influenced by the presence of intramolecular C – H···F hydrogen bonding between the Cp* and BF moieties [82].

Reactions of **8.44** with nucleophiles proceed with retention of the Fe–B bond to generate the corresponding heteroatom-substituted boryl complexes $CpFe(CO)_2B(Mes)ER_n$ [ER_n = SPh, **8.49**; ER_n = $OC_6H_4{}^tBu$-4, **8.50**; ER_n = O^tBu, **8.51**] (Fig. 19), thereby allowing further comparison of these boryl ligand systems as a function of the B-substituent [62, 89]. The decreasing π donor ability of the substituents ER_n (O^tBu > $OC_6H_4{}^tBu$-4 > SPh > Br) is

Table 5 Comparison of the structural parameters and spectroscopic data for the complexes $Cp^*Fe(CO)_2B(2,4,6\text{-}Me_3C_6H_2)X$

	8.54	**8.46**	8.53	8.52
X	I	Br	Cl	F
$\nu(CO)/cm^{-1}$	2015, 1969	2006, 1961	1996, 1937	1989, 1931
$d(Fe-B)/Å$	–	1.972(2)	1.985(2)	2.017(3)
δ_B/ppm	110.7	113.2	112.1	90.4

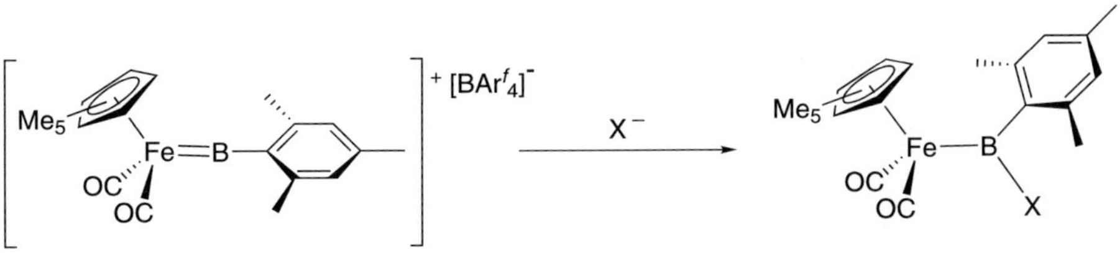

8.52: X = F
8.53: X = Cl
8.46: X = Br
8.54: X = I

Scheme 9 Synthesis of mesityl(halo)boryl complexes via halide addition to a cationic terminal borylene complex

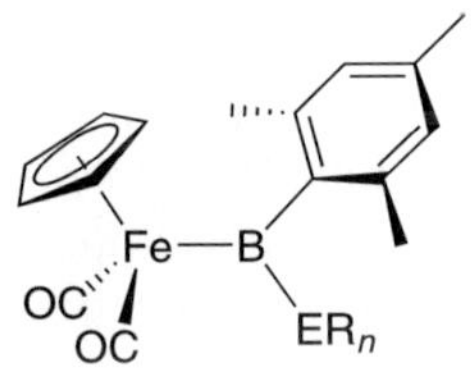

8.49: ER_n = SPh
8.50: ER_n = $OC_6H_4{}^tBu$-4
8.51: ER_n = O^tBu

Fig. 19 Asymmetrically substituted mesitylboryl complexes **8.49–8.51**

reflected in successive downfield shifts in the ^{11}B NMR resonances for these compounds (δ_B 72.9, 80.3, 103.5, 111.4 for **8.51, 8.50, 8.49** and **8.35**, respectively). In addition, the significant elongation of the Fe–B lengths for **8.50** and **8.51** [2.040(2) and 2.056(2) Å, respectively] compared to **8.44** [1.964(5) Å] is a reflection of the differences in ligand bulk and the π acceptor properties of the boryl ligands. The sequential reduction in π acceptor properties of the – B(Mes)ER_n ligand is also evidenced by the successively lower carbonyl stretching frequencies for **8.44, 8.49, 8.50** and **8.51**, respectively (2016, 1962; 2000, 1935; 1997, 1933 and 1987, 1918 cm^{-1}, respectively). The fact that alkoxoboryl ligands are poorer π acceptors than their aryloxo counterparts is reflected in the longer Fe–B bond and lower carbonyl stretching frequencies for **8.51** compared to **8.52**, a phenomenon also observed for related osmium systems [71, 72].

Ferrocenyl(bromo)boryl complexes have also been reported, the reaction of dibromoborylferrocene with one equivalent of $[(\eta^5\text{-}C_5R_5)Fe(CO)_2]^-$ yielding the compounds $(\eta^5\text{-}C_5R_5)Fe(CO)_2B(Fc)Br$ (R = H, **8.63**; R = Me, **8.64**, Fig. 20) [92]. The Fe–B distances for **8.63** and **8.64** are similar [1.997(2) Å for **8.63**; 1.972(3) and 1.985(3) Å for **8.64**], and are comparable to other haloboryl complexes [e.g. 1.964(5) 1.962(4), 1.972(2) and 2.005(10) Å, for **8.44–8.47**, re-

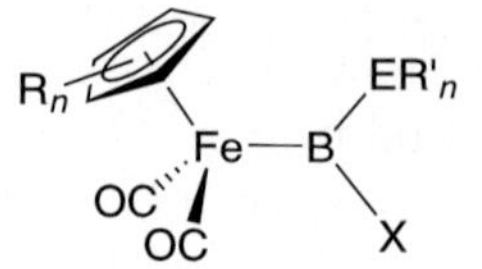

8.60: $R_n = H_5$, $ER'_n = Si(SiMe_3)_3$, X = Cl
8.61: $R_n = H_5$, $ER'_n = Ge(SiMe_3)_3$, X = Cl
8.62: $R_n = H_5$, $ER'_n = Si(SiMe_3)_3$, X = F
8.63: $R_n = H_5$, ER'_n = Fc, X = Br
8.64: $R_n = Me_5$, ER'_n = Fc, X = Br
8.65: $R_n = H_5$, ER'_n = Cym, X = Br
8.66: $R_n = Me_5$, ER'_n = Cym, X = Br

Fig. 20 Asymmetric haloboryl systems **8.60–8.66**

spectively] [62, 88, 90]. The torsion angle between the boryl ligand θ (Fig. 14) is found to be 15.2° for **8.63** and 88.8° and 89.0° for **8.64**, indicating a much greater possibility for Fe→B π backbonding in **8.63** than **8.64**. This is supported by the CO stretching frequencies for these complexes (2015, 1955 cm^{-1} for **8.63**; 1995, 1934 cm^{-1} for **8.64**), and is presumably due to the smaller steric demands of the Cp ligand compared to Cp*. The Fe→B π backbonding in these compounds has also been probed by monitoring the bending of the boryl moiety toward the ferrocenyl iron centre. This interaction, characterized by the "dip angle" β (Fig. 21), is typically greater for more strongly Lewis acidic boron centres [93]. Thus, β is found to be smaller in these boryl complexes than in $FcBBr_2$ (18°), and smaller in **8.63** (4.2°) than in **8.64** (8.0 and 7.6°). This in turn implies a less Lewis acidic boron centre in **8.63**, consistent with greater back-bonding from the $CpFe(CO)_2$ fragment. Ferrocenyl(bromo)boryl complexes also prove to be useful starting materials for the synthesis of unusual heteronuclear bridged borylenes by oxidative addition of the pendant B–Br bond to Pd(0) [94].

The cymantrene-functionalized complexes $(\eta^5\text{-}C_5R_5)Fe(CO)_2B(Cym)Br$ (R = H, **8.65**; R = Me, **8.66**, Fig. 20), display similar ^{11}B NMR chemical shifts

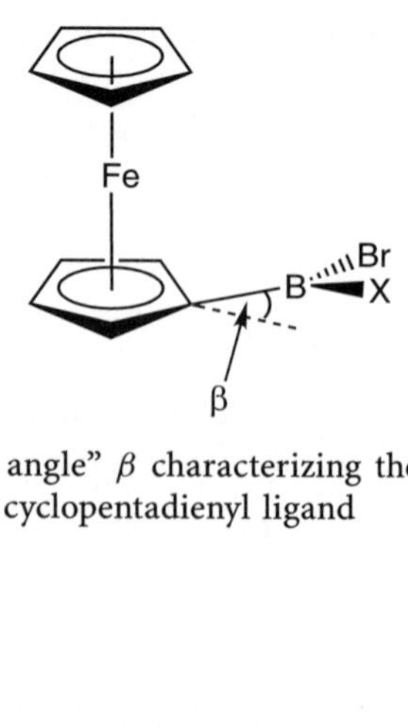

Fig. 21 Definition of the "dip angle" β characterizing the bending of a ferrocenylboryl moiety out of the plane of the cyclopentadienyl ligand

to ferrocenyl analogues **8.63** and **8.64** (δ_B 102.2 for **8.65**, 106.1 for **8.66**, 99.1 for **8.63**, 103.0 for **8.64**); **8.66** exhibits an essentially identical orientation of the boryl ligand with respect to the metal fragment as found in **8.64** (torsion angle $\theta = 88.7°$ for **8.66**; 88.8° and 89.0° for **8.64**), the Fe – B distances in these compounds also being very similar [1.983(4) Å for **8.66** 1.972(3) and 1.985(3) Å for **8.64**] [95]. The structural and spectroscopic data for **8.66** therefore indicate the presence of a weak Fe→B π interaction in this compound.

(iv) Aryloxy(halo)boryl Complexes

Given the structural and reaction studies carried out on iron amino- and aryl(halo)boryl complexes, research effort has also been directed at analogous aryloxy(halo)boryl systems. The complexes $(\eta^5\text{-}C_5R_5)Fe(CO)_2B(OMes)X$ (R = H, X = Cl, **8.55**; R = H, X = Br, **8.56**; R = Me, X = Cl, **8.57**) have been synthesized from the reaction of the corresponding haloborane with one equivalent of $Na[(\eta^5\text{-}C_5R_5)Fe(CO)_2]$ (Fig. 22) [75]. Boryl complexes **8.55** and **8.57** both feature relatively short Fe–B distances [1.977(4) Å for each] with the torsion angle (θ) for **8.57** [87.6°] being consistent with the possibility for π overlap between the boron-centred *p* orbital and the metal-based HOMO-2.

Substitution of the chloride substituent in **8.55** with para-*tert*-butyl phenoxide and thiophenolate, respectively, yields $CpFe(CO)_2B(OMes)ER_n$ [$ER_n = OC_6H_4{}^tBu\text{-}4$, **8.58**; ER_n = SPh, **8.59**, Fig. 22], as characterized by upfield shifts in the ^{11}B NMR resonances (**8.55**: δ_B 61.5; **8.58**: δ_B 47.4; **8.59**: δ_B 69.1) [62, 89]. Complex **8.59** containing mixed oxygen/sulfur donor substituents has a significantly longer Fe–B distance [2.034(4) Å] than chloroboryl precursor **8.55**, presumably reflecting, at least in part, the greater steric requirements of the SPh substituent compared to Cl.

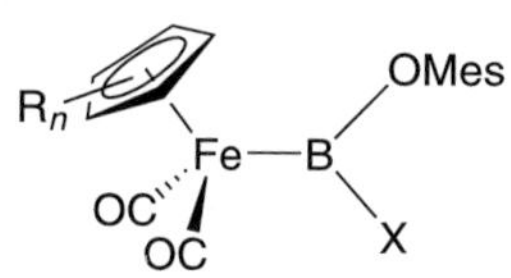

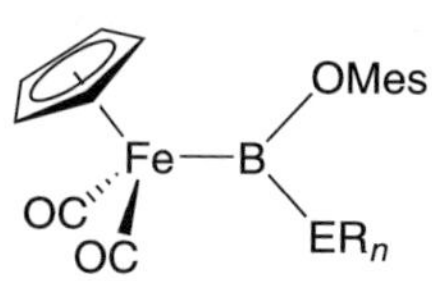

8.55: $R_n = H_5$, X = Cl
8.56: $R_n = H_5$, X = Br
8.57: $R_n = Me_5$, X = Cl

8.58: $ER_n = OC_6H_4{}^tBu\text{-}4$
8.59: ER_n = SPh

Fig. 22 Mesityloxyboryl complexes **8.55–8.59**

(v) Silyl- and Germyl(halo)boryl Complexes

The use of steric bulk has allowed the synthesis of boryl complexes featuring substituents derived from the heavier group 14 elements viz. $CpFe(CO)_2B(ER_n)Cl$ [$ER_n = Si(SiMe_3)_3$, **8.60**; $ER_n = Ge(SiMe_3)_3$, **8.61**,

Fig. 20] [61]. Noteworthy are the very low field ^{11}B NMR chemical shifts for these compounds ($\delta_B = 141.2$ and 139.0 for **8.60** and **8.61**, respectively), indicating a very deshielded boron centre in each case. Further reaction of **8.60** with TlF yields the structurally authenticated fluoroboryl complex $CpFe(CO)_2B\{Si(SiMe_3)_3\}F$ (**8.62**), the ^{11}B resonance showing a diagnostic upfield shift on going from **8.60** to **8.62** (δ_B 141.2 to δ_B 113.2), reflecting the substitution of chloride with the better π-donating fluoride substituent. The Fe–B distances for **8.60–8.62** are relatively short [1.964(8), 1.985(11) and 1.983(9) Å, respectively], with torsion angles implying that any π backbonding must originate in the HOMO-2 of the $[CpFe(CO)_2]^+$ fragment.

6.1.3
Dialkyl- and Diarylboryl Complexes

Half-sandwich iron complexes featuring dialkyl- or diarylboryl ligands have also been investigated. Thus, the complex $Cp^*Fe(CO)_2$(9-BBN) was synthesized as part of a broad-ranging study of ligand influences on CH activation, although no structural data were forthcoming [37]. The highly Lewis acidic nature of boranes containing the pentafluorophenyl substituent has led to the use of the bis(pentafluorophenyl)boryl ligand in an attempt to probe the upper limits of Fe→B π bonding. In common with the diphenylboryl complex $CpFe(CO)_2B(C_6H_5)_2$ (**8.69**), the structurally authenticated compounds $(\eta^5\text{-}C_5R_5)Fe(CO)_2B(C_6F_5)_2$ (R = H, **8.67**; R = Me, **8.68**) (Fig. 23) [63, 96], are found to be photolytically sensitive [66, 97]. The ^{11}B NMR resonances for **8.67–8.69** are similar ($\delta_B = 121.5$, 121.3 and 121, respectively), appearing at the relatively low fields expected for a boron centre with very poorly π-donating substituents. The Fe–B distances measured for **8.67** [1.964(4), 1.965(5) Å] are much shorter than that found in **8.69** [2.034(4) Å], with the orientation of the boryl ligand also being more appropriate for an Fe→B π interaction involving the HOMO of the organometallic fragment [torsion angles $\theta = 28.4$, 27.9 and 75° for **8.67** and **8.69**, respectively]. The corresponding Fe–B distance and angle for the sterically more encumbered system **8.68** are

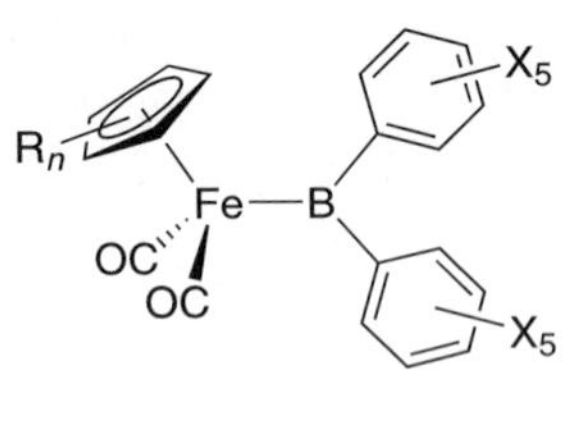

8.67: $R_n = H_5$, X = F
8.68: $R_n = Me_5$, X = F
8.69: $R_n = H_5$, X = H

Fig. 23 Diarylboryl complexes **8.67–8.69**

1.968(5) Å and 41.3(3)°, respectively. Paradoxically, the carbonyl stretching frequencies measured for **8.67** and **8.68** (2014, 1968 and 2004, 1954 cm^{-1} for **8.67** and **8.68**, respectively), do not imply such a strong Fe→B π interaction, being *red-shifted* compared to those reported for **8.69** (2021, 1951 cm^{-1}).

6.1.4
Base-Stabilized Complexes

The base-stabilized BH_2 complex $Cp^*Fe(CO)_2(BH_2{\cdot}PMe_3)$ (**8.72**, Fig. 24), has been synthesized via two routes: photolytic methane elimination and salt elimination. The upfield ^{11}B NMR chemical shift for this complex (δ_B –25.1) is indicative of a four-coordinate boron centre [98]. Likewise, the long Fe–B bond measured for the ethyltetramethylcyclopentadienyl derivative $(\eta^5\text{-}C_5Me_4Et)Fe(CO)_2(BH_2{\cdot}PMe_3)$ [**8.73**, 2.195(14) Å] and the very low carbonyl stretching frequencies (1932, 1869 cm^{-1}), are as expected for a four-coordinate base-stabilized system [c.f. 2.129(3) Å and 1958, 1899 cm^{-1} for $Cp^*Fe(CO)_2(BCl_2 \cdot 4pic)$, **8.42**].

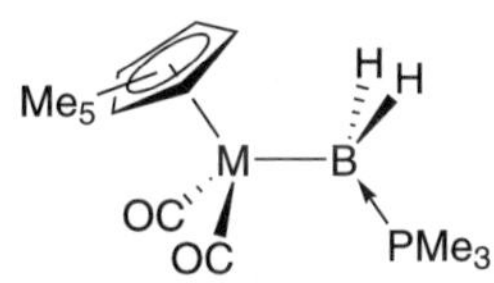

8.72: M = Fe
8.103: M = Ru

Fig. 24 The base-stabilized iron and ruthenium dihydridoboryl systems **8.72** and **8.103**

6.2
Ruthenium Compounds

Ruthenium complexes featuring boryl ligands are somewhat less numerous than their lighter homologues (see Table 6), although a number of interesting σ-borane and boratrane systems have also been reported [99–102]. In general, these systems can be produced by two routes: salt elimination and oxidative addition reactions and conform to two main types: three-legged piano-stool complexes similar to the $CpFe(CO)_2$ systems discussed above, and five- and six-coordinate complexes with square pyramidal or octahedral coordination geometries [12].

Table 6 Selected spectroscopic and structural parameters for ruthenium boryl complexes

Compound	d(Ru – B) (Å)	δ^{11}B	ν(CE) (cm^{-1})[a]	Refs.
$Ru(Bcat)Cl(CO)(PPh_3)_2$ (**8.79**)	–	–	1944, 1935[b]	[103]
$Ru(B\text{-}1,2\text{-}O_2C_{10}H_6)Cl(CO)(PPh_3)_2$ (**8.80**)	–	–	1950, 1933[b]	[103]
$Ru[B(3\text{-}Mecat)]Cl(CO)(PPh_3)_2$ (**8.81**)	–	–	–	[103]
$Ru\{B\text{-}1,2\text{-}(NH)_2C_6H_4\}Cl(CO)(PPh_3)_2$ (**8.82**)	–	–	1923	[103]
$Ru(B\text{-}1\text{-}S\text{-}2\text{-}NHC_6H_4)Cl(CO)(PPh_3)_2$ (**8.83**)	–	–	1937, 1921, 1907[b]	[103]
$Ru(Bcat)Cl(CS)(PPh_3)_2$ (**8.84**)	–	–	1292	[103]
$Ru(B1\text{-}S\text{-}2\text{-}NHC_6H_4)Cl(CS)(PPh_3)_2$ (**8.85**)	–	–	1275	[103]
Ru(Bcat)Cl(CN-*p*-tolyl)$(PPh_3)_2$ (**8.86**)	–	–	2070, 2020, 1989, 1962[b]	[103]
$Ru(Bcat)_2(CO)_2(PPh_3)_2$ (**8.87**)	2.100(3), 2.095(4)	48.2	2057, 1975	[105]
$Ru(Bcat)_2$(CO)(CN-*p*-tolyl)$(PPh_3)_2$ (**8.88**)	2.093(3), 2.086(3)	50.0	1974, 2141 (CN)	[105]
$Ru(dmpe)_2(H)Bpin$ (**8.89**)	–	50	–	[181]
$Ru(depe)_2(H)Bpin$ (**8.90**)	–	50	–	[181]
$Cp^*Ru(CO)_2[B(3,5\text{-}Me_2cat)]$ (**8.91**)	–	48	2012, 1952	[37]
$Cp^*Ru(CO)_2Bpin$ (**8.92**)	–	44.7	2002, 1940	[37]
$Cp^*Ru(CO)_2[B(4\text{-}Methiocat)]$ (**8.93**)	–	78	2007, 1948	[37]
$Cp^*Ru(CO)_2BBN$ (**8.94**)	–	117	1994, 1931	[37]
$Cp^*Ru(CO)_2BCy_2$ (**8.95**)	–	120	1984, 1921	[37]
$Cp^*Ru(CO)_2BMe_2$ (**8.96**)	–	118.7	[c]	[37]
$CpRu(CO)_2B(NMe_2)B(NMe_2)Br$ (**8.97**)	2.173(3)	63.3	2005, 1945	[40]
$CpRu(CO)_2B(NMe_2)Cl$ (**8.98**)	–	50.3	2019, 1956	[80]
$CpRu(CO)_2B(NMe_2)Br$ (**8.99**)	–	48.1	2022, 1960	[80]
$CpRu(CO)_2B(Cl)N(SiMe_3)B(Cl)N(SiMe_3)_2$ (**8.100**)	2.115(2)	60.3	2018, 1955	[80]
$Cp'Ru(CO)_2B(Fc)Br$ (**8.101**)	–	90.1	2021, 1958	[92]
$Cp'Ru(CO)_2BCl_2$ (**8.102**)	–	81.0	2021, 1958	[60]
$Cp^*Ru(CO)_2(BH_2 \cdot PMe_3)$ (**8.103**)	2.243(8)	–28.9	1951, 1883	[98]

[a] E = O, S, *N*-*p*-tolyl
[b] Solid-state splitting
[c] Not given

6.2.1 Catecholboryl and Related Complexes

Systematic appraisal of structure and bonding in ruthenium(II) and osmium(II) boryl systems has been aided by the syntheses by Roper and coworkers of a number of compounds featuring a range of B-substituents and ancillary ligands. Reaction of the ruthenium hydride $RuHCl(CE)(PPh_3)_3$ with a borane, X_2BH, gives rise to a number of coordinatively unsaturated boryl systems $Ru(BX_2)Cl(CE)(PPh_3)_2$ [E = O, X_2 = cat, **8.79**; X_2 = 1,2-$O_2C_{10}H_6$,

8.80; X_2 = 3-Mecat, **8.81**; X_2 = 1,2-$(NH)_2C_6H_4$, **8.82**; X_2 = 1-*S*-2-NHC_6H_4, **8.83**; E = S, X_2 = cat, **8.84**; X_2 = 1-*S*-2-NHC_6H_4, **8.85**; E = N *p*-tolyl, X_2 = cat, **8.86**] containing five-coordinate metal centres which have been examined spectroscopically (Scheme 10) [103]. Interestingly, Bcat complex **8.79** has been shown to insert ethyne into the Ru–B bond to give a borylalkenyl complex that is stabilized by intramolecular coordination of one of the catecholate oxygen atoms to the osmium centre. The resulting six-coordinate ruthenium complex features a five-membered chelate ring and can undergo further substitution chemistry at the boron centre with a range of alcohols [104].

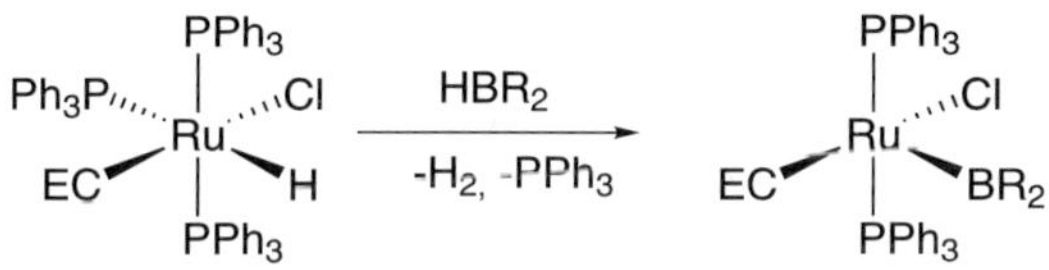

8.79: E = O, R_2 = cat
8.80: E = O, R_2 = 1,2-$O_2C_{10}H_6$
8.81: E = O, R_2 = 3-Mecat
8.82: E = O, R_2 = 1,2-$(NH)_2C_6H_4$
8.83: E = O, R_2 = 1-S-2-NHC_6H_4
8.84: E = S, R_2 = cat
8.85: E = S, R_2 = 1-S-2-NHC_6H_4
8.86: E = N p-tolyl, R_2 = cat

Scheme 10 The synthesis of ruthenium boryl complexes **8.79–8.86**

Related ruthenium(II) bis(catecholboryl) complexes can be formed by oxidative addition of B_2cat_2 to $Ru(CO)(CE)(PPh_3)_3$ yielding $Ru(Bcat)_2(CO)(CE)(PPh_3)_2$ [E = O, **8.87**; E = *N*-*p*-tolyl, **8.88**] [105]. The structures of **8.87** and **8.88** (together with that of its osmium analogue **8.119**) feature the metal in an approximately octahedral environment, with pairs of mutually *cis* triphenylphosphine and Bcat ligands; the two carbonyls (or carbonyl and isocyanide ligands in the case of **8.88** and **8.119**) occupy the axial positions. The Bcat ligands are arranged face-to-face, with B–Ru–B angles of less than 90°[75.43(14)° and 75.57(12)° for **8.87** and **8.88**, respectively]; the accompanying P–Ru–P angles for **8.87** and **8.88** are 109.51(3)° and 104.00(3)°, respectively. As such, the geometries of the P_2RuB_2 fragments resemble closely those of square planar *cis*-bis(phosphine)bis(boryl)platinum(II) complexes (vide infra) for which steric factors have largely been held responsible. Thus, the narrow B–M–B and wide P–M–P angles are thought to be due to the bulky three-dimensional nature of the phosphine ligands, with the flat, essentially two-dimensional geometry of the Bcat ligand allowing close approach in a face-to-face fashion.

Ruthenium boryl complexes containing catecholate, pinacolate and dithiolate substituents have been analyzed as potential alkane functionalization

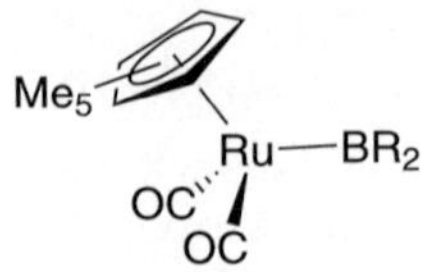

8.91: R_2 = 3,5-Me_2cat
8.92: R_2 = pin
8.93: R_2 = 4-Methiocat
8.94: R_2 = BBN
8.95: R_2 = Cy_2

Fig. 25 Ruthenium boryl systems studied in the functionalization of alkanes (**8.91–8.95**)

reagents. The compounds $Cp^*Ru(CO)_2BX_2$ [X_2 = 3,5-Me_2cat, **8.91**; X_2 = pin, **8.92**; X_2 = 4-Methiocat, **8.93**], along with the alkyl-substituted boryl systems $Cp^*Ru(CO)_2BX_2$ [BX_2 = 9-BBN, **8.94**; X_2 = Cy_2, **8.95**; X_2 = Me_2 **8.96**] were studied for their efficacy in the activation of alkanes (Fig. 25) [37]. The differing ^{11}B NMR chemical shifts for these complexes reflect the π donor ability of the substituents: low field shifts are reported for alkyl substituted boryl ligands (δ_B 117 and 120 for **8.91** and **8.95**, respectively), with successively higher field resonances being observed for the thiolate-substituted complex **8.93** (δ_B 78) and alkoxy/aryloxy substituted complexes **8.91** and **8.92** (δ_B 48 and 45, respectively). Although not structurally characterized, these systems were further analyzed via the electronic effects of the boryl substituents on carbonyl stretching frequencies. The electron-donating ability of the boryl substituents (alkyl > pinacolato > dithiolato > catecholato) is reflected in the lowest carbonyl stretching frequencies being measured for **8.95** (1984, 1921 cm^{-1}) and the highest for **8.91** (2012, 1952 cm^{-1}). With regard to alkane (n-pentane) CH activation chemistry, it was found that the highest yield of 1-functionalized product was obtained with dialkoxyboryl ligands. Furthermore, it was postulated that differences in reactivity between these boryl systems are related to the differential reactivity of the intermediate systems $Cp^*Ru(CO)BX_2$ towards C–H bonds, rather than the efficiency of intermediate generation [37].

6.2.2
Aminoboryl and Diboran(4)yl Systems

The diboran(4)yl system $CpRu(CO)_2B(NMe_2)B(NMe_2)Br$ (**8.97**) was the first structurally authenticated ruthenium boryl complex to be reported (Fig. 26), [40] and has a structure similar to the related iron complex $CpFe(CO)_2B(NMe_2)B(NMe_2)Cl$ (**8.32**). The Ru–B distance is 2.173(3) Å and this, taken with the orientation of the boryl ligand with respect to the ruthenium fragment [torsion angle θ = 103.4°] and carbonyl stretching

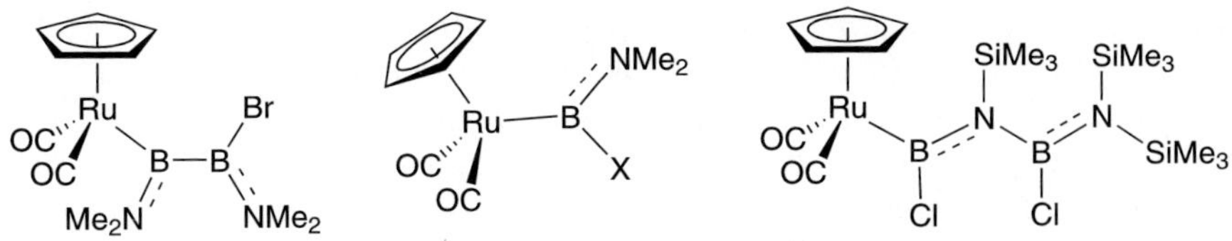

Fig. 26 Ruthenium boryl and related complexes bearing amino substituents

frequencies of 2005, 1945 cm^{-1} indicates little or no Ru→B π backbonding. In related chemistry, reaction of Na[$CpRu(CO)_2$] with the dihaloboranes Me_2NBX_2 (X = Cl, Br) gives rise to the amino(halo)boryl complexes $CpRu(CO)_2B(NMe_2)X$ (X = Cl, **8.98**; X = Br, **8.99**) (Fig. 26). These complexes, which have been characterized spectroscopically, give rise to ^{11}B NMR chemical shifts typical for aminoboryl systems (δ_B 50.3 and 48.1, for **8.98** and **8.99**, respectively) and carbonyl stretching frequencies (2019, 1956 cm^{-1} for **8.98**; 2022, 1960 cm^{-1} for **8.99**] consistent with minimal Ru–B π interaction [80]. By contrast, the analogous reaction of Na[$CpRu(CO)_2$] with $(Me_3Si)_2NBCl_2$ does not yield the corresponding aminoboryl system, but instead gives rise to $CpRu(CO)_2B(Cl)N(SiMe_3)B(Cl)N(SiMe_3)_2$ (**8.10**), presumably formed from two molecules of the borane by elimination of Me_3SiCl (Fig. 26). The Ru–B distance in **8.100** [2.115(2) Å], although shorter than that found in **8.97**, is still not significantly shorter than the sum of the respective covalent radii (2.12 Å) [106].

6.2.3 Haloboryl Complexes

The first ruthenium dihaloboryl $Cp'Ru(CO)_2BCl_2$ (**8.102**) has recently been spectroscopically characterized, although no structural data were reported [60]. As with the related iron complex $CpFe(CO)_2BCl_2$ (**8.35**), **8.102** exhibits the relatively downfield-shifted ^{11}B resonance (δ_B 81.0) and high frequency carbonyl stretching bands (2012, 1958 cm^{-1}), expected for a boryl system featuring poorly π-donating substituents. The ferrocenyl(bromo)boryl complex $Cp'Ru(CO)_2B(Fc)Br$ (**8.101**) has also been synthesized recently and has spectroscopic properties similar to those of the closely related iron systems **8.63** and **8.64** [92].

6.2.4 Base-Stabilized Complexes

As with iron, Shimoi and co-workers have reported the synthesis of half-sandwich metal complexes containing the $BH_2{\cdot}PMe_3$ ligand. The Ru–B distance for the base-stabilized boryl complex, $Cp^*Ru(CO)_2(BH_2{\cdot}PMe_3)$ (**8.103**; Fig. 24) is long [2.243(8) Å], as expected for a σ only interaction, and this is mirrored in the low carbonyl stretching frequencies (1951, 1883 cm^{-1}) [98]. The ^{11}B NMR chemical shift for **8.103** (δ_B –28.9) is markedly upfield compared to that for other Ru boryls, reflecting the four-coordinate boron centre in the system.

6.3 Osmium Compounds

A range of five- and six-coordinate osmium boryl complexes has been synthesized making use of similar approaches to those reported above for ruthenium. In particular, the reaction of phenylosmium(II) precursors with boranes, which proceed via elimination of benzene, has been shown to be a useful entry point into octahedral and square pyramidal osmium boryl complexes. Significant further chemistry has been reported on these systems, including substitution at both boron and metal centres, which has shed light on fundamental issues of structure/bonding and reactivity [12].

6.3.1 Catecholboryl and Related Complexes

A number of osmium boryl complexes featuring different boryl substituents and ancillary ligands have been reported, allowing for a systematic study of these systems. The first such compounds were reported in 1997, the reaction of $Os(Ph)Cl(CE)(PPh_3)_2$ with X_2BH yielding the coordinatively unsaturated complexes $Os(BX_2)Cl(CE)(PPh_3)_2$ [E = O, X_2 = cat, **8.104**; X_2 = 3-Mecat, **8.105**; X_2 = 1,2-$(NH)_2C_6H_4$, **8.106**; X_2 = 1-*S*-2-NHC_6H_4, **8.107**; E = S, X_2 = cat, **8.108**] (Scheme 11), which were characterized spectroscopically. This approach mirrors that reported by the same authors for related ruthenium boryls synthesized from hydridoruthenium(II) precursors [103].

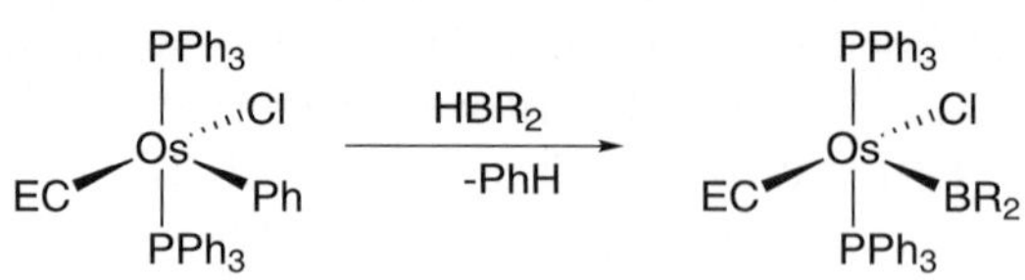

8.104: E = O, R_2 = cat
8.105: E = O, R_2 = 3-Mecat
8.106: E = O, R_2 = 1,2-$(NH)_2C_6H_4$
8.107: E = O, R_2 = 1-S-2-NHC_6H_4
8.108: E = S, R_2 = cat

Scheme 11 The synthesis of five-coordinate osmium(II) boryl complexes **8.104–8.108**

Table 7 Selected spectroscopic and structural parameters for osmium boryl complexes

Compound	d(Os – B) (Å)	δ^{11}B	ν(CE) (cm^{-1})[a]	Refs.
$Os(Bcat)Cl(CO)(PPh_3)_2$ (**8.104**)	2.019(3)	–	1923	[103, 107]
$Os(B\text{-}1,2\text{-}O_2\text{-}3\text{-}MeC_6H_3)Cl(CO)(PPh_3)_2$ (**8.105**)	–	–	1923	[103]
$Os\{B\text{-}1,2\text{-}(NH)_2C_6H_4\}Cl(CO)(PPh_3)_2$ (**8.106**)	–	–	1898	[103]
$Os(B\text{-}1\text{-}S\text{-}2\text{-}NHC_6H_4)Cl(CO)(PPh_3)_2$ (**8.107**)	–	–	1924, 1906, 1894[b]	[103]
$Os(Bcat)Cl(CS)(PPh_3)_2$ (**8.108**)	–	–	1302	[103]
$[Os(Bcat)(CO)(CH_3CN)_2(PPh_3)_2]SbF_6$ (**8.109**)	2.094(5)	–	1941, 1925	[107]
$[Os(Bcat)(CO)_2(CH_3CN)(PPh_3)_2]SbF_6$ (**8.110**)	–	–	2052[b], 2040, 1983	[107]
$[Os(Bcat)(CO)(CN\text{-}p\text{-}tolyl)(CH_3CN)(PPh_3)_2]SbF_6$ (**8.111**)	2.135(10)	–	1978 (CO), 2149 (CN)	[107]
$Os(Bcat)(\eta^2\text{-}O_2CH)(CO)(PPh_3)_2$ (**8.112**)	–	–	1925	[107]
$Os(Bcat)(\eta^2\text{-}O_2N)(CO)(PPh_3)_2$ (**8.113**)	–	–	1928	[107]
$Os(Bcat)(\eta^2\text{-}S_2CNEt_2)(CO)(PPh_3)_2$ (**8.114**)	2.073(7)	–	1913	[107]
$Os(Bcat)(o\text{-}tolyl)(CO)(PPh_3)_2$ (**8.115**)	–	26.5	1903	[105, 109]
cis-$Os(Bcat)(o\text{-}tolyl)(CO)_2(PPh_3)_2$ (**8.116**)	2.155(7)	46.0	1997, 1937	[105, 109]
cis-$Os(Bcat)(o\text{-}tolyl)(CO)(CN\text{-}p\text{-}tolyl)(PPh_3)_2$ (**8.117**)	–	47.2	1941 (CO), 2097 (CN)	[105]
$Os(Bcat)_2(CO)_2(PPh_3)_2$ (**8.118**)	–	43.9	–	[105]
$Os(Bcat)_2(CO)(CN\text{-}p\text{-}tolyl)(PPh_3)_2$ (**8.119**)	2.104(6), 2.093(6)	45.6	1964 (CO), 2141 (CN)	[105]
$Os(Bcat)H(CO)_2(PPh_3)_2$ (**8.120**)	–	45.2	2009, 1971	[105]
trans-$Os(Bcat)(o\text{-}tolyl)(CO)_2(PPh_3)_2$ (**8.121**)	2.132(7)	43.3	2038, 1941	[105, 109]
trans-$Os(Bcat)I(CO)_2(PPh_3)_2$ (**8.122**)	2.090(3)	36.6	2064, 1970	[105]
cis-$Os(Bcat)I(CO)_2(PPh_3)_2$ (**8.123**)	2.145(5)	44.3	2020, 1958	[105]
$Os(BCl_2)Cl(CO)(PPh_3)_2$ (**8.124**)	–	52.5	1937	[110]
$Os\{B(OH)_2\}Cl(CO)(PPh_3)_2$ (**8.125**)	2.056(3)	–	1917	[71]

Table 7 (continued)

Compound	d(Os – B) (Å)	δ^{11}B	ν(CE) (cm^{-1})[a]	Refs.
Os{B(OMe)$_2$}Cl(CO)(PPh$_3$)$_2$ (**8.126**)	–	–	1908	[71]
Os{B(OEt)$_2$}Cl(CO)(PPh$_3$)$_2$ (**8.127**)	2.081(5)	–	1906	[71]
Os{B[1,2-(NMe)$_2$C$_6$H$_4$]}Cl(CO)(PPh$_3$)$_2$ (**8.128**)	2.082(10)	–	1885	[71]
Os{B[1,2-(NMe)$_2$C$_2$H$_4$]}Cl(CO)(PPh$_3$)$_2$ (**8.129**)	–	–	1890	[71]
Os{B(1,2-O$_2$C$_2$H$_4$)}Cl(CO)(PPh$_3$)$_2$ (**8.130**)	2.043(4)	27.8	1912	[72]
Os{B(1,2-O$_2$C$_3$H$_6$)}Cl(CO)(PPh$_3$)$_2$ (**8.131**)	2.062(9)	23.5	1904	[72]
Os{B(1,2-O$_2$C$_2$H$_4$)}Cl(CO)$_2$(PPh$_3$)$_2$ (**8.132**)	2.170(9)	45.0	2031, 1945	[72]
Os{B(OEt)NHC$_9$H$_6$N}Cl(CO)(PPh$_3$)$_2$ (**8.133**)	2.072(3)	–	–	[110]
Os{BCl(OC$_5$H$_4$N)}Cl(CO)(PPh$_3$)$_2$ (**8.134**)	–	63.0	1911	[111]
Os{BCl(OC$_5$H$_4$N)}I(CO)(PPh$_3$)$_2$ (**8.135**)	2.039(4)	62.3	1919	[111]
Os{B(OEt)(OC$_5$H$_4$N)}Cl(CO)(PPh$_3$)$_2$ (**8.136**)	–	47.5	1895	[111]
Os{B(NH^{n}Bu)(OC$_5$H$_4$N)}Cl(CO)(PPh$_3$)$_2$ (**8.137**)	2.075(3)	46.0	1887	[111]
Os{BF(OC$_5$H$_4$N)}Cl(CO)(PPh$_3$)$_2$ (**8.138**)	–	48.8	1921, 1904 [b]	[111]
Os{BCl(NHC$_5$H$_4$N)}Cl(CO)(PPh$_3$)$_2$ (**8.139**)	–	60.9	1906	[112]
Os{BCl(NHC$_5$H$_4$N)}Br(CO)(PPh$_3$)$_2$ (**8.140**)	2.047(7), 2.053(8)	60.2	1908	[112]
Os{BCl(NHC$_5$H$_4$N)}H(CO)(PPh$_3$)$_2$ (**8.141**)	2.104(3)	73.1	1877	[112]
Os{B(OH)(NHC$_5$H$_4$N)}Cl(CO)(PPh$_3$)$_2$ (**8.142**)	2.085(3)	–	1876	[112]
[Os{B(OH)(NHC$_5$H$_4$N)}(CO)(MeCN)(PPh$_3$)$_2$]SbF$_6$ (**8.143**)	2.105(4)	48.7	1919	[112]
[Os{B(OEt)(NHC$_5$H$_4$N)}(CO)$_2$(PPh$_3$)$_2$]SbF$_6$ (**8.144**)	–	54.6	2022, 1948	[112]
Os{B(OEt)(OC$_2$H$_4$OH)}Cl(CO)(PPh$_3$)$_2$ (**8.145**)	2.107(4)	–	1916	[72]

[a] E = O, S, N-p-tolyl
[b] Solid-state splitting

Catecholboryl complex **8.104** has subsequently been structurally authenticated and computationally evaluated; due to the presence of a labile chloride ligand at the osmium centre, **8.104** is also found to be a versatile substrate for further substitution chemistry [107, 108]. The solid-state structure of **8.104** shows the metal in a square pyramidal geometry with the boryl ligand in the apical position, consistent with its strong *trans* influence. As a result of the lack of a competing ligand *trans* to the boryl, the Os–B distance for **8.104** is short [2.019(3) Å c.f. 2.10 Å for the sum of the respective covalent radii] [11]. Abstraction of the chloride ligand from **8.104** using $Ag[SbF_6]$, in the presence of acetonitrile yields the bis(acetonitrile) complex $[Os(Bcat)(CO)(CH_3CN)_2(PPh_3)_2]SbF_6$ (**8.109**), which can subsequently be reacted with CO or *p*-tolyl isocyanide to yield the complexes $[Os(Bcat)(CO)(CE)(CH_3CN)(PPh_3)_2]SbF_6$ [E = O, **8.110**; E = N *p*-tolyl, **8.111**], via selective substitution of the acetonitrile ligand *trans* to the boryl (Scheme 12).

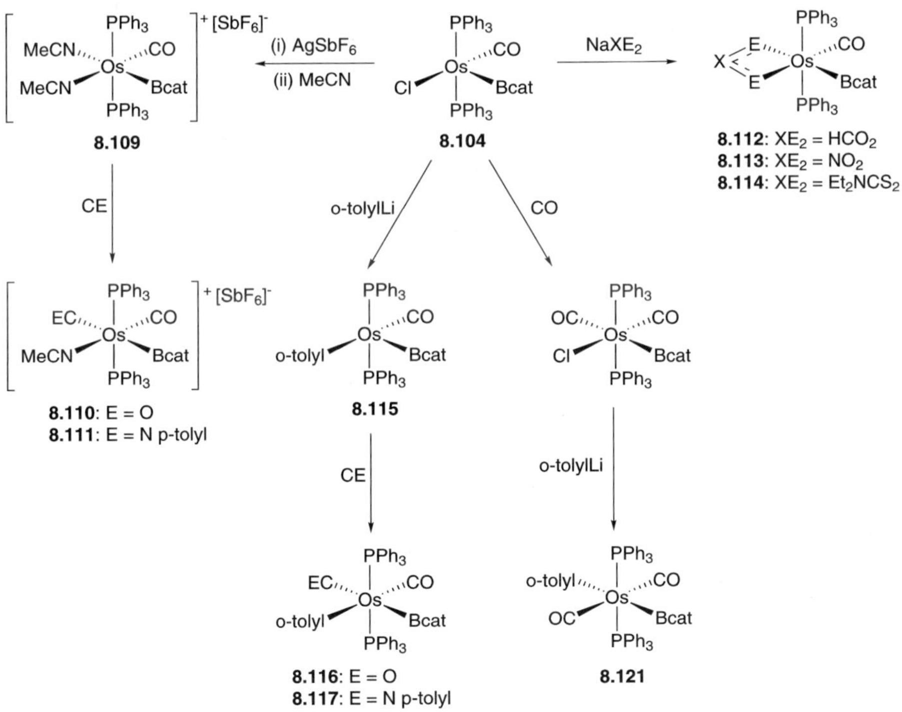

Scheme 12 Osmium boryl complexes generated from catecholboryl system **8.104**

Complex **8.109** has a distorted octahedral geometry with *trans* triphenylphosphine and *cis* acetonitrile ligands. The Os–B distance for this complex [2.094(5) Å] is longer than that found in **8.104**, a reflection both of the presence of a competing *trans* ligand and of the net positive charge. The Os–N distance for the acetonitrile ligand *trans* to the Bcat ligand [2.168(3) Å] is longer than that *trans* to the CO ligand [2.101(3) Å] illustrating the greater *trans* influence of Bcat compared to CO. The structure of **8.111** is that of a distorted octahedron, and the Os–B length [2.135(10) Å] is similar to that of **8.109**. Reaction of **8.104** with bidentate anions yields the complexes Os(Bcat)(η^2-XE_2)(CO)(PPh_3)$_2$ [XE_2 = O_2CH, **8.112**; XE_2 = O_2N, **8.113**; XE_2 = S_2CNEt_2, **8.114**] (Scheme 12). The structure of **8.114** in the solid state reveals a distorted octahedral geometry, with the dithiocarbamate ligand occupying coordination sites *trans* to both carbonyl and boryl ligands. Here too the stronger *trans* influence of the boryl ligand (vs. CO) is in evidence, with the Os–S distance *trans* to Bcat [2.5209(14) Å] being significantly longer than that *trans* to CO [2.4709(13) Å]. The Os–B distance in **8.114** [2.073(3) Å] is similar to that of cationic systems **8.109** and **8.111**, but significantly longer than that measured for five-coordinate Bcat precursor **8.104**, presumably due to coordination of one arm of the dithiocarbamate ligand *trans* to the boryl.

In an attempt to investigate further the mechanisms involved in metal-catalyzed hydroboration and arene borylation, osmium boryl compounds containing ancillary aryl ligands have been targeted. Reaction of **8.104** with *o*-tolyllithium yields the five-coordinate system Os(Bcat)(*o*-tolyl)(CO)(PPh_3)$_2$ (**8.115**), which, on reaction with CO or *p*-tolylisocyanide yields the coordinatively saturated complexes *cis*-Os(Bcat)(*o*-tolyl)(CO)(CE)(PPh_3)$_2$ (E = O, **8.116**; E = CN *p*-tolyl, **8.117**) (Scheme 12) [105, 109]. The distorted octahedral geometry of **8.116** in the solid state features mutually *cis* *o*-tolyl and Bcat ligands, with a *trans* triphenylphosphine arrangement. The Os–C distance *trans* to the boryl ligand [1.964(7) Å] is significantly longer than that *trans* to *o*-tolyl [1.903(6) Å], revealing a stronger *trans* influence for Bcat compared to the (known strong σ donor) *o*-tolyl ligand. The Os–B distance for **8.116** is relatively long [2.155(7) Å] compared, for example to **8.104** and **8.114**, as expected for a boryl ligand *trans* to CO.

Facile reductive elimination of o-tolylBcat from complexes **8.116** and **8.117** generates the four-coordinate intermediates Os(CO)$_2$(PPh_3)$_2$ and Os(CO)(CN-*p*-tolyl)(PPh_3)$_2$, respectively, which have been found to oxidatively add B_2cat_2 to give Os(Bcat)$_2$(CO)$_2$(PPh_3)$_2$ (**8.118**) and Os(Bcat)$_2$(CO)(CN-*p*-tolyl)(PPh_3)$_2$ (**8.119**). In the case of **8.116**, similar reactivity towards HBcat generates Os(Bcat)H(CO)$_2$(PPh_3)$_2$ (**8.120**) (Scheme 13) [105, 109]. B–C reductive elimination and B–H oxidative addition have been proposed as two key steps in metal-catalyzed borylation reactions [2–5, 9, 10], and osmium systems such as **8.116–8.120** have provided a fruitful testing ground on which to probe this fundamental chemistry. The structure of **8.119** is qualitatively similar to those of the ruthenium bis(boryl) systems **8.87** and **8.88**, with Os–B distances of 2.093(6) and 2.104(6) Å.

An isomeric form of **8.114**, namely *trans*-Os(Bcat)(*o*-tolyl)(CO)$_2$(PPh_3)$_2$ (**8.121**, featuring a *trans* arrangement of Bcat and *o*-tolyl fragments) has been

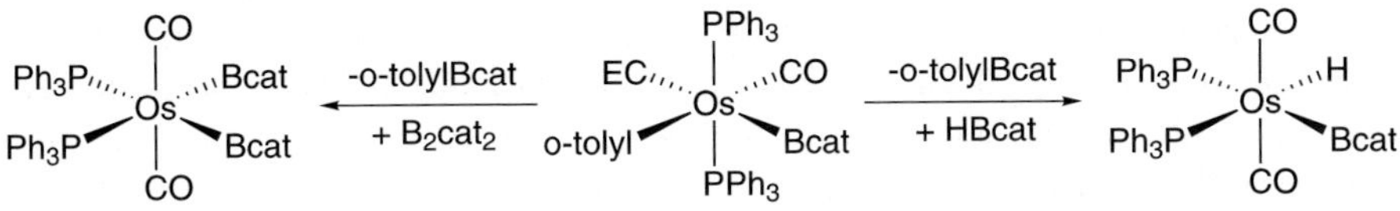

8.118: E = O
8.119: E = N p-tolyl
8.116: E = O
8.117: E = N p-tolyl
8.120: E = O

Scheme 13 Generation of osmium boryl complexes from oxidative addition of B_2cat_2 and HBcat

generated by the reaction of *cis*-Os(Bcat)Cl(CO)$_2$(PPh$_3$)$_2$ with *o*-tolyllithium (Scheme 12), and features an Os–B distance [2.132(7) Å] very similar to that measured for **8.116**. The strong *trans* influence of the Bcat ligand compared to CO is reflected in an Os–C distance for **8.118** [2.277(5) Å] which is much longer than that in **8.116** [2.202(6) Å]. Unlike **8.116**, **8.121** does not readily reductively eliminate o-tolylBcat; this process occurs only under photolytic conditions for **8.121**, presumably due to the necessity for initial isomerization to **8.116** in which the boryl and aryl ligands are mutually *cis* [109].

Further reaction of *trans*-Os(Bcat)(*o*-tolyl)(CO)$_2$(PPh$_3$)$_2$ (**8.121**) with I_2 yields *trans*-Os(Bcat)I(CO)$_2$(PPh$_3$)$_2$ (**8.122**), along with *trans*-Os(*o*-tolyl) I(CO)$_2$(PPh$_3$)$_2$ [105]. Synthesis of the corresponding *cis* isomer of **8.122** (i.e. **8.123**) via an alternative route allows comparison of the molecular structures of these complexes. Both **8.122** and **8.123** possess almost octahedral symmetry around the metal centre, with the phosphine ligands being mutually *trans* in each case (Fig. 27). As expected, the Os–B distance for **8.123** [2.145(5) Å] is lengthened in comparison to that in **8.122** [2.090(3) Å], due to the Bcat ligand in **8.123** competing with a stronger σ-donor *trans* ligand (i.e. CO, rather than I). The Os–I distance in **8.122** [2.8346(2) Å] is much longer than that in **8.123** [2.7883(3) Å] due to the higher *trans* influence of the Bcat ligand compared to CO. Furthermore, comparison of the two distinct Os–CO distances in **8.123** [*trans* to Bcat: 1.968(5) Å; *trans* to I: 1.867(4) Å] and the Os–CO distances in **8.122** [both *trans* to CO: 1.938(3), 1.943(3) Å] is consistent not only with the strong *trans* influence of the Bcat ligand, but also with its appreciable π acidity.

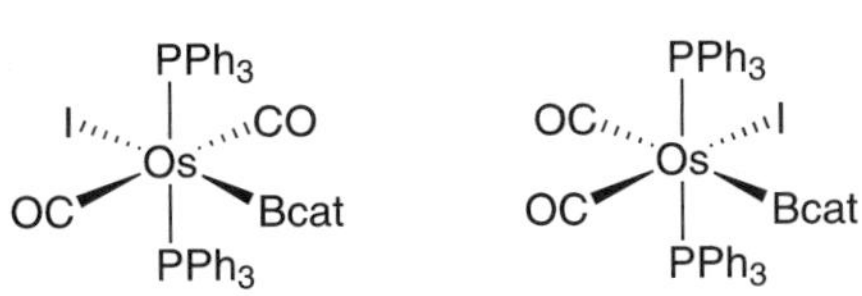

8.122 **8.123**

Fig. 27 *Trans*- and *cis*-Os(Bcat)I(CO)$_2$(PPh$_3$)$_2$ (**8.98** and **8.99**)

The dichloroboryl complex Os(BCl$_2$)Cl(CO)(PPh$_3$)$_2$ (**8.124**) [71, 72, 110–112] has proved to be a versatile substrate for further derivatization of the boron centre. Reactions of **8.124** with ROH yield the hydroxy- and alkoxyboryl compounds $Os[B(OR)_2]Cl(CO)(PPh_3)_2$ (R = H, **8.125**; R = Me, **8.126**; R = Et, **8.127**) while those with chelating bis(amines) give rise to the diaminoboryls $Os\{B[1,2\text{-}(NMe)_2C_6H_4]\}Cl(CO)(PPh_3)_2$ (**8.128**) and $Os\{B[1,2\text{-}(NMe)_2C_2H_4]\}Cl(CO)(PPh_3)_2$ (**8.129**) (Scheme 14) [71]. Reaction of these complexes with CO leads to six-coordinate systems via ligand coordination *trans* to the boryl ligand. Differences in spectroscopic and structural properties of these systems compared to the parent dichloroboryl complex **8.124** and the corresponding Bcat system **8.104** have been rationalized by a "competitive π-bonding model" analogous to that used for Fischer carbenes [12,71]. Thus, the weaker π-donor nature of aryloxy compared to alkoxy substituents renders the Bcat ligand a better π acceptor with respect to the osmium centre than $B(OH)_2$ or $B(OEt)_2$, leading to a shorter Os–B bond length for **8.104** [2.019(3) vs. 2.056(3) and 2.081(5) Å for **8.123** and **8.125**]; consistent with these observations, the carbonyl stretching frequency for **8.104** is slightly higher than those of **8.123** and **8.125** [1923 vs. 1917 and 1906 cm^{-1}]. The corresponding carbonyl stretching frequency for dichloroboryl system **8.122** is 1937 cm^{-1}. In a similar vein, the Os–B bond length for complex **8.126**, which features an *N,N′*-dimethylphenylenediaminoboryl ligand (two anilino substituents at boron) is markedly longer [2.082(10) Å] than that measured for catecholboryl complex **8.104**, and the corresponding carbonyl stretching frequency is red-shifted by 38 cm^{-1} [71].

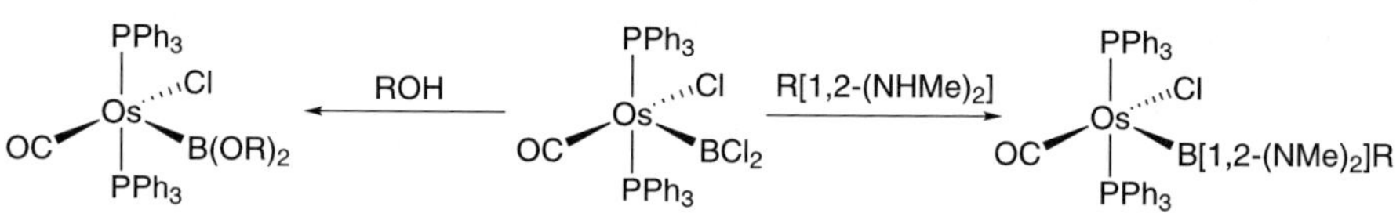

8.125: R = H
8.126: R = Me
8.127: R = Et

8.124

8.128: R = C_6H_4
8.129: R = C_2H_4

Scheme 14 Generation of osmium alkoxy- and aminoboryl complexes via the reaction of dichloroboryl system **8.124** with nucleophiles

Interestingly, diethoxyboryl complex **8.127**, also proves to be a useful substrate for boron-centred substitution chemistry in the presence of added trimethylsilyl chloride. Thus, the cyclic dialkoxyboryl systems $Os\{B(1,2\text{-}O_2C_2H_4)\}Cl(CO)(PPh_3)_2$ (**8.130**) and $Os\{B(1,2\text{-}O_2C_3H_6)\}Cl(CO)(PPh_3)_2$ (**8.131**) (Scheme 15) have been synthesized from **8.127** in the presence of two equivalents of Me_3SiCl [72]. The use of a single equivalent of silane leads to the formation of tethered mono-substituted complexes (vide infra). The longer Os–B distances measured for **8.130** and **8.131** [2.043(4) and 2.062(9) Å, respectively] compared to **8.104** are also consistent with the poorer π acceptor properties of alkoxyboryl ligands compared to their aryloxy counterparts. This trend is further reflected in the lower carbonyl stretching frequencies for

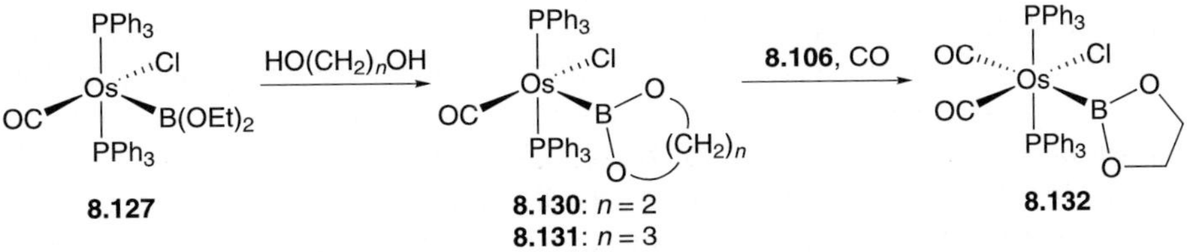

Scheme 15 Generation of bidentate osmium boryl complexes from reaction of **8.127** with diols

8.130 and **8.131** (1912 and 1904 cm^{-1}, respectively) than **8.104** (1923 cm^{-1}). The differences in bond lengths and carbonyl stretching frequencies between **8.130** and **8.131** imply that the five-membered cyclic boryl ligand is a marginally better π acceptor than the six-membered cyclic boryl ligand.

Reaction of **8.130** with CO yields the coordinatively saturated species $Os\{B(1,2\text{-}O_2C_2H_4)\}Cl(CO)_2(PPh_3)_2$ (**8.132**, Scheme 15), allowing direct comparison between otherwise identical five-coordinate and CO-ligated six-coordinate complexes. The molecular structure of **8.132** reveals an approximately octahedral geometry, with the boryl ligand situated *trans* to CO. The Os–B distance measured for **8.132** [2.179(7) Å] is significantly longer than that for **8.130** [2.043(4) Å], where there is no competing σ-donor, π-acceptor ligand in the *trans* position [72].

6.3.2
Tethered Systems

A number of tethered osmium boryl complexes have been developed recently in which one of the boryl substituents features an additional pendant donor which coordinates to the metal centre as a "tether". For systems resulting from reactions with bifunctional donors such as 2-aminopyridine two possibilities exist, depending on which donor atom interacts with the boron centre and which with the metal. Thus, tethered boryl complexes result from coordination of the anionic donor at boron, with the neutral donor tether coordinated at osmium; the reverse coordination possibility leads to the formation of intramolecular base-stabilized borylene complexes.

The first such system to be reported was an 8-aminoquinoline-stabilized compound formed from the reaction of the base-stabilized borylene $Os(=BNHC_9H_6N)Cl_2(CO)(PPh_3)_2$ with ethanol, thereby yielding $Os\{B(OEt)NHC_9H_6N\}Cl(CO)(PPh_3)_2$ (**8.133**) [110]. A more widely applicable route to such systems involves the use of dichloroboryl complex **8.124** as the precursor. Reaction of **8.100** with 2-hydroxypyridine gives rise to the pyridine tethered complex $Os\{BCl(OC_5H_4N)\}Cl(CO)(PPh_3)_2$ (**8.134**), which can further be converted to $Os\{BCl(OC_5H_4N)\}I(CO)(PPh_3)_2$ (**8.135**) and $Os\{B(ER_n)(OC_5H_4N)\}Cl(CO)(PPh_3)_2$ [ER_n = OEt, **8.136**; ER_n = NH^nBu, **8.137**; ER_n = F, **8.138**] by reaction at the osmium and boron centres, respectively (Scheme 16) [111]. The ^{11}B NMR chemical shifts for **8.134** and **8.135** ($\delta_B = 63.0, 62.3$, respectively) are downfield compared with those for **8.136–8.138** ($\delta_B = 47.5, 46.0$ and 48.8 respectively) due to the poorer π-donating ability of chloride compared to alkoxo, amino or fluoro substituents. This is also reflected in the differing values for the carbonyl stretching frequencies for these complexes [e.g. 1911, 1895 and 1887 cm^{-1} for **8.134, 8.136, 8.137**, respectively]. The solid-state structures of **8.135** and **8.137** allow for further analysis of the bonding characteristics of these tethered boryl systems which, as with related non-tethered complexes, is consistent with a "competitive π-bonding model". Thus, the Os–B distance in **8.135** [2.039(4) Å] is much shorter than that for **8.137** [2.075(3) Å], the chloroboryl ligand in **8.135** being a better π acceptor than the amino-substituted boryl ligand in **8.137**. The B–O distance for **8.135** [1.417(5) Å] is correspondingly shorter than that in **8.137** [1.445(4) Å], a reflection of the poorer π-donating ability of the Cl compared to NH^{n}Bu.

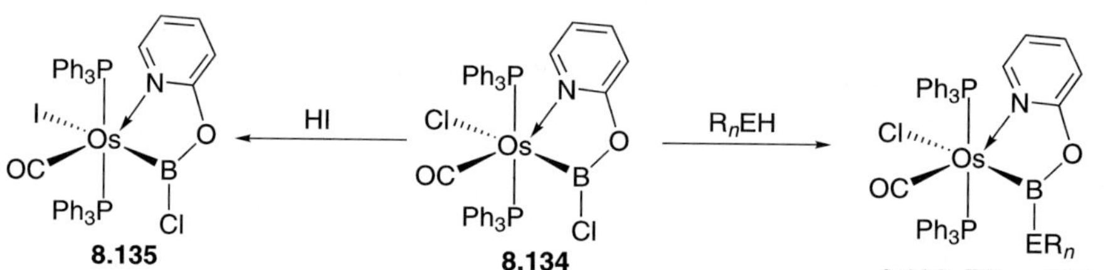

8.135 **8.134**

8.136: ER$_n$ = OEt
8.137: ER$_n$ = NH^{n}Bu
8.138: ER$_n$ = F

Scheme 16 Substitution at either osmium or boron centres in tethered boryl complex 8.134

The analogous 2-aminopyridine-tethered boryl complex Os[BCl(NHC$_5$H$_4$N)]Cl(CO)(PPh$_3$)$_2$ (**8.139**) has also been prepared from **8.124** [112]. Further reaction yields the related systems Os{BCl(NHC$_5$H$_4$N)}X(CO)(PPh$_3$)$_2$ [X = Br, **8.140**; X = H, **8.141**; X = OH, **8.142**] and [Os{B(OR)(NHC$_5$H$_4$N)}L(CO)(PPh$_3$)$_2$][SbF$_6$] [R = H, L = MeCN, **8.143**; R = Et, L = CO, **8.144**] (Scheme 17). Interestingly, and in contrast to **8.134**, **8.139** is found to be more inert to nucleophilic attack at the boron centre, presumably due to the better π-donating ability of amino over aryloxy substituents rendering the boron less electrophilic in the amino-tethered boryl system [111]. The Os–B distances for **8.140** [2.047(7) and 2.053(8) Å] and the corresponding compound **8.135** derived from 2-hydroxypyridine [2.039(4) Å] are also an indication of greater E→B (E = N, O) π-donation in the amino case; this is additionally reflected in the differences in B–Cl lengths for these complexes [1.811(7) and 1.812(9) Å for **8.140**; 1.773(4) Å for **8.135**]. Furthermore, the lengthening of the Os–B distance in the cationic hydroxyboryl complex **8.143** [2.105(4) Å] compared to neutral system **8.142** [2.085(3) Å]

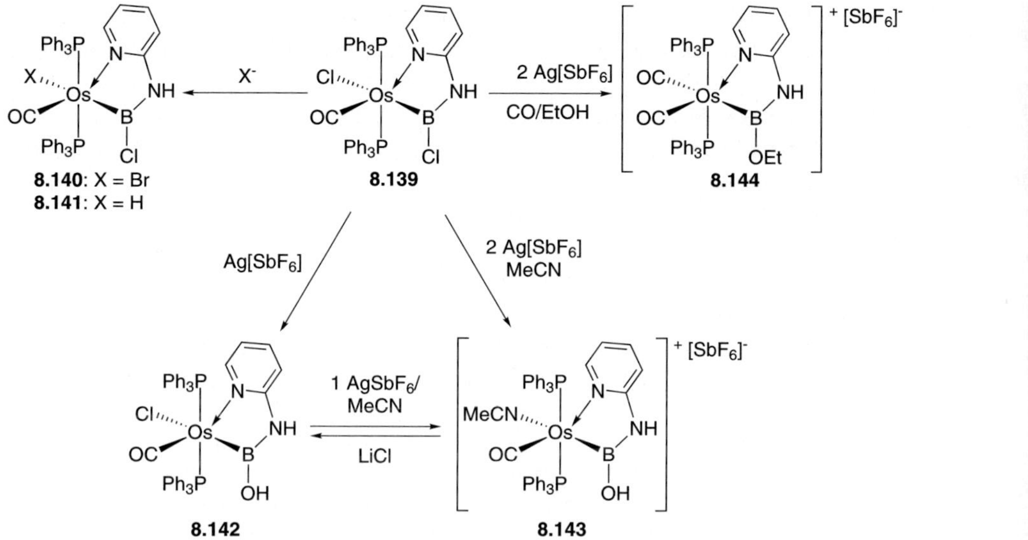

Scheme 17 Tethered aminopyridyl systems **8.139–8.144**

suggests the participation of an Os→B π interaction in the bonding of these compounds.

Finally, the ethanediolate-tethered boryl system Os[B(OEt)(OC_2H_4OH)] Cl(CO)(PPh_3)$_2$ (**8.145**) prepared from the corresponding diethoxyboryl complex and ethane-1,2-diol in the presence of a single equivalent of trimethylsilyl chloride, features an approximately octahedral coordination geometry and an Os–B distance [2.107(4) Å] towards the longer end of the range measured for tethered osmium boryl complexes [72]. The molecular structure also reveals a hydrogen-bonding interaction between **8.145** and an ethanol molecule in the solid state.

7 Group 9 (Cobalt, Rhodium and Iridium)

Complexes of the group 9 metals, especially rhodium and iridium, represent one of the more numerous families on which a systematic appraisal of structure/bonding properties for the boryl ligand can be based. In part, this reflects the involvement of such systems not only in earlier work on metal-catalyzed hydroboration chemistry [2–5, 35, 113–123], but in more recent studies of diboration [124, 125], and the activation of C–H bonds in both saturated [9, 10, 126–135] and unsaturated hydrocarbons [9, 10, 50, 51, 127, 129, 134, 136–159].

The range of group 9 mono-, bis- and tris(boryl) complexes discussed below provides a useful structural basis on which to discuss the *trans* influence of the boryl ligand, together with more specific issues such as geometric preferences in 5-coordinate systems, and the structural and chemical significance of residual B···H interactions in boryl hydride complexes. Experimental reports have been complemented by a number of quantum chemical studies of group 9 boryl complexes, targeting aspects of both structural and reaction chemistry [14, 15, 50, 51, 113, 118–121, 128, 134, 135, 138, 151–153, 160–162].

Of related interest, but not strictly of relevance to the material described here, are a number of M→B donor/acceptor complexes featuring a group 9 metal as the donor atom, reported principally by Hill and co-workers [163–167].

7.1 Cobalt

In comparison to its heavier congeners, rhodium and iridium and its predecessor iron, the coordination chemistry of cobalt with boryl and related ligand systems represents a numerically small area (Table 8). Although boryl and borylene systems stabilized by B-coordination of an external base have been reported by Fehlner, Puddephatt and Shimoi [168–171], and a number of cobalt-containing species were reported by Schmid and Nöth [1], structurally characterized three-coordinate ligand systems are restricted to just four examples [160, 161]. The paramagnetic 17-electron bis(boryl) systems $(Me_3P)_3Co(Bcat)_2$ (**9.1**), $(Me_3P)_3Co[B(4\text{-Mecat})]_2$ (**9.2**) and $(PhMe_2P)_3Co(Bcat)_2$ (**9.3**) have been synthesized by a common synthetic route (Scheme 18) involving B–B oxidative addition to Co(0) phosphine precursors [160, 161]. Mirroring chemistry reported earlier for rhodium [14], the 18-electron mono(boryl) system $(Me_3P)_4Co(Bcat)$ (**9.4**) has been prepared from $(Me_3P)_4CoMe$ and B_2cat_2.

Table 8 Selected data for structurally characterized cobalt boryl complexes

Compound	$d(M-B)$ (Å)	Other data	Refs.
$(Me_3P)_4Co(Bcat)$ (**9.4**)	1.949(2)	δ_B 52.5	[160, 161]
$(Me_3P)_3Co(Bcat)_2$ (**9.1**)	1.945(11)	B···B 2.185 Å	[160, 161]
	1.970(11)	B–Co–B 67.9(4)°	
$(Me_3P)_3Co[B(4\text{-Mecat})]_2$[a] (**9.2**)	1.947(6)	B···B 2.211, 2.192 Å	[160, 161]
	1.966(5)	B–Co–B 68.8(2), 68.1(2)°	
	1.954(6)		
	1.960(6)		
$(PhMe_2P)_3Co(Bcat)_2$ (**9.3**)	1.948(3)	B···B 2.271 Å	[160, 161]
	1.955(3)	B–Co–B 71.2(1)°	

[a] Two crystallographically independent molecules

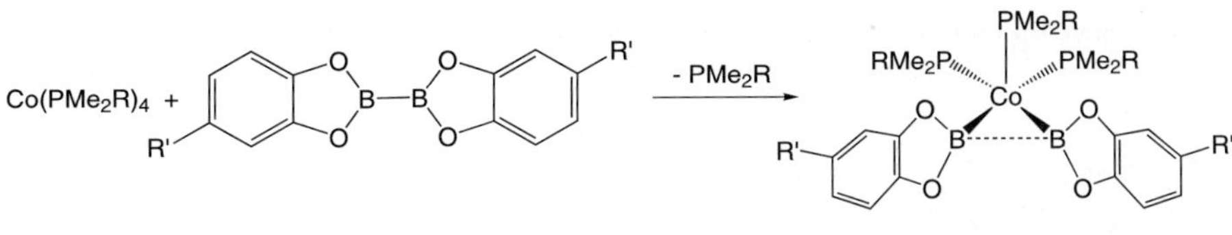

9.1: R = Me, R' = H
9.2: R = Me, R' = Me
9.3: R = Ph, R' = H

Scheme 18 Syntheses of the 17-electron cobalt(II) boryls **9.1–9.3** by B – B oxidative addition

Bis(boryl) complexes **9.1–9.3** have each been shown in the solid state to feature a distorted square pyramidal coordination environment similar to that observed for 16-electron rhodium(III) species of the type $(R_3P)_2RhCl(BX_2)_2$ (vide infra), with the important difference that in **9.1–9.3** both boryl ligands occupy basal coordination sites [15, 115, 172]. Of significant structural interest are the acute B–Co–B angles [67.9(1)–71.2(1)° c.f. 77.8(7)° for *cis*-$(Ph_3P)_2Pt(Bcat)_2$] and relatively short B···B separations [2.185–2.271 Å c.f. 1.678(3) Å for B_2cat_2 and 2.552 Å in *cis*-$(Ph_3P)_2Pt(Bcat)_2$] which attest to a degree of residual B···B interaction. Although the boron centres remain trigonal planar [sum of angles = 359.4°(mean) for **9.1**], the relative orientation of the two boryl ligands is such that the two π systems face one another, providing a geometric basis for overlap of the boron p_z orbitals. Consistent with this, molecular orbital analysis using DFT methods reveals a weak B···B interaction and a contribution to boryl ligand binding from a three-centre CoB_2 interaction.

9.1–9.3 also represent rare examples of paramagnetic boryl complexes [173, 174]; ESR studies of **9.2** reveal that the unpaired electron occupies an orbital (the SOMO) with significant Co d_{z^2} character, with no measurable spin density at boron [161]. Indeed, further studies have suggested that the antibonding character of this SOMO with respect to the apical ligand is a key factor in the boryl ligands occupying basal coordination sites [15]. Occupation of the apical site by the weaker σ donor PMe_3 ligand (c.f. Bcat) means that the SOMO is less elevated in energy and the complex more stable than it would be with the boryl ligand in the apical site. The differences in ligand positions compared to 16-electron distorted square pyramidal systems are thus explained, since with one electron fewer, this orbital is unoccupied and the Bcat ligand can then occupy the apical site typical of strong σ donors [15].

7.2 Rhodium

Rhodium boryl complexes have been the subject of intense research effort, reflecting not only fundamental issues of structure and bonding but also their

Table 9 Selected data for structurally characterized rhodium boryl complexes

Compound	d(M–B) (Å)	Other data	Refs.
$(Me_3P)_4Rh(Bcat)$ **(9.5)**	2.047(2)	δ_B 35.2	[176, 177]
$(Me_3P)_3Rh(Cl)_2(Bcat)$-mer,cis **(9.7)**	2.913(2)	Rh–Cl 2.536(1) Å trans to Bcat Rh–Cl 2.412(1) Å trans to PMe_3 δ_B 40.1	[178]
$Cp^*Rh(H)_2(SiEt_3)Bpin$ **(9.16)**	2.038(5)	B···H 1.74(4) Å H–Rh–B 55.7(16)° δ_B 40	[162]
$Cp^*Rh(H)(Bpin)P(p\text{-tol})_3$ **(9.19)**	2.029(4)	B···H 2.23(3) Å H–Rh–B 79(1)° δ_B 42.3	[128]
$(^iPr_3P)_2RhH(Cl)(Bcat)$ **(9.9)**	1.965(2) 1.973(7) [b]	B···H 2.01(2), 2.004(10) Å [b] H–Rh–B 70.9(8), 68.5(4)° [b] δ_B 37.7	[14, 114]
$(^iPr_3P)_2RhH(Cl)(Bpin)$ **(9.10)**	1.981(2) 1.985(4) [b]	B···H 2.02(2), 2.013(5) Å [b] H–Rh–B 70.0(8), 67.8(2)° [b] δ_B 31.8	[14, 127]
$(Ph_3P)_2RhCl(Bcat)_2$ [c] **(9.11)**	1.954(4) [e] 2.008(4) [f]	B–Rh–Cl 118.9(1) [e], 162.1(1)° [f] δ_B38.4	[115]
$(Ph_3P)_2RhCl(Bcat)_2$ [d] **(9.11)**	1.956(8) [e] 2.008(7) [f]	B–Rh–Cl 117.5(2) [e], 163.4(2)° [f]	[115, 172]
$(Ph_3P)_2RhCl(B\text{-}4\text{-}Mecat)_2$ **(9.12)**	1.906(13) [e] 2.034(12) [f]	B–Rh–Cl 118.9(3) [e], 160.2(4)° [f] δ_B 40.3	[172, 179]
$(Et_3P)_2RhCl(Bcat)_2$ **(9.13)**	1.973(2) [e] 1.994(2) [f]	B–Rh–Cl 132.3(1) [e], 152.3(1)° [f] δ_B 39.7	[172]
$Cp^*Rh(H)_2(Bpin)_2$ [a] **(9.17)**	2.055(7)- 2.081(6)	B···H 1.57(5)–1.70(6) Å H–Rh–B 49(2)–54(2)° δ_B 40.4	[128]
$Cp^*Rh(Bpin)_2PEt_3$ [a] **(9.20)**	2.028(1)- 2.033(1)	δ_B 41.3	[128]
$(Me_3P)_3Rh(Bcat)_3$ **(9.6)**	2.055(4) 2.053(4) 2.061(4)	δ_B 46.8	[176, 177]
$Cp^*Rh(H)(Bpin)_3$ [a] **(9.18)**	2.056(3)- 2.078(3)	B···H 1.53(3), 1.69(3) Å H–Rh–B 48(1)–54(1)° δ_B 39.9	[128]
(dippe)Rh(μ-H)$_2$(μ-Bcat)-Rh(H)(dippe) **(9.21)**	2.057(8) [g] 2.444(9) [h]	δ_B 35.0	[113]

[a] Two crystallographically independent molecules
[b] Determined from neutron diffraction data
[c] As the tetra dichloromethane solvate
[d] As the tris 1,2-dichloroethane solvate
[e] For the apical boryl ligand
[f] For the basal boryl ligand
[g] For the Rh(III)–B bond
[h] For the Rh(I)–B bond

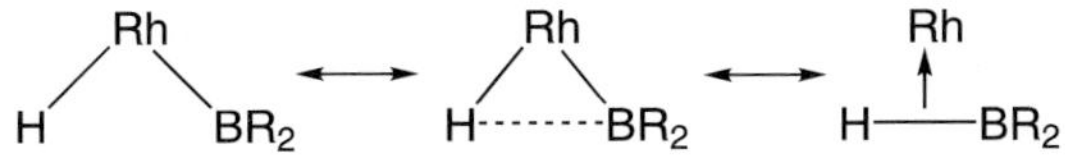

Scheme 19 The continuum of electronic structure between limiting descriptions of rhodium complexes as boryl hydrides or as σ boranes

implication as catalytic intermediates in important organic transformations (notably hydroboration and CH functionalization processes). From a structural standpoint, boryl compounds of rhodium have emerged as one of the key proving grounds for establishing the fundamental properties of the boryl ligand e.g. its strong σ donor properties and high *trans* influence (Table 9). More recently much interest has focused on complexes containing both hydride and boryl ligands, and the relative merits of their description as boryl hydrides or as σ-borane complexes (Scheme 19). In particular relatively short B$\cdots$H contacts and narrow H–Rh–B angles have been taken as indications of residual BH interaction.

7.2.1 Complexes Without Ancillary Hydride Ligands

A range of five- and six-coordinate rhodium(I) and rhodium(III) mono-, bis- and tris(boryl) systems have been reported by Marder and co-workers, which provide a useful platform to discuss the fundamental properties of boryl ligands, notably their *trans* influence (σ donor properties) and potential as π-acceptors. The structural features of several of these systems were reviewed in 1998 [11] and these (along with new systems) are assessed in the light of more recent theoretical studies [15, 175].

The 18-electron mono-boryl complex $(Me_3P)_4Rh(Bcat)$ (**9.5**) was initially reported in 1997 from the reaction of $(Me_3P)_4RhMe$ with B_2cat_2 (Scheme 20) [176], a route subsequently adopted for the corresponding cobalt complex $(Me_3P)_4Co(Bcat)$ (**9.4**) [160, 161]. Both compounds adopt the trigonal bipyramidal structure which is prevalent for 18-electron ML_5 systems, with the boryl ligand occupying one of the apical coordination sites. Theoretical studies have shown that in such trigonal bipyramidal systems, strong σ-donor ligands have a preference for the apical coordination site, while π-acceptors prefer the equatorial sites [178]. Thus, superficially, the overall geometries of **9.4** and **9.5** might be considered as evidence for the Bcat ligand having strong σ donor and weak π-acceptor properties. That said, there is some indication of shortening of the Co–B and Rh–B bonds [1.949(2) and 2.047(2) Å, respectively] compared to the bond lengths expected on the basis of the respective covalent radii (2.00 and 2.09 Å) [11]. Computational studies on model systems related to rhodium complex **9.5** have shown that the alternative structure, featuring an equatorial boryl ligand, lies some 14 kcal mol^{-1} higher in energy than the axial isomer [15]. This in turn is related to the form

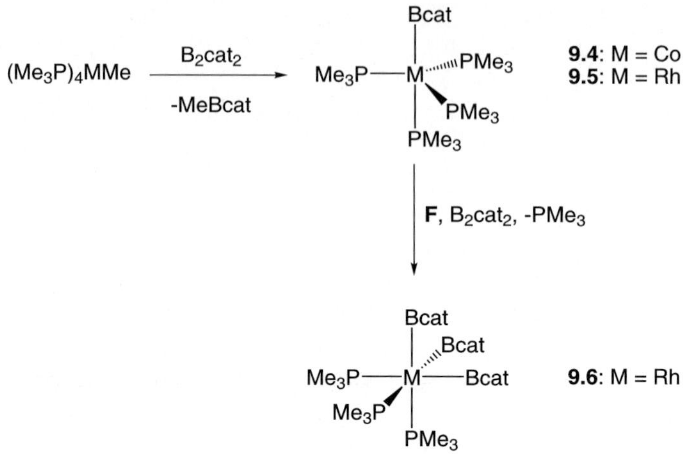

Scheme 20 Formation of cobalt and rhodium mono- and tris(boryl) systems **9.4–9.6** from methylmetal precursors

of the HOMO, which is shown to have significant σ^* character with respect to the equatorial ligands. Thus, occupation of an equatorial site by the stronger σ-donor ligand (i.e boryl rather than phosphine) will cause significant destabilization through elevation of the HOMO.

An octahedral coordination environment has been deduced crystallographically for the rhodium(III) mono(boryl) species *mer,cis*-$(Me_3P)_3Rh(Cl)_2Bcat$ (**9.7**), which along with its dibromo counterpart has been synthesized by the reaction of $(Me_3P)_3RhCl$ with ClBcat (or BrBcat), followed in the latter case by halide redistribution [177]. Further evidence for the strong *trans* influence of the Bcat ligand is obtained from the fact that it is located *trans* to the weakest donor ligand (i.e. chloride) and that the Rh–Cl distance *trans* to Bcat [2.536(1) Å] is significantly longer than that *trans* to PMe_3 [2.412(1) Å]. The effect of the ligand *trans* to the Bcat moiety on the Rh–B interaction can also be addressed: Rh–B distances of 2.013(2) and 2.047(2) Å for **9.7** and **9.5** reflect the presence of *trans* chloride and PMe_3 ligands, respectively.

A range of five-coordinate bis(boryl) systems have also been synthesized primarily via oxidative addition of diboron(4) precursors to rhodium(I) centres (Scheme 21). These complexes typically adopt distorted square pyramidal geometries with one apical and one basal boryl ligand [15, 115, 172]. As such, they offer the opportunity to compare the bonding characteristics of boryl ligands with/without a competing *trans* ligand. These ML_5 systems typically have formal electron counts of 16, for which both square bipyramidal and Y-shape distorted trigonal bipyramidal structures are conceivable. That the former structural model offers the best description for boryl systems **9.11–9.13** is consistent with studies of the related 16-electron five-coordinate

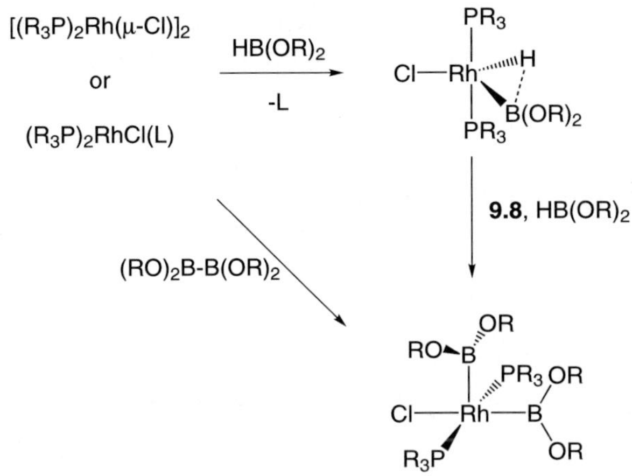

9.8: R = Ph, $(OR)_2$ = cat
9.9: R = ^{i}Pr, $(OR)_2$ = cat
9.10: R = ^{i}Pr, $(OR)_2$ = pin

9.11: R = Ph, $(OR)_2$ = cat
9.12: R = Ph, $(OR)_2$ = 4-Mecat
9.13: R = Et, $(OR)_2$ = cat

Scheme 21 Synthetic routes to the crystallographically characterized rhodium boryl hydride and bis(boryl) complexes **9.8–9.13**

osmium boryl complexes $(Ph_3P)_2OsCl(CO)(BX_2)$ (see Sect. 6.3) [12], and reflects the strong σ donor properties of the boryl ligand. That the 16-electron rhodium boryl hydrides **9.9** and **9.10** adopt structures better described as intermediate between square pyramidal and the alternative trigonal bipyramidal limit, is thought to reflect the presence of weak B···H interactions (vide infra) [14, 15, 114, 127]. In the cases of compounds **9.11–9.13** the possibility of any residual interaction between the two boryl ligands (of the type found in the 17-electron cobalt systems **9.1–9.3**) can be eliminated by consideration that the two BO_2 planes are essentially perpendicular.

Inspection of the Rh–B bond lengths for the bis(triphenylphosphine) systems $(Ph_3P)_2RhCl(Bcat)_2$ (**9.11**) and $(Ph_3P)_2RhCl(B\text{-}4\text{-}Mecat)_2$ (**9.12**) reveals that in each case the bond length involving the apical boryl ligand [1.955 (mean), 1.906(13) Å, respectively] is significantly shorter (2.6% for **9.11**, 6.3% for **9.12**) than that involving the corresponding basal ligand [2.008 (mean), 2.034(12) Å]. Such a trend might be expected simply on the basis of the lack of a competing *trans* σ donor, the influence of which on Rh–B bond lengths has already been noted (vide supra). Alternatively, it has been suggested that the degree of π back-bonding available to the apical boryl ligand in such an environment would also be augmented [11, 172]. As such, the apical Rh–B bond lengths are significantly shorter (8.8% for **9.12**) than the sum of the relevant covalent radii (2.09 Å) [11]. The corresponding bis(triethylphosphine) complex **9.13** has a geometry which is slightly further distorted toward Y-shaped trigonal bipyramidal (as indicated by smaller differences between the "apical" and "basal" Rh–B distances and Cl–Rh–B angles); interestingly attempts to form the analogous PMe_3 complex by phosphine exchange from **9.11**, lead instead to the formation of the six-coordinate bis(boryl) *cis,mer*-$(Me_3P)_3RhCl(Bcat)_2$ (**9.14**), presumably on steric grounds [172]. Exchange chemistry can also be effected on the boryl ligands themselves. Thus, reaction of **9.11** with $B_2(4\text{-Mecat})_2$ in dichloromethane solution generates a mixture of three species: **9.11**, **9.12** and the mixed boryl complex $(Ph_3P)_2RhCl(B\text{-}4\text{-Mecat})(Bcat)$, **9.15**. A σ-bond metathesis process has been suggested as a possible mechanism for this exchange [179].

Tris(boryl) complexes of group 9 metals have been shown to be implicated in arene borylation processes [136, 137]; notwithstanding this, only one structurally characterized Rh(III) tris(boryl) complex has been reported to date [176]. $(Me_3P)_3Rh(Bcat)_3$ (**9.6**) is generated according to the chemistry outlined in Scheme 20, either directly from $(Me_3P)_4RhMe$ and two equivalents of B_2cat_2, or from the mono(boryl) complex $(Me_3P)_4Rh(Bcat)$ (**9.5**) and a single equivalent of the diboron reagent. The corresponding reaction with $(Me_3P)_4Co(Bcat)$ (**9.4**) does not appear to go to completion [161]. The fac coordination geometry provides further evidence for the strong *trans* influence of the Bcat ligand.

7.2.2 Complexes Containing Ancillary Hydride Ligands

Reflecting their historical origins in the studies of CH functionalization and hydroboration, respectively, crystallographically characterized rhodium boryl hydrides can be sub-divided into two broad categories: (i) half-sandwich formally Rh(III) or Rh(V) systems featuring a three- or four-legged piano-stool geometry; and (ii) Rh(III) systems featuring bis(phosphine) ancillary ligand sets and a distorted trigonal bipyramidal coordination geometry.

The complex $Cp^*Rh(H)_2(SiEt_3)Bpin$ (**9.16**: Fig. 28) offers a number of distinct potential bonding situations, for example a description as a formally Rh(V) silyl(boryl)bis(hydride) or as a Rh(III) system featuring a sigma-bound borane, silane or dihydrogen ligand. Crystallographic, computational and reactivity studies reported by Hartwig and co-workers are consistent with a significant residual BH interaction [162]. Thus, one of the hydride ligands located crystallographically is measurably distorted towards the Bpin moiety; the two Rh–H distances differ by 0.3 Å, and one of the H–Rh–B angles is significantly more acute than the other [55.7(16) vs. 67.9(14)°]. Although the shorter B···H separation [1.74(4) Å] is still markedly (ca. 0.4 Å) longer than the range of distances reported for σ-borane complexes [25–27, 59, 100, 101],

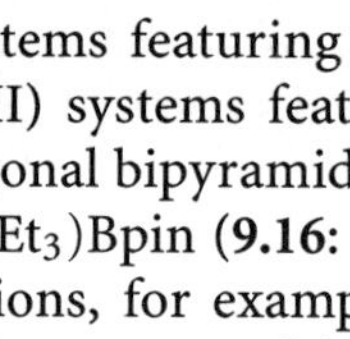

Fig. 28 Rhodium complex **9.16**: a system containing boryl, silyl and hydride ligands

it is however, much shorter than that measured for $(^{i}Pr_3P)_2RhH(Cl)Bpin$, [**9.10**, 2.013(5) Å] by neutron diffraction, for which a residual B···H interaction has also been inferred (vide infra) [14]. Consistent with the idea of a structurally and chemically significant B···H interaction, addition of phosphine to **9.16** leads primarily to the elimination of borane (rather than silane) and the calculated potential energy surface for shortening of the B···H distance is significantly flatter than that for Si···H shortening.

A range of related formally Rh(V) species, of direct relevance to alkane functionalization chemistry has been synthesized from $Cp^*Rh(\eta^4\text{-}C_6Me_6)$ by B–H oxidative addition processes (Scheme 22) [128]. In common with the mixed silyl boryl complex **9.16**, bis- and tris(boryl) species $Cp^*Rh(H)_2(Bpin)_2$, **9.17**, and $Cp^*Rh(H)(Bpin)_3$, **9.18**, also eliminate HBpin on addition of phosphine, in this case yielding $Cp^*Rh(H)(Bpin)P(p\text{-tol})_3$, **9.19**, and $Cp^*Rh(Bpin)_2$ PEt_3, **9.20**, respectively. Furthermore, both **9.17** and **9.18** not only convert arenes and alkanes to the corresponding boronate esters, but can also be observed spectroscopically in the corresponding catalytic reactions using $Cp^*Rh(\eta^4\text{-}C_6Me_6)$ and HBpin. As in the case of **9.16**, NMR, crystallographic and quantum chemical data for **9.17** and **9.18** are consistent with the presence of structurally significant B–H bond character. Thus, the asymmetric positioning of each of the two hydride ligands with respect to the neighbouring boryl groups in **9.17** [short B···H distances: 1.57(7)–1.70(6) Å, long B···H distances: 2.13(5)–2.30(6) Å; narrow H–Rh–B angles: 49(2)–53(2)°, wide H–Rh–B angles: 72(2)–79(2)°] is consistent with significant residual B–H bonding. Computational studies are also consistent with B–H interac-

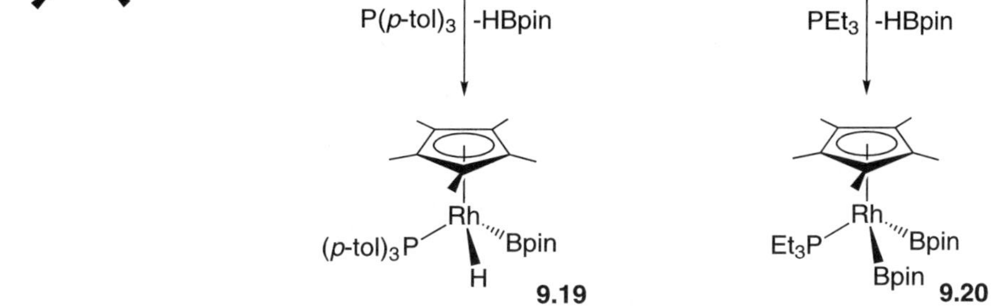

Scheme 22 Interplay between formally Rh(III) and Rh(V) boryl complexes: formation of bis- and tris(boryl) complexes **9.17** and **9.18** via B–H oxidative addition, and their subsequent reactions with phosphine via B – H reductive elimination

tions between each of the boryl ligands and one of the two adjacent hydride ligands. Similar studies on tris(boryl) complex **9.18** also imply the presence of a measurable B–H interaction between the lone hydride ligand and one of the adjacent boryl moieties [short B···H distances: 1.55(3), 1.69(3) Å, long B···H distances: 1.99(3), 2.00(3) Å; narrow H–Rh–B angles: 48(1), 54(1)°, wide H–Rh–B angles: 66(1)–69(1)°]. Furthermore, the residual B–H interactions in compounds **9.17** and **9.18** are not only of relevance to their ground-state structures, but have also been shown computationally: (i) to facilitate the dissociation of borane thereby forming the reactive intermediate which brings about C–H bond cleavage; and (ii) to stabilize the transition state for C–H bond cleavage. These factors, together with the thermodynamic incentive of strong B–C bond formation and the existence of a formally unoccupied orbital on the boryl ligand (which provides rapid kinetics for B–C reductive elimination) are thought to be key factors in catalytic alkane functionalization by these rhodium-based systems. Phosphine-substituted system **9.19** features a much longer Rh–H distance [2.23(3) Å] and a much wider H–Rh–B angle [79(1)°] than either **9.17** or **9.18** and has been formulated as a conventional Rh(III) boryl hydride. A similar description as a Rh(III) bis(boryl) complex pertains for **9.20**.

Residual B···H interactions between mutually *cis* boryl and hydride ligands have also been shown to be relevant in bis(phosphine) rhodium(III) systems of the type $(R_3P)_2RhH(Cl)[B(OR)_2]$ (Scheme 21). Such systems have been shown to be of relevance in the functionalization of organic molecules via a number of approaches, including hydroboration and CH functionalization [127]. The gross structural features of $(^iPr_3P)_2RhH(Cl)Bcat$ (**9.9**) were first established in 1991 [114], namely a distorted trigonal bipyramidal coordination geometry featuring apical phosphine ligands and a torsion angle between the ClRhB and BO_2 planes of 14.9°. Unfortunately the location of the hydride ligand could not be established in this initial crystallographic study, with its presence being inferred from the Cl–Rh–B angle [137.5(2)°] and from NMR data in solution. More recent work from the group of Marder has shown that the related complex $(^iPr_3P)_2RhH(Cl)Bpin$ (**9.10**), which is formed readily from $(^iPr_3P)_2RhCl(N_2)$ and HBpin at 140 °C, functions as a precursor to the active catalytic species in the rhodium-catalyzed borylation of aryl and benzyl CH bonds [127, 134]. As a consequence, a combination of low-temperature neutron and X-ray diffraction experiments and DFT calculations have been brought to bear to provide accurate structural data for the key compounds **9.9** and **9.10** [14]. These studies, which have provided the first accurate location of a hydride ligand in a metal boryl hydride species, also show that the gross structures of both compounds lie approximately midway between the limiting Y-shaped trigonal bipyramidal and distorted square planar structures typically observed for five-coordinate d^6 systems [15, 180]. Several key structural parameters for the superficially similar systems **9.9** and **9.10** are, however, found to differ markedly. Thus, the torsion angle between ClRhB and BO_2 planes for **9.9** (14.9°) reflects a near co-planar alignment, while that for **9.10** (81° by X-ray; 82° by neutron diffraction) reflects a near perpendicular orientation; the Cl–Rh–B angles also differ markedly between the two compounds [137.9(5) and 117.73(4)° for **9.9** and **9.10**, respectively]. Despite these differences, the H–Rh–B angles [68.5(4) and 67.8(2)° for **9.9** and **9.10**] and B···H separations [2.004(10) and 2.013(5) Å] measured from the neutron data, are remarkably similar. In comparison with known σ-borane systems and even with the half-sandwich rhodium systems **9.17** and **9.18**, however, the B···H contacts measured for **9.9** and **9.10** attest to a modest interaction between the hydride and boryl ligands, and consequently to a predominant description as rhodium(III) boryl hydrides.

Interestingly, the difference in ligand orientation between Bpin and Bcat systems is found to be mainly steric in origin. In the case of Bpin system **9.10** the perpendicular orientation allows for interaction of boryl and hydride ligands via the formally vacant BO_2 π^* orbital (to which the B $2p_z$ orbital is a major contributor) in a manner reminiscent of conventional boron-containing Lewis acids. In the case of Bcat system **9.9**, however, the near coplanar arrangement of ClRhB and BO_2 planes means that the B···H interaction involves the perpendicular BO_2 σ^* as the acceptor orbital (Fig. 29) [154].

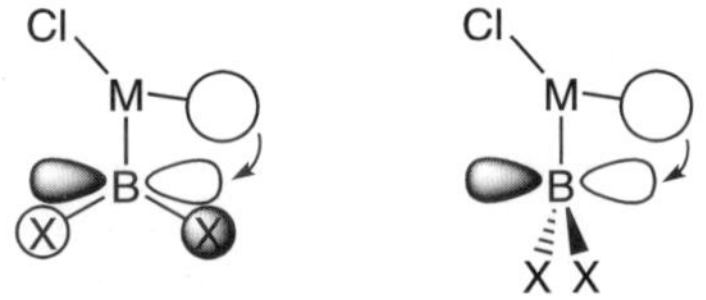

Fig. 29 B···H interactions featuring (*left*) the BX_2 σ^* MO as the acceptor orbital (as in Bcat complex **9.9**) and (*right*) the BO_2 π^* as acceptor (as in Bpin complex **9.10**)

Finally, recent work has examined in detail the *kinetics* of the fundamental B–H oxidative addition step which leads to the formation of rhodium(III) [and ruthenium(II)] boryl hydrides. Conversion of fac-(triphos)Rh(H)$_3$ into fac-(triphos)Rh(H)$_2$(Bpin) via sequential H_2 reductive elimination/HBpin oxidative addition was induced by laser flash photolysis and kinetic data determined from UV measurements. Thus, an extremely high second-order rate constant was determined for the reaction of the 16-electron intermediate (triphos)Rh(H) with HBpin in THF (2.3×10^8 mol^{-1} s^{-1} dm^3) [181].

7.2.3 Semi-Bridging Ligands

Rhodium also provides a very rare crystallographically characterized example of a boryl ligand adopting a semi-bridging mode of coordination

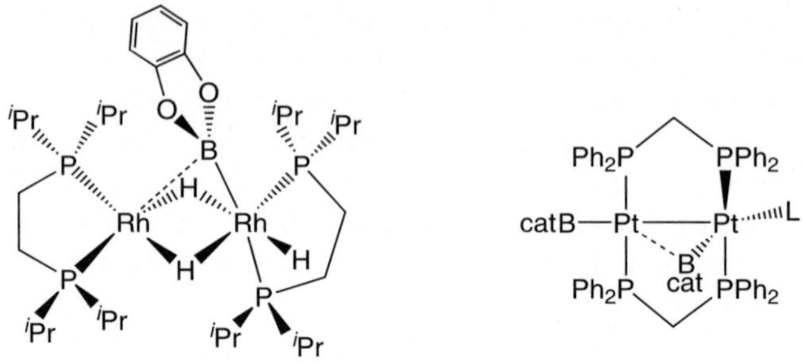

9.21

10.31: L = PPh_3
10.32: L = κ^1-dppm

Fig. 30 Structures of dinuclear rhodium and platinum complexes featuring semi-bridging Bcat ligands

between two metal centres [76, 113, 182]. Here too an auxiliary interaction involving the Bcat ligand has been identified, in this case involving a second rhodium centre. Reaction of $[(dippe)Rh(\mu\text{-}H)]_2$ with HBcat initially generates the mixed valence Rh(I)/Rh(III) boryl hydride $(dippe)Rh(\mu\text{-}H)_2(\mu\text{-}Bcat)Rh(H)(dippe)$ (**9.21**: Fig. 30). The structure and bonding in **9.21** have been analyzed with the aid of DFT calculations carried out on model systems [113]. Starting from an initial description in which the semi-bridging Rh(I)···B interaction is ignored, the two metals can be described as a distorted square planar d^8 Rh(I) centre and a distorted octahedral d^6 Rh(III) boryl hydride. The interaction between the two rhodium centres is exclusively via the bridging hydride ligands and the Rh(I)···B interaction can then be described in terms of a Lewis acid/base interaction with the filled d_{z^2} orbital at the square planar Rh(I) centre acting as the donor component and the formally vacant p orbital at boron as the acceptor. Indeed, this type of metal Lewis acid/borane Lewis base interaction is reminiscent of that found in group 9 metal boratranes [163–165, 167] and in recently reported examples of late transition-metal complexes of PBP ambiphilic ligands [166].

7.3 Iridium Complexes

As with rhodium and platinum, iridium boryl systems have proved to be a valuable structural proving ground on which to establish fundamental properties of the BX_2 ligand (Table 10). In a manner also mimicking rhodium, a significant body of work has emerged more recently due to the implication of iridium boryls in hydrocarbon activation [10, 126, 129, 136–148, 150, 151, 155–159]

Table 10 Selected data for structurally characterized iridium boryl complexes

Compound	*d*(M–B) (Å)	Other data	Refs.
$(Me_3P)_3IrH(Cl)(Bcat)$-mer (**9.22**)	2.023(10)	Ir–P 2.351(2) Å trans to H; 2.304(2), 2.307(2) Å δ_B 32.8	[117]
$(Me_3P)_3IrCl(Br)(Bcat)$-mer[a] (**9.24**)	2.035(8) 2.023(8)	Ir–P 2.268(2), 2.274(2) Å trans to Cl; 2.352(2), 2.357(2), 2.357(2), 2.341(2) Å	[178]
$(Me_3P)_3Ir(Br)_2(Bcat)$-mer[a] (**9.25**)	2.012(16) 2.026(16)	Ir–P 2.266(3), 2.277(3) Å trans to Br; 2.357(4) 2.366(4), 2.354(4), 2.345(4) Å δ_B 22.5	[178]
$(Ph_3P)_2(OC)IrH(Cl)(Bcat)$-trans[b] (**9.26**)	2.045(5)	Ir–P 2.330(1), 2.335(1) Å δ_B 30.8	[116]
$(Me_3P)_3Ir(H)_2$(9-BBN)-fac (**9.27**)	2.093(7)	Ir–P 2.345 Å trans to boryl; 2.288, 2.304 Å δ_B 106.2	[35]
$(Me_3P)_3IrCl[B(F)(biph)]^+[BPh_4]^-$ (**9.28**)	2.00(1)	Ir–P 2.348(3) Å trans to boryl; 2.432(3) Å trans to aryl 2.333(4) Å trans to chloride	[173, 174]
$Cp^*Ir(H)_3Bpin$ (**9.30**)	2.047(4)	δ_B 36.0	[129]
$(Me_3P)_3Ir(Cl)(Bcat)_2$-mer (**9.23**)	2.080(6) 2.024(6)	Ir–P 2.399(1) Å trans to Bcat; 2.322(1), 2.329(1) Å δ_B 41.7, 32.6	[183]
$(Et_3P)_2Ir(Cl)(Bcat)_2$-*trans* (**9.31**)	1.991(6)[c] 2.004(6)[d]	B–Rh–Cl 132.9(2)[c], 150.4(2)°[d] δ_B 30.1	[172]
$[(tbbpy)(cod)Ir(Bpin)_2]^+[OTf]^-$ (**9.33**)	2.06(2) 2.12(2)	δ_B 34	[136]
$(\eta^6$-toluene$)Ir(Bcat)_3$ (**9.34**)	2.018(5) 2.024(5) 2.036(4)	B–Ir–B 80.3(2), 83.1(2), 84.4(2)° δ_B 37.9	[137]
$(\eta^6$-mesitylene$)Ir(Bcat)_3$ (**9.35**)	2.016(3)	B–Ir–B 81.6(1)° δ_B 37.8	[137]
$(tbbpy)(coe)Ir(Bpin)_3$ (**9.36**)	2.055(7) 2.057(6) 2.207(6)	B–Ir–B 81.7(2), 84.0(2), 84.9(2)° δ_B 37	[136]
fac-$(Ph_3P)_2(OC)Ir(BF_2)_3$ (**9.37**)	2.066(10) 2.083(6) 2.088(5)	B–Ir–B 80.7(2), 82.4(4), 86.2(5)° δ_B 32.4	[185]

[a] Two crystallographically independent molecules
[b] As the dichloromethane solvate
[c] For the apical boryl ligand
[d] For the basal boryl ligand

7.3.1 Mono(boryl) Complexes

A number of octahedral iridium mono(boryl) complexes were reported in the period prior to 1998. These have been reviewed in detail previously [11] and are briefly examined below, together with more recent additions. With the exception of the 9-BBN derivative fac-$(Me_3P)_3Ir(H)_2$(9-BBN) (**9.27**), the structures of the iridium(III) mono(boryl) complexes **9.22**, **9.24–9.26** and the closely related bis(boryl) complex mer-$(Me_3P)_3Ir(Cl)(Bcat)_2$ (**9.23**) all feature a meridional arrangement of the three neutral donors [i.e. containing mer-$(R_3P)_3Ir$ or $(R_3P)_2(OC)Ir$ fragments] (Fig. 31). Such an arrangement of the coordination sphere is consistent with the ligand *trans* influences being ordered along the lines of H^-, $BX_2 > PR_3 > Cl^-$, Br^- [35, 116, 117, 178, 183]. A further direct comparison of the *trans* influences of the Bcat, H^-, PMe_3, Cl^- and Br^- ligands can be made by consideration of the structural parameters for mer-$(Me_3P)_3IrH(Cl)(Bcat)$ (**9.22**), **9.23**, and mer-$(Me_3P)_3IrBr(X)(Bcat)$ (**9.24**: X = Cl; **9.25** X = Br). Thus, the Ir–P distance associated with the PMe_3 ligand *trans* to the hydride in **9.22** is 2.351(2) Å, while that measured for PMe_3 *trans* to Bcat in the otherwise identical coordination sphere of **9.23** is 2.399(1) Å. The corresponding Ir–P distances in **9.24** and **9.25**, for PMe_3 *trans* to chloride or bromide are reported to be 2.271 (mean) and 2.272 Å (mean), respectively [117, 178, 183]. Additionally, the Ir–P distances in **9.22** and **9.23** for the remaining pair of PMe_3 ligands (which are *trans* to each other) are 2.305 (mean) and 2.326 Å (mean), respectively. Thus, the available data on these systems is consistent with Bcat not only being a significantly stronger σ-donor ligand than halide or PMe_3 ligands, but also a stronger σ donor than hydride. A similar conclusion can be reached by consideration of the Ir–P

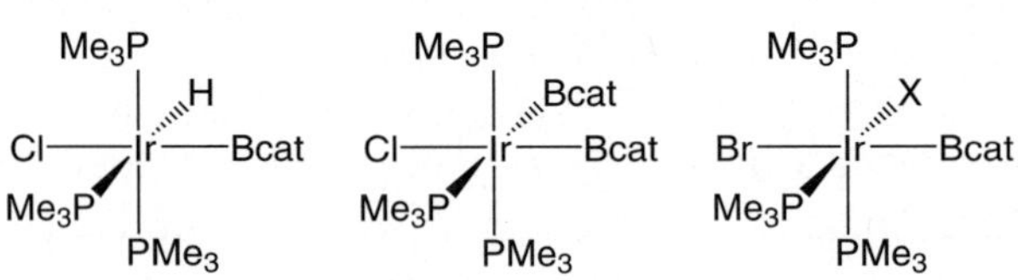

9.22 **9.23** **9.24**: X = Cl **9.25**: X = Br

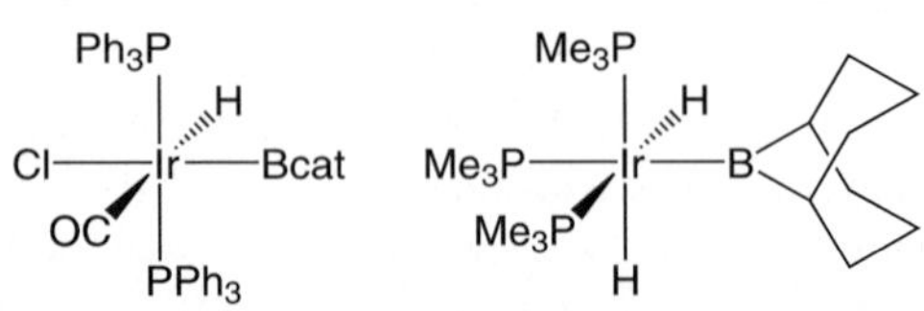

9.26 **9.27**

Fig. 31 Crystallographically characterized octahedrally coordinated iridium(III) boryl complexes 9.22–9.27

bond lengths for the PMe_3 ligands *trans* to the boryl ligand (2.345 Å) and to the two hydrides [2.296 Å (mean)] in **9.27** [35]. Recent theoretical studies carried out on square planar platinum(II) boryl complexes have sought to further rank the *trans* influence of a range of boryl systems with respect to other ligands known to have strong σ-donor capabilities, e.g. alkyl and silyl ligands (vide infra) [175].

The Ir–B bond lengths in the octahedral iridium(III) mono(boryl) systems **9.22** and **9.24–9.27** fall within the range 2.012(16)–2.093(7) Å. In general, variation in this distance can be correlated with the expected properties of the ligand *trans* to the boryl group. Thus, in the bis(boryl) system **9.23** Ir–B distances of 2.080(6) and 2.024(6) Å have been determined for the Bcat ligands *trans* to PMe_3 and chloride, respectively [183]. Furthermore, comparison of the bond lengths measured for Bcat ligands *trans* to a halide, i.e. **9.22** [2.023(10) Å], **9.23** [2.024(6) Å], **9.24** [2.029 Å (mean)], **9.25** [2.019 Å (mean)] and **9.26** [2.045(4) Å] with (i) the sum of the covalent radii (2.10 Å) [11]; and (ii) the Ir–B distance measured for the effectively σ-only bond in *trans,cis*-$(Me_3P)_2(OC)Ir(Br)_2(2\text{-}B_5H_8)$ [184], in which the boron ligand is also *trans* to a halide [2.071(14) Å], indicates a degree of shortening which might be consistent with a modest Ir→B π interaction.

Iridium mono(boryl) species featuring a metal oxidation state of greater than three are relatively uncommon. The Ir(IV) species $(Me_3P)_3IrCl[B(F)(biph)]^+[BPh_4]^-$ (**9.28**) produced by oxidation of the Ir(III) complex $(Me_3P)_3IrCl(biph)$ (biph = biphenyl-2,2′-diyl) by $[NO]^+$ and insertion of a BF moiety from the $[BF_4]^-$ counter-ion, also has a mer-octahedral geometry and is one of only two cationic iridium boryl systems [136, 173, 174]. Interestingly, the Ir–P distances in this case for the phosphines *trans* to the aryl and boryl ligands are 2.432(3) and 2.348(3) Å, respectively. Whether this somewhat counter-intuitive ordering is due to the relative *trans* influences of the aryl and boryl ligands, or is related to the geometric constraints of the chelating biphenylboryl unit is not clear. The iridium(V) boryl species $Cp^*Ir(H)_2(Bpin)_2$, **9.29**, and $Cp^*Ir(H)_3Bpin$, **9.30**, have been synthesized by Kawamura and Hartwig using both oxidative addition and salt-elimination approaches [129]. These systems are of relevance as potential intermediates in the iridium-catalyzed functionalization of alkanes [133], and indeed both are found to react with *n*-octane at 200 °C to give a ca. 50% yield of (*n*-octyl)Bpin after 48 h. The crystal structure of **9.30** was obtained, although in contrast to the structures of the related rhodium(V) species **9.17** and **9.18**, location of the hydride ligands was not possible. Nevertheless, the authors favour a description as an iridium(V) boryl tris(hydride), rather than as an iridium(III) σ-borane, σ-dihydrogen or borohydride complex, principally on the basis of sharp hydride ^1H NMR resonances (even at low temperatures), and an ^{11}B NMR signal which shows no ^1H coupling and a chemical shift (δ_B 36.0) reminiscent of other iridium boryl systems (see Table 10). Additionally, the Ir–B distance [2.047(7) Å] is also within the range expected for iridium boryl complexes (Table 10), with correspondingly longer contacts being expected for σ-borane or borohydride ligands [25–27, 33, 59, 100, 101].

7.3.2 Bis(boryl) Complexes

Bis(boryl) complexes of iridium have also been isolated, primarily by exploiting the oxidative addition of diborane(4) species to iridium(I) precursors. Thus, reaction of B_2cat_2 with $(Et_3P)_3IrCl$ leads to the isolation of five-coordinate *trans*-$(Et_3P)_2IrCl(Bcat)_2$ (**9.31**), which can be compared to the six-coordinate complex mer-$(Me_3P)_3IrCl(Bcat)_2$ (**9.23**) discussed above. As in the case of the corresponding rhodium(III) complexes *trans*-$(Et_3P)_2RhCl(Bcat)_2$ (**9.13**) and mer-$(Me_3P)_3RhCl(Bcat)_2$ (**9.14**), the differences in steric constraints between PMe_3 and PEt_3 ligands are clearly important in determining whether a five- or six-coordinate geometry is prevalent. In contrast to the rhodium systems, however, Marder and co-workers report NMR evidence for the formation of the tris(triethylphosphine) complex $(Et_3P)_3IrCl(Bcat)_2$ (**9.32**) in solution, with the expected high lability of the phosphine *trans* to the Bcat ligand being responsible for the crystallization of the bis(triethylphosphine) complex **9.31** [172]. The structure of **9.31** is isomorphous with its rhodium analogue **9.13** discussed above, and the relatively small differences between the Ir–B distances [1.991(6), 2.004(6) Å] and B–Ir–Cl angles [132.9(2), 150.4(2)°] for the two Bcat ligands reflect a structure which is similarly intermediate between square pyramidal and Y-shaped trigonal bipyramidal [15, 172].

The structure of a further iridium bis(boryl) complex, the cationic species $[(tbbpy)(cod)Ir(Bpin)_2]^+[OTf]^-$ (**9.33**), together with that of a related neutral tris(boryl) system (vide infra), has been reported as part of a study to elucidate the mechanism and intermediates of arene borylation chemistry [136]. **9.33** features an approximately octahedral coordination geometry with mutually *cis* pairs of bipyridyl, diene and Bpin ligands. The two Ir–B bond lengths [2.06(2) and 2.12(1) Å] are effectively identical within the standard 3σ limit.

7.3.3 Tris(boryl) Complexes

Tris(boryl) complexes have been shown to have a central role in the iridium-catalyzed borylation of arenes [136, 137]. The syntheses of the half-sandwich iridium(III) systems (η^6-toluene)$Ir(Bcat)_3$, **9.34**, and (η^6-mesitylene)$Ir(Bcat)_3$, **9.35**, were reported by Marder and co-workers in 1993, by exploiting the reaction of (η^5-Ind)Ir(cod) with excess HBcat using the arene as a solvent. The organic products of the reactions were found not only to include species resulting from the hydroboration or hydrogenation of the cod or indenyl ligands, but also compounds derived from borylation of the solvent [137]. Ir–B bond lengths for **9.34** [2.018(5)–2.036(4) Å] and **9.35** [2.016(3) Å] are within the range expected for iridium(III) boryl linkages (Table 10) and the B–Ir–B angles [80.3(2)–84.4(2)° for **9.34**; 81.6(1)° for **9.35**] are as expected for three-legged piano-stool complexes of this type. There is no hint, based on the B···B distances or on the relative orientations of the BO_2 ligand planes of any intramolecular B···B interactions. The relatively weak nature of the Ir(III) arene interaction in these complexes is evidenced by ready reaction of **9.35** with three equivalents of triethylphosphine which generates fac-$(Et_3P)_3Ir(Bcat)_3$. More recently, the complex (tbbpy)(coe)Ir$(Bpin)_3$ (**9.36**) has been reported by Ishiyama, Miyaura, Hartwig and co-workers to be a kinetically and chemically viable intermediate in the room-temperature borylation of arenes [136, 138]. As with all known examples of group 9 tris(boryls), the strong *trans* influence of the boryl ligand enforces a facial coordination geometry (Fig. 32). **9.36** is shown to be produced in high yield from the reaction of $[(cod)Ir(OMe)]_2$, tbbpy, coe and HBpin in cyclohexane-d_{12}, although in much lower yield using B_2pin_2 as the boron source. Both stoichiometric and catalytic arene borylation reactions are thought to feature the intermediate (tbbpy)Ir$(Bpin)_3$, formed by reversible dissociation of coe from **9.36**, as the active species in C–H bond cleavage. The viability of an alternative intermediate, namely the Ir(I) monoboryl complex (tbbpy)Ir(Bpin), was discounted on the basis of its much lower concentration in the catalytic borylation system and of theoretical studies showing that the barrier to reductive elimination of B_2pin_2 from (tbbpy)Ir$(Bpin)_3$ is higher than the barrier to C–H activation [138].

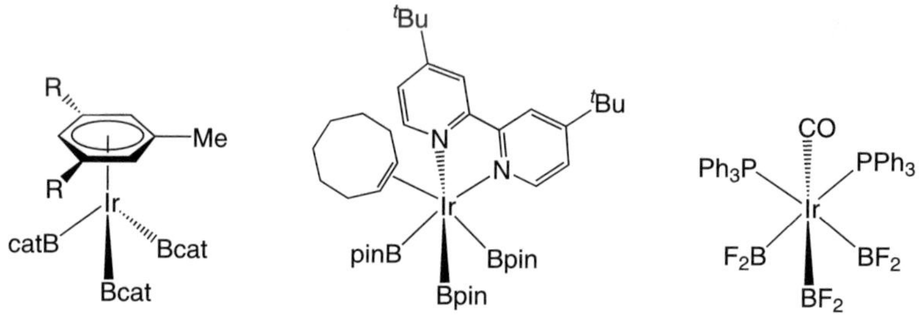

9.34: R = H
9.35: R = Me

9.36

9.37

Fig. 32 Crystallographically characterized iridium tris(boryl) complexes **9.34–9.37**

One final crystallographically characterized iridium tris(boryl) system has been reported to date. Norman and co-workers, as apart of a study of the coordination chemistry of the difluoroboryl ligand, have synthesized fac-$(Ph_3P)_2(OC)Ir(BF_2)_3$ (**9.37**) from *trans*-$(Ph_3P)_2(OC)IrCl$ and B_2F_4 [185]. Although the mechanism for the formation of **9.37** has not been definitively established, BF_2Cl has been suggested as a possible co-product, with the formation of **9.37** from the iridium(I) chloride starting material proceeding via a combination of oxidative addition and σ-bond metathesis steps. Superficially at least, the formation of **9.37** under these conditions mirrors the reaction of $(Me_3P)_4RhMe$ with two equivalents of B_2cat_2 to generate fac-$(Me_3P)_3Rh(Bcat)_3$ (**9.6**) and MeBcat [176, 177]. The structure of **9.37** shows the expected fac orientation of the boryl ligand set and Ir–B bond lengths within the precedented range (Table 10). A related iridium mono(difluoroboryl) complex was reported by Bergman and co-workers, although no structural data were forthcoming [186].

8
Group 10 (Nickel, Palladium and Platinum)

The boryl chemistry of group 10 is dominated by platinum, with mono- and bis(boryl) platinum(II) systems typically being synthesized by oxidative addition of B – E bonds to low valent metal precursors. The relevance of such oxidative addition processes to the platinum (and to a lesser extent palladium) catalyzed addition of B – E linkages across unsaturated organic bonds has been a powerful driving force for the investigation of group 10 boryl complexes. Thus, for example, the platinum-catalyzed diboration of alkynes and other unsaturated C – E bonds by diboron(4) reagents [6, 187–198], has led to the isolation of a wide range of geometrically *cis* bis(boryl) complexes. As with the group 9 systems discussed above, a range of complementary quantum chemical studies have sought to clarify issues of structure/bonding and reactivity [113, 153, 175, 199–205].

8.1
Nickel and Palladium Complexes

Isolated boryl complexes of nickel are essentially unknown, there being no reports in the literature of crystallographically authenticated examples, despite the implication of nickel catalysts in the addition of B – Si bonds to unsaturated substrates [206–208]. Attempts to form Ni – B bonds by exploiting the salt elimination chemistry used widely for group 8 systems, and organonickel anions such as K[CpNi(CO)] lead instead to boryloxycarbyne species, resulting from attack at the boron electrophile by the oxygen of the ancillary carbonyl ligand [44]. Such behaviour, indicative of alternative (non-metal based) sources of nucleophilic reactivity for organometallic anions, has also been observed for group 6 and group 8 systems [42–44, 209].

Examples of palladium boryl systems are slightly more common (four crystallographically characterized examples, Table 11) and their reactivity, for example with respect to alkyne insertion, has been investigated [210–213]. The complex *cis*-(dmpe)Pd($SnMe_3$)[B{N(Me)CH_2}$_2$] (**10.1**: Fig. 33) was

Table 11 Selected data for structurally characterized palladium boryl complexes

Compound	d(M–B) (Å)	Other data	Refs.
(dmpe)Pd($SnMe_3$)[B{N(Me)CH_2}$_2$]-cis (**10.1**)	2.077(6)	Pd – P 2.338(2) Å trans to boryl 2.280(2) Å trans to $SnMe_3$ δ_B 46.9	[208]
(Me_3P)$_2$PdCl[B{N(Me)CH_2}$_2$]-*trans* (**10.2**)	2.037(9)	δ_B 40.0	[207]
(Cy_3P)$_2$PdBr(BNSiMe_3)-*trans* (**10.4**) [a]	1.958(3) 1.967(3)	Pt – Br 2.553(1), 2.522(1) Å B – N 1.251(4), 1.262(3) Å δ_B 22.0	[211]
Cp*Fe(μ_2-CO)$_2$(μ_y-BCl_2)Pd(PCy_3) [a] (**10.6**)	2.062(4) [b] 2.090(4) [b] 2.095(4) [c] 2.078(4) [c]	Fe – B – Pd 75.2(1), 75.0(1) δ_B 72.2	[76]

[a] Two crystallographically independent molecules
[b] For the Pd–B bond
[c] For the Fe–B bond

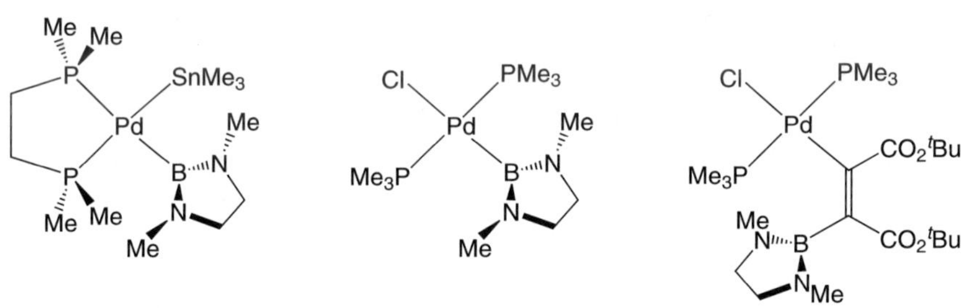

10.1 **10.2** **10.3**

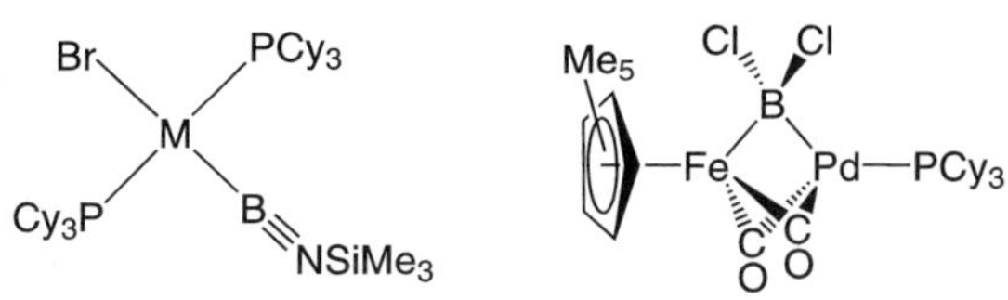

10.4: M = Pd
10.5: M = Pt

10.6

Fig. 33 Crystallographically characterized palladium boryl and related complexes **10.1**–**10.6**

isolated by Tanaka and co-workers in 1996, as part of a study designed to elucidate the mechanism of syn addition of boron-tin bonds to alkynes catalyzed by Pd(0) complexes [211]. **10.1** was synthesized by the reaction of (dmpe)$PdMe_2$ with $Me_3SnB\{N(Me)CH_2\}_2$ with concomitant generation of $SnMe_4$ and $MeB\{N(Me)CH_2\}_2$. Structurally, the square-planar Pd(II) centre features a *cis* disposition of the boryl and stannyl ligands, with the relevant *trans* Pd–P distances [2.338(2) and 2.280(2) Å, respectively] reflecting the greater *trans* influence of the boryl ligand compared to stannyl [175, 211]. The related complex *trans*-$(Me_3P)_2PdCl[B\{N(Me)CH_2\}_2]$ (**10.2**) has been reported in a follow-up study, and the product of alkyne insertion into the Pd–B bond (**10.3**) has also been structurally characterized in one case. Reaction of $CpPd(\eta^3\text{-}C_3H_5)$ with $Cl[B\{N(Me)CH_2\}_2]$ in the presence of two equivalents of trimethylphosphine proceeds via oxidative addition of the B–Cl bond to the formation of **10.2**. In common with the much wider range of related square planar platinum(II) complexes containing halide and boryl ligands (vide infra), and as expected on the basis of the respective ligand *trans* influences, a *trans* configuration of the phosphine ligands is observed. The Pd–B distances measured for both **10.1** and **10.2** [2.077(6), 2.037(9) Å, respectively] are significantly shorter than the sum of the covalent radii (2.16 Å) [11], and this taken together with the near perpendicular alignment of the metal coordination plane and that of the boryl ligand, is suggestive of moderate Pd–B π interactions [11, 210]. A similar *trans* coordination geometry is also observed for the iminoboryl complex $(Cy_3P)_2PdBr(BNSiMe_3)$ (**10.4**), which has been synthesized from $Pd(PCy_3)_2$ and $(Me_3Si)_2NBBr_2$ in a manner analogous to that used for the corresponding platinum complex **10.5**. Both systems are discussed in the section on platinum complexes (below) which offers a greater basis for comparative structural analysis [214].

Finally, the compound $Cp^*Fe(\mu_2\text{-}CO)_2(\mu_2\text{-}BCl_2)Pd(PCy_3)$ (**10.6**) reported in 2005 by Braunschweig and co-workers represents a very rare example of a metal complex exhibiting a bridging coordination mode of the boryl ligand [76, 113, 182]. As might be expected, given the increased coordination number at boron, and the bridging mode of coordination adopted by the BCl_2 ligand, the Fe–B bond length [2.087 Å (mean)] is markedly longer than that found in the parent compound $CpFe(CO)_2BCl_2$ [1.942(3) Å] [74]. The Pd–B distance [2.076 Å (mean)], on the other hand, is similar to those reported for Pd(II) boryl complexes, despite the fact that the formal oxidation state of palladium in **10.6** is zero, *and* the boryl ligand is four-coordinate. These observations led the authors to conclude that the Pd–B interaction is a strong one, and that **10.6** therefore represents a unique example of a symmetrically bridging boryl ligand [76]. Such a description contrasts with the highly asymmetric semi-bridging interaction of the boryl ligand with the two adjacent metal centres found in the rhodium complex (dippe)$Rh(\mu\text{-}H)_2(\mu\text{-}Bcat)Rh(H)$(dippe) (**9.21**) and the platinum systems $Pt_2(PPh_3)(\mu_2\text{-}dppm)_2(Bcat)(\mu_2\text{-}Bcat)$ (**10.31**) and $Pt_2(\kappa^1\text{-}dppm)(\mu_2\text{-}$

dppm)$_2$(Bcat)(μ_2-Bcat) (**10.32**) [113, 182]. An alternative description of the bonding in **10.6** is as a σ-borane complex, i.e. a Pd–B bond consisting of a Lewis acid/base interaction in which the Pd(PCy$_3$) fragment functions as the donor and the boryl ligand as the acceptor component [16–165, 167, 268]. In related work, a number of structurally interesting borylene complexes featuring bonds between boron and palladium have also been reported from reactions involving Pd(PCy$_3$)$_2$ [94, 216]. These are discussed elsewhere in this volume.

8.2
Platinum Complexes

Compared to the lighter elements of group 10, platinum boryl complexes have received considerable attention, not least because of their implication in the diboration of unsaturated organic substrates [6, 187–195]. Much of this chemistry has been reviewed previously [11], and mechanistic aspects of both stoichiometric and catalytic reactions of platinum bis(boryl) systems with alkynes have been thoroughly investigated. Platinum(II) boryl complexes also offer useful structural and spectroscopic probes with which to interrogate the fundamental properties of the boryl ligand (Table 12). Thus, for example $^1J_{\mathrm{Pt-P}}$ coupling constants offer a way of comparing the σ donor properties of boryl and other ligands *trans* to the phosphine of interest. Such studies have also been complemented by recent computational studies of fundamental boryl ligand properties in platinum(II) systems [175].

8.2.1
Mono(boryl) Complexes

Crystallographically characterized platinum(II) mono(boryl) complexes have been reported featuring both *cis* and *trans* square planar coordination geometries at the metal centre (Fig. 34) [92, 182, 214, 217, 218]. In addition, a single example of a T-shaped, three-coordinate bis(phosphine) platinum(II) boryl system has been reported [219]. The adoption of either *cis* or *trans* geometries for systems of the type (R$_3$P)$_2$PtY(BX$_2$) can be rationalized in terms of the relative *trans* influences of the phosphine, boryl and Y ligands. Given the ordering BX$_2$ > PR$_3$ established above for rhodium and iridium systems, complexes with the remaining ligand (Y) of weaker *trans* influence than PR$_3$ consequently adopt *trans* geometries [e.g. *trans*-(Ph$_3$P)$_2$PtCl(Bcat), **10.7**]. For systems featuring a Y ligand of stronger *trans* influence than PR$_3$, a *cis* geometry is therefore rationalized {e.g. *cis*-(Ph$_3$P)$_2$Pt(EMe$_3$)[B(NMe$_2$)$_2$]; **10.8**: E = Ge, **10.9**: E = Sn} [217, 218].

The complexes **10.8**, **10.9** and *cis*-(Ph$_3$P)$_2$Pt(EMe$_3$)[B(N^{i}Pr)$_2$C$_6$H$_4$-1,2] (**10.10**) have been synthesized by Nöth and Habereder via the oxidative addition of boron-tin or boron-germanium bonds to Pt(0) precursors [217].

Table 12 Selected data for structurally characterized platinum boryl complexes

Compound	d(M–B) (Å)	Other data	Refs.
$(Cy_3P)_2PtBr(BNSiMe_3)$-*trans* (**10.5**)	1.960(3)	Pt–Br 2.552(1) Å B–N 1.260(4) Å δ_B 25.9	[211]
$(Ph_3P)_2Pt(GeMe_3)[B(NMe_2)_2]$-*cis* (**10.8**)	2.139(6)	Pt–P 2.377(1) Å trans to boryl 2.304(1) Å trans to germyl δ_B 46.7 $^1J_{Pt-P}$ 1200 Hz trans to boryl 2220 Hz trans to germyl	[214]
$(Ph_3P)_2Pt(SnMe_3)[B(NMe_2)_2]$-*cis* (**10.8**)	2.136(4)	Pt–P 2.369(1) Å trans to boryl 2.303(1) Å trans to stannyl δ_B 45.6 $^1J_{Pt-P}$ 1392 Hz trans to boryl 2681 Hz trans to stannyl	[214]
$(Ph_3P)_2Pt(EMe_3)[B(N^iPr)_2C_6H_4$-1,2]-*cis* (**10.10**)	2.085(6)	Pt–P 2.376(1) Å trans to boryl 2.301(1) Å trans to stannyl δ_B 44.7 $^1J_{Pt-P}$ 1560 Hz trans to boryl 2448 Hz trans to stannyl	[214]
$(Ph_3P)_2PtCl(Bcat)$-*trans* (**10.11**)	2.008(8)	Pt–Cl 2.445(1) Å δ_B 28.7	[215]
$(Ph_3P)_2PtCl[B(NMe_2)Cl]$-*trans* (**10.12**)	2.084(3)	Pt–Cl 2.470(2) Å δ_B 36.4	[182]
$(Cy_3P)_2PtBr[B(Fc)Br]$-*trans* (**10.13**)	1.996(3)	Pt–Br 2.614(1) Å δ_B 82	[182]
$[(Cy_3P)_2Pt\{B(Fc)Br\}]^+[BAr^f_4]^-$ (**10.14**)	1.966(3)	P–Pt–P 162.96(3)° No ^{11}B signal observed	[216]
$[(Cy_3P)_2PtBr\{BFc(4\text{-pic})\}$-*trans*$]^+[BAr^f_4]^-$ (**10.15**)	2.014(5)	Pt–Br 2.606(1) Å No ^{11}B signal observed	[216]
$(Ph_3P)_2Pt(Bcat)_2$-*cis*[a] (**10.16**)	2.040(6) 2.058(6)	B···B 2.552 Å B–Pt–B 77.1(2)° δ_B 47.0 $^1J_{Pt-P}$ 1639 Hz	[188]
$(Ph_3P)_2Pt(Bcat)_2$-*cis*[b] (**10.16**)	2.08(2) 2.07(2)	B···B 2.61 Å B–Pt–B 77.9(7)° No ^{11}B signal observed $^1J_{Pt-P}$ 1608 Hz	[189]
$(Ph_3P)_2Pt(4\text{-}^tBuBcat)_2$-*cis* (**10.17**)	2.045(11) 2.046(13)	B···B 2.554 Å B–Pt–B 77.2(4)° δ_B 50.1 $^1J_{Pt-P}$ 1621 Hz	[188]
$(Ph_3P)_2Pt(BcatCl_4)_2$-*cis*[b] (**10.18**)	2.04(2) 2.03(2)	B···B 2.616 Å B–Pt–B 80.0(8)° δ_B 51.5 $^1J_{Pt-P}$ 1608 Hz	[215]

Table 12 (continued)

Compound	d(M–B) (Å)	Other data	Refs.
$(Ph_3P)_2Pt(Bpin)_2$-*cis*[b] (**10.19**)	2.076(6)	B···B 2.537 Å	[190]
	2.078(6)	B–Pt–B 75.3(3)°	
		δ_B 46.0	
		$^1J_{Pt-P}$ 1516 Hz	
$(Ph_3P)_2Pt(Bped\text{-}S)_2$-*cis* (**10.20**)	2.054(4)	B···B 2.525 Å	[187]
	2.070(3)	B–Pt–B 75.5(2)°	
		δ_B 48.2	
		$^1J_{Pt-P}$ 1578 Hz	
$(Ph_3P)_2Pt(Btart\text{-}R,R)_2$-*cis* (**10.21**)	2.052(7)	B···B 2.451 Å	[187]
	2.067(5)	B–Pt–B 73.0(4)°	
		δ_B 48.1	
		$^1J_{Pt-P}$ 1634 Hz	
$(Ph_3P)_2Pt(Bthiocat)_2$-*cis*[b] (**10.22**)	2.056(4)	B···B 2.612 Å	[215]
	2.075(4)	B–Pt–B 78.5(2)°	
		δ_B 72	
		$^1J_{Pt-P}$ 1600 Hz	
$(Ph_3P)_2Pt(BF_2)_2$-*cis* (**10.23**)	2.052(6)	B···B 2.591 Å	[185,
	2.058(6)	B–Pt–B 78.2(3)°	223]
		δ_B 42.3	
		$^1J_{Pt-P}$ 1607 Hz	
$(Ph_3P)_2Pt[B(OMe)_2]_2$-*cis* (**10.24**)	2.098(4)	B···B 2.494 Å	[11]
	2.100(4)	B–Pt–B 72.9(2)°	
$(Ph_3P)_2Pt[B(NMe_2)Cl]_2$-*cis* (**10.25**)	2.076(3)	B···B 2.537 Å	[182]
	2.084(3)	B–Pt–B 75.2(1)°	
		δ_B 50.8	
		$^1J_{Pt-P}$ 1434 Hz	
$(Ph_3P)(Cy_3P)Pt(Bcat)_2$-*cis* (**10.26**)	2.044(4)	B···B 2.445 Å	[182]
	2.050(4)	B–Pt–B 73.3(2)°	
		δ_B 48.8	
		$^1J_{Pt-P}$ 1713, 1564 Hz	
$(dppe)Pt(Bcat)_2$-*cis* (**10.27**)	2.048(8)	B···B 2.667 Å	[188]
	2.058(8)	B–Pt–B 81.0(3)°	
		δ_B 48.9	
		$^1J_{Pt-P}$ 1454 Hz	
$(dppb)Pt(Bcat)_2$-*cis* (**10.28**)	2.031(8)	B···B 2.514 Å	[188]
		B–Pt–B 76.5(4)°	
		δ_B 48.9	
		$^1J_{Pt-P}$ 1589 Hz	
$(dppb)Pt(BF_2)_2$-*cis* (**10.29**)	2.044(6)	B···B 2.671 Å	[185]
	2.047(6)	B–Pt–B 81.5(1)°	
		δ_B 43.9	
		$^1J_{Pt-P}$ 1526 Hz	

Table 12 (continued)

Compound	d(M–B) (Å)	Other data	Refs.
$(Et_3P)_2Pt[\{B(NMe_2)\}_2(\eta^5\text{-}C_5H_4)\text{-}(\eta^7\text{-}C_7H_6)Cr]$-*cis* (**10.30**)	2.108(3) 2.096(3)	B···B 2.601 Å B–Pt–B 76.4(1)° δ_B 60.6, 67.0 $^1J_{Pt\text{–}P}$ 1139, 1186 Hz	[224]
$Pt_2(PPh_3)(\mu_2\text{-dppm})_2(Bcat)(\mu_2\text{-}Bcat)$ (**10.31**)	2.04(2)^c 2.12(2)^d 2.49(2)^e	Pt–Pt 2.772(1) Å ^{11}B resonances not reported	[182]
$Pt_2(\kappa^1\text{-dppm})(\mu_2\text{-dppm})_2(Bcat)(\mu_2\text{-}Bcat)$ (**10.32**)	2.06(2)^c 2.14(2)^d 2.51(2)^e	Pt–Pt 2.769(1) Å ^{11}B resonances not reported	[182]

[a] As the benzene-d_6 solvate
[b] As the toluene solvate
[c] For the Pt(II)–B bond associated with the terminal Bcat ligand
[d] For the Pt(0)–B bond associated with the semi-bridging Bcat ligand
[e] For the Pt(II)–B bond associated with the semi-bridging Bcat ligand

10.8: $NR_2 = NMe_2$; E = Ge
10.9: $NR_2 = NMe_2$; E = Sn
10.10: $NR_2 = (NPr)_2C_6H_4\text{-}1,2$; E = Sn
10.11: BX(X') = Bcat; Y = Cl; R = Ph
10.12: BX(X') = $B(NMe_2)Cl$; Y = Cl; R = Ph
10.13: BX(X') = B(Fc)Br; Y = Br; R = Cy

Fig. 34 Crystallographically characterized square planar platinum(II) mono(boryl) systems **10.8–10.13**

Thus, for example reaction of $Me_3GeB(NMe_2)_2$ with $(Ph_3P)_2Pt(C_2H_4)$ leads to the formation of **10.8**; the corresponding bis(germyl) and bis(stannyl) aminoboranes, however, are shown to be unreactive towards oxidative addition chemistry. As with the related palladium complex **10.1**, the relative *trans* influences of diaminoboryl and trimethylgermyl/stannyl ligands can be established on the basis of structural data [211, 217]. Thus, the Pt–P distances for the PPh_3 ligands *trans* to the boryl moiety in **10.8, 10.9** and **10.10** are 2.377(1), 2.369(1) and 2.376(1) Å, respectively, whereas the corresponding Pt–P distances *trans* to the germyl/stannyl ligand are 2.304(1), 2.303(1) and 2.301(1) Å. A similar conclusion can be reached from an analysis of the corresponding $^1J_{Pt\text{–}P}$ coupling constants, which are respectively 1200, 1392 and 1560 Hz (*trans* to boryl) and 2220, 2681 and 2448 Hz (*trans* to germyl/stannyl). Similar coupling constants for the PPh_3 ligands *trans* to Bcat (1761 Hz) and to $SnMe_3$ (2371 Hz) have been reported by Smith [192]. Finally, in each of **10.8–10.10** the boryl ligand is orientated approximately perpendicular with respect to the metal coordination plane, as might be expected on the basis of both steric and electronic effects (vide infra). Related *cis*-silyl(boryl)platinum(II) complexes have been reported by Ozawa and co-workers and their reactivity with respect to the silylboration of alkynes investigated [220]. Interestingly, very similar $^1J_{PtP}$ coupling constants are reported for the phosphine ligands *trans* to the boryl and silyl fragments in these systems (ca. 1350–1470 Hz).

Oxidative addition of boron–halogen bonds represents a second synthetic route used to give access to platinum(II) mono(boryl) species. In this case the *trans* geometry is observed in all structurally characterized systems, with no examples of chelating phosphine halo(boryl) complexes having been reported. $(Ph_3P)_2PtCl(Bcat)$ (**10.11**), reported in 1998, represented the first example of a platinum(II) boryl featuring a *trans* arrangement of the phosphine ligands; in the interim a number of related systems have been reported viz. $(Ph_3P)_2PtCl[B(NMe_2)Cl]$ (**10.12**) and $(Cy_3P)_2PtBr[B(Fc)Br]$ (**10.13**) [92, 178, 182]. Comparison of the Pt–Cl and Pt–Br distances in these systems [2.445(1), 2.470(2) and 2.6141(3) Å, respectively] with the corresponding distances in *trans*-$(Ph_3P)_2PtCl(H)$ and *trans*-$(Ph_3P)_2PtBr(H)$ [2.36(1) and 2.535(1) Å, respectively] provides further evidence for the stronger *trans* influence of the boryl ligand over hydride [221, 222]. Furthermore, the Pt–Br distance in **10.13** can additionally be compared to the corresponding Pt–Br distances in the complexes *trans*-$(Ph_3P)_2PtBr(X)$ [2.502(1), 2.563(1), 2.478(1) and 2.531(1) for X = vinyl, alkyl, alkynyl and aryl ligands] [223–226]. Thus, convincing structural evidence is also obtained for the stronger *trans* influence of the boryl ligand compared to these carbon-donor ligands. An in-depth computational evaluation of the *trans* influence of the boryl ligand, both comparatively with respect to carbon-, silicon- and tin-based systems, and as a function of the boryl substituent, has recently been reported by Zhu, Lin and Marder [175]. These authors used DFT to probe the electronic structure of complexes of the type *trans*-$(Me_3P)_3PtCl(Y)$ thereby deducing a hierarchy for the *trans* influences of the ligand Y: $BMe_2 > SiMe_3 > BH_2 > SnMe_3 > B[N(H)CH_2]_2 > Bpin > B(OCH_2)_2 > Bcat \sim BCl_2 \sim BBr_2 \sim SiH_3 > CH_2CH_3 > CH{=}CH_2 > H \sim Me > Ph > SiCl_3 > SnCl_3 > C{\equiv}CH$ [175]. The Pt–Cl bond length was used as a thermodynamic probe of the (ground state) *trans* influence, and was found to correlate well with the percentage of platinum orbital character in the Pt–B bond calculated by NBO methods. This in turn was related to the s/p composition of the boron orbital contributing to the Pt–B bond. A greater p contribution to the boron hybrid orbital implies (i) a higher energy for that orbital; (ii) better covalent overlap with the Pt-based orbitals; and hence (iii) that the boryl ligand will be a stronger σ donor (higher *trans* influence). Somewhat counter-intuitively this also implies a longer Pt–B bond length for ligands which have a high *trans* influence. Variation of *trans* influence as a function of the boryl substituent (i.e. X in BX_2) was rationalized using Bent's rule. Substituents X with greater electronegativities will utilize boron orbitals with greater *p* character in forming the B–X bonds; the Pt–B bond will therefore feature a smaller boron *p* component (greater *s* character) and the BX_2 ligand will be a weaker σ donor.

Chemical evidence for a strong boryl *trans* influence is provided by the reaction of **10.13** with $Na[BAr^f_4]$ which proceeds by abstraction of the platinum-bound bromide ligand to generate the T-shaped cationic boryl complex $[(Cy_3P)_2Pt\{B(Fc)Br\}]^+[BAr^f_4]^-$ (**10.14**; Scheme 23) [219]. Intriguingly, and in contrast to the five-coordinate rhodium complexes **9.11** and **9.12** discussed above, there is little shortening of the Pt–B bond in the absence of a competing *trans* ligand, presumably on steric grounds. Thus, the Pt–B distance in **10.14** [1.966(4) Å] is only marginally shorter than that found in the precursor **10.13** [1.996(3) Å]. Despite this, there is no structural, spectroscopic or computational evidence for any agostic CH interaction in the position *trans* to the boryl ligand—an observation consistent with previous computational studies of 14-electron d^8 $(R_3P)_2Rh(BX_2)$ systems, and thought to be another consequence of the strong boryl *trans* influence [134, 219]. Interestingly, **10.13** undergoes a bromide migration reaction on addition of 4-picoline to generate the base-stabilized borylene complex $[(Cy_3P)_2PtBr\{BFc(4\text{-pic})\}]^+[BAr^f_4]^-$ (**10.15**) which features a *trans* arrangement of borylene and bromide ligands [219].

The orientation of the boryl ligand in complexes of the type $(R_3P)_2Pt(hal)(BX_2)$ is typically such that the BX_2 plane lies approximately perpendicular to that defined by the metal coordination environment (e.g. 85.0° for **10.13**) [92, 182, 218]. Such an alignment is consistent not only with steric factors, but also with maximum overlap between the filled Pt d_{xy} orbital and the formally vacant boron-centred p orbital (Fig. 35). Further evidence for significant Pt→B back-bonding can be obtained by considering the Pt–B bond lengths for **10.11–10.14** [2.008(8), 2.084(3), 1.996(3) and 1.966(4) Å, respectively]. Comparison of the Pt–B bond length for **10.11**, for example, with the sum of the covalent radii deduced for the $[(Ph_3P)_2PtCl]$ and Bcat fragments (2.15 Å) is consistent with a substantial π component; furthermore

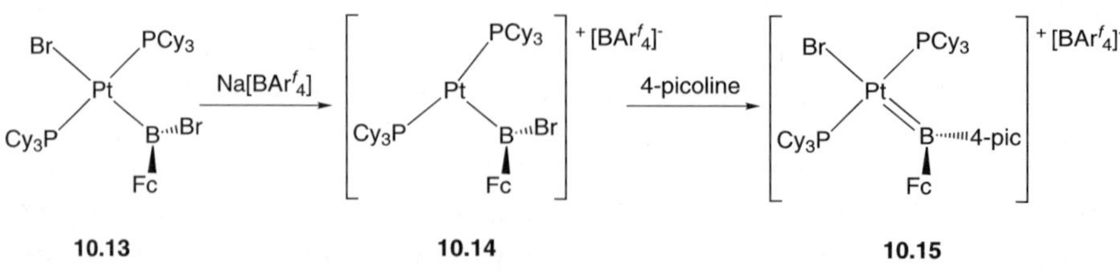

10.13 **10.14** **10.15**

Scheme 23 Interconversion of square planar boryl complex **10.13**, T-shaped boryl **10.14** and base-stabilized borylene system **10.15**

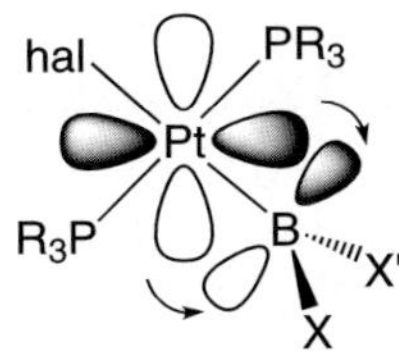

Fig. 35 Pt→B back-bonding resulting from overlap between the filled Pt d_{xy} orbital and the formally vacant boron-centred p orbital

comparison of the Pt–B distances for **10.11** and **10.13**, for example, reveals the expected bond contraction for the more π acidic B(Fc)Br ligand [11, 92, 218].

Useful verification of computational data comes in the form of the structures of the iminoboryl palladium and platinum complexes *trans*-$(Cy_3P)_2MBr(BNSiMe_3)$ (**10.4**: M = Pd; **10.5**: M = Pt), formed (along with Me_3SiBr) from the reaction of $M(PCy_3)_2$ with $(Me_3Si)_2NBBr_2$ [214]. These two complexes are unique in featuring a coordinated RN≡B ligand, short M–B distances [1.963 (mean) and 1.960(3) Å, respectively, c.f. **10.1**, **10.2** and **10.11–10.14**] and relatively short *trans* Pt–Br linkages [2.538 (mean) and 2.552(1) Å]. These observations are entirely consistent with the likely s/p composition of the boron hybrid orbital involved in the Pt–B bond. Since the boron centre is effectively linear, a greater s orbital contribution to this boron hybrid would be expected (i.e. approximately sp hybridized c.f. approximately sp^2 for a three-coordinate boryl), with the consequence not only of a shorter M–B bond, but also of weaker σ donor properties and hence a lower *trans* influence (shorter *trans* M–Br bond) [175, 214].

8.2.2 Bis(boryl) Complexes

A large number of square planar platinum(II) bis(boryl) complexes have been reported which are typically the products of B–B oxidative addition chemistry (see Fig. 36). All of the crystallographically characterized species (**10.16–10.30**) feature a pair of phosphine donor co-ligands and the *cis* configuration of the boryl ligands expected on the basis of the relative ligand *trans* influences [11, 182, 185, 187–190, 224, 227, 228]. Further evidence for the high boryl ligand *trans* influence is provided by the small Pt–P coupling constants {e.g. 1639 [188] or 1608 Hz [189] for $(Ph_3P)_2Pt(Bcat)_2$ (**10.16**), c.f. 3677 and 1765 Hz for *cis*-$(Ph_3P)_2PtCl_2$ and *cis*-$(Ph_3P)_2Pt(CH_2CH_2CH{=}CH_2)_2$, respectively [229, 230]} and relatively long Pt–P bonds found in these compounds {e.g. 2.363 (mean), 2.263 (mean), 2.302 Å (mean) for **10.16**, *cis*-$(Ph_3P)_2PtCl_2$ and *cis*-$(Ph_3P)_2Pt(CH_2CH_2CH{=}CH_2)_2$, respectively [63, 189, 229, 230]}. Structurally there is little diversity between each of the complexes **10.16–10.30**, with Pt–B distances in the range 2.031(8)-2.108(3) Å typically being shorter than the expected sum of the covalent radii (ca.

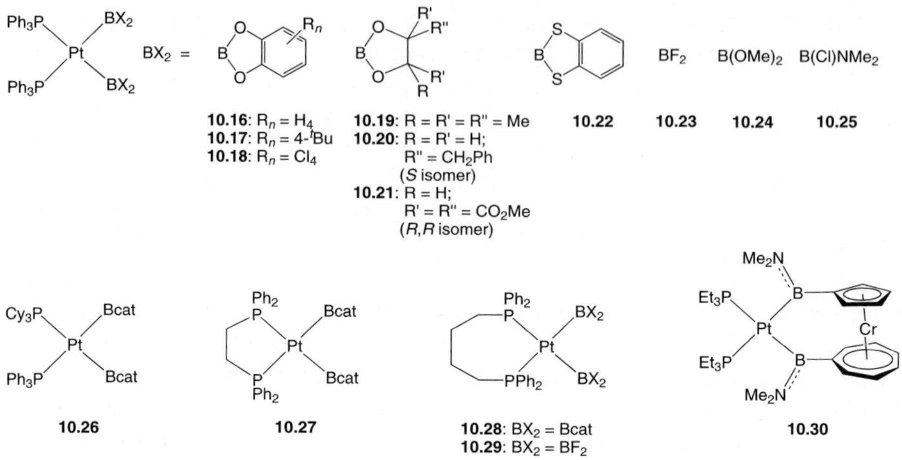

Fig. 36 Crystallographically characterized square planar platinum(II) bis(boryl) complexes **10.16–10.30**

2.14 Å) [11]. Few statistically significant trends can be established in terms of the variation in the Pt–B distance as a function of the boryl substituent. Thus, although the Pt–B bonds in the bis(dialkoxy)boryl complex $(Ph_3P)_2Pt[B(OMe)_2]_2$ [**10.24**: 2.098(4), 2.100(4) Å] are longer than those found in the bis(diaryloxy)boryl system $(Ph_3P)_2Pt(Bcat)_2$ [**10.16**: 2.040(6), 2.058(6) Å], there is little significant difference between the bond lengths in **10.16**, $(Ph_3P)_2Pt(4\text{-}^tBuBcat)_2$ [**10.17**: 2.045(11), 2.046(13) Å], $(Ph_3P)_2Pt(Bpin)_2$ [**10.19**: 2.076(6), 2.078(6) Å], $(Ph_3P)_2Pt(4\text{-Bped-}S)_2$ [**10.20**: 2.054(4), 2.070(3) Å], $(Ph_3P)_2Pt[B(NMe_2)Cl]_2$ [**10.25**: 2.076(3), 2.084(3) Å] and $(Ph_3P)_2Pt(BF_2)_2$ [2.052(6), 2.058(6) Å] [11, 182, 185, 187, 188, 190, 227].

Other notable structural features common to each of the complexes **10.16–10.30** are the acute B–Pt–B angles and correspondingly wide P–Pt–P angles. Thus, the B–Pt–B angles fall within the range 72.9(2)–81.5(1)°, with the value measured for **10.30** [76.4(1)°] which features a chelating boryl ligand, not differing from those reported for bis(monodentate) boryl systems [228]. Such acute B–Pt–B angles, together with the near perpendicular alignments of boryl and metal coordination planes typically found in these systems (e.g. torsion angles of 81.1, 87.4° for the two BF_2 ligands in **10.23**) allows for the possibility of residual B···B interaction of the type found in the cobalt(II) bis(boryl) systems **9.1–9.3** (Scheme 18). However, in contrast to **9.1–9.3**, the B···B contacts found in platinum complexes **10.16–10.30** are relatively long [2.445–2.671 c.f. 2.185–2.271 Å for **9.1–9.3**] [160, 161], and the B···B distances in **10.16** and **10.23** [2.552, 2.591 Å, respectively] are put in appropriate context by consideration of the B–B distances in B_2cat_2 and B_2F_4 [1.68, 1.67 Å] [185, 188, 227]. It seems likely, therefore, that the origins of the relatively acute B–Pt–B angles are principally steric in origin, with the trigonal planar (and effectively two-dimensional) boryl ligands being constrained to occupy a relatively compressed volume of the platinum coordination sphere by the bulkier phosphine ligands. Thus, a comparison of the bis(difluoroboryl) system **10.29**, with the isoelectronic complex $(dppb)_2Pt(NO_2)_2$ which contains the planar nitrito ligand, reveals similar structural properties, with due allowance made for the differing electronegativities of the B, F, N and O atoms [185]; the P–Pt–P angles for the two complexes are very similar [95.70(7), 94.95(11)°]. In general the P–Pt–P angles for bis(monodentate) phosphine complexes **10.16–10.26** are significantly greater than 90° [e.g. 107.14(4)° for **10.16**], consistent with the greater steric requirement of the phosphine ligands compared to boryl; narrowing of the P–Pt–P angle is observed as expected for related chelating phosphines [e.g. 85.36(6), 100.72(9)° for **10.27** and **10.28**]. Theoretical studies on model bis(phosphine) bis(boryl) platinum(II) complexes have been shown to accurately reproduce the structural features noted above, although little specific attention has been paid to the origins of the B···B separation [199, 200].

8.2.3
Semi-Bridging Ligands

Substitution of monodentate phosphine ligands has been exploited using *cis*-$(Ph_3P)_2Pt(Bcat)_2$ (**10.16**) as the starting material, to generate a range of related systems of the type $(R_3P)_2Pt(Bcat)_2$ [R_3 = Me_3, Et_3, Me_2Ph, $MePh_2$, $(OEt)_3$] [182]. With the sterically demanding tricyclohexyl phosphine, the mixed donor system **10.26** results, whereas the smaller more π acidic ligand $P(OMe)_3$ brings about reductive elimination of B_2cat_2 and the formation of Pt(0) species. The reaction of **10.16** with dppm, on the other hand leads to the formation of a major product formulated as either *cis*-(dppm)$Pt(Bcat)_2$ or dimeric *cis,cis*-$Pt_2(\mu_2$-dppm)$(Bcat)_4$ on the basis of spectroscopic data. Interestingly, the latter reaction also gives rise to two unusual semi-bridging boryl systems as minor co-products (see Fig. 30). Both $Pt_2(PPh_3)(\mu_2$-dppm$)_2$(Bcat)$(\mu_2$-Bcat) (**10.31**) and $Pt_2(\kappa^1$-dppm$)(\mu_2$-dppm$)_2$(Bcat)$(\mu_2$-Bcat) (**10.32**) feature a $Pt(I)_2$ core [with Pt–Pt separations of 2.772(1), 2.769(1) Å, respectively] and highly asymmetric interactions of the bridging Bcat ligand with the two platinum centres [182]. Thus, in each case the shorter Pt–B bond to the bridging boryl ligand is marginally longer than that found for the terminally bound Bcat moiety [2.12(2), 2.04(2) and 2.14(2), 2.06(2) Å for **10.31** and **10.32**, respectively], while the remaining bridging interaction is significantly longer [2.49(2), 2.51(2) Å] [182]. Recent DFT studies on model complexes related to **10.31** and **10.32** propose a bonding model comprising an approximately tetrahedral 18-electron Pt(0) tris(phosphine) boryl anion linked to T-shaped 14-electron Pt(II) bis(phosphine) boryl cation via a Pt→Pt donor/acceptor interaction. The dimeric structure is further stabilized by a Pt→B donor/acceptor interaction utilizing the d_{xz} orbital of the Pt(II) centre and the formally vacant *p* orbital of the Bcat ligand. Quantitatively this description resembles that proposed for the related dirhodium complex **9.21**, with the difference that the donor component of the semi-bridging interaction in this case utilizes the metal d_{z2} orbital [113].

9
Group 11 (Copper, Silver and Gold)

Although a number of studies have reported the use of group 11 reagents as catalysts in a range of borylation reactions [231–236], and copper(I) boryl complexes, for example, have been postulated in a number of reaction schemes leading to the formation of C–B bonds [231–235], structurally characterized group 11 complexes are a very recent development. Indeed, to date only one crystallographical example of a boryl complex of copper, silver or gold has been described [232].

Ito, Kawakami and Sawamura recently described the borylation of allylic carbonates by B_2pin_2, catalyzed by bis(phosphine)copper(I) alkoxides. It was proposed that bis(phosphine)copper(I) boryl species formed by alkoxide/boryl σ-bond metathesis are key intermediates in the catalytic cycle [231]. Making use of related *N*-heterocyclic carbene stabilized precursors, Sadighi and co-workers have very recently isolated the thermally labile copper(I) boryl complex (IPr)CuBpin (**11.1**) together with the products of oxygen atom, styrene and aldehyde insertion into the Cu–B bond (**11.2–11.5**; Scheme 24) [232, 233, 237]. The structure of **11.1** in the solid state reveals an approximately linear Cu(I) coordination geometry [∠B–Cu–C 168.1(2)°] and a Cu–B distance [2.002(3) Å] which is somewhat shorter than the sum of the expected covalent radii [2.05 Å] [106]. Yet further evidence for the

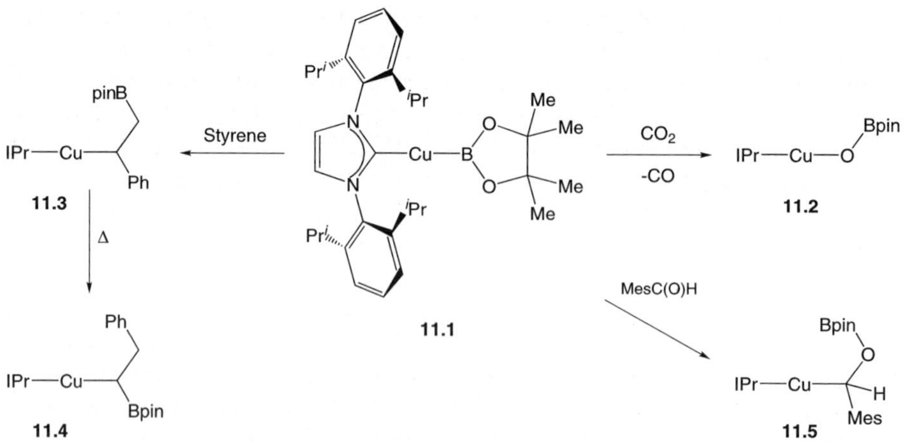

Scheme 24 Cu – B insertion reactions of copper(I) boryl complex **11.1**

Table 13 Selected data for structurally characterized copper boryl complexes

Compound	d(M – B) (Å)	Other data	Refs.
(IPr)CuBpin (**11.1**)	2.002(3)	Cu – C 1.937(2) Å C – Cu – B 168.1(2)° δ_B 41.7	[228]

strong *trans* influence of the boryl ligand is provided by a Cu–C distance for the *N*-heterocyclic carbene co-ligand [1.937(2) Å] which is markedly longer than found in related derivatives of the type (carbene)CuX [e.g. 1.811(7) and 1.887(5) for (IPr)CuCl and (IPr)CuMe, respectively] [238].

10 Outlook

The chemistry of boryl systems continues to be an exciting and rapidly developing field of synthetic chemistry, driven primarily by the development of novel methodologies for introducing functionality into organic molecules, but also encompassing more fundamental studies of structure and bonding. This is amply demonstrated by two significant papers in the area that appeared during the production of this review. Hartwig and co-workers have reported that the catalytic functionalization of methyl CH bonds by B_2pin_2 can be effected by ruthenium complexes at 150 °C. Thus, the well-known Ru(III) dimer $(Cp^*RuCl_2)_2$ reacts with octane under such conditions to generate 1-octylBpin in 98% yield. Somewhat surprisingly, these ruthenium systems are much less effective at functionalizing arene CH bonds, an observation ascribed to the ready formation of the stable $[Cp^*Ru(\eta^6\text{-arene})]^+$ cations [239]. Although the active catalyst in the alkane activation cycle is not known, Hartwig reports the structure of the bridged boryl complex, $(Cp^*Ru)_2(\mu\text{-H})_3\{\mu\text{-B}[1,2\text{-}(NMe)_2C_6H_4]\}$(**12.1**, Fig. 37) from the reaction of a hydridoruthenium precursor with $HB[1,2\text{-}(NMe)_2C_6H_4]$. It is possible that an analogous (but more reactive) Bpin complex might feature in alkane functionalization processes. The structure of **12.1** in the solid state (Fig. 37) represents the first example of a symmetrically bridged homodinuclear transition-metal boryl complex. Both the ^{11}B resonance (at δ_B 43.9) and the crystallographically located hydrogen atom positions are consistent with a metal boryl system (as opposed to a σ-borane or hydroborate). The boryl ligand sits on a crystallographic mirror plane, with the structural parameters associated with the Ru_2B unit [d(Ru–B) = 2.307(6), d(Ru–Ru) = 2.455(1) Å, $\angle$(Ru–B–Ru) = 64.3(2)°] therefore necessarily being consistent with a symmetrically bridging boryl ligand.

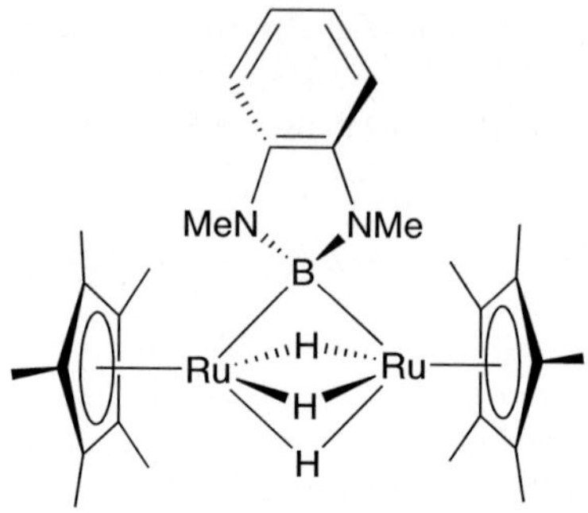

12.1

Fig. 37 A symmetrically bridged homodinuclear ruthenium boryl complex

Perhaps the most significant fundamental development in synthetic boron chemistry in 2006 was reported by Segawa, Yamashita and Nozaki in Science [240, 241]. Using a synthetic approach previously developed for the corresponding gallium systems [242], these authors reported the isolation of a boryllithium species **12.2** which reacts towards organic substrates as a *nucleophilic* boryl anion (Scheme 25). The crystallographic characterization of **12.2** as the dme adduct reveals a dimeric structure with terminal B–Li contacts [2.291(6) Å] which are significantly (ca. 8.5%) longer than the sum of the covalent radii of boron and lithium (2.11 Å) [106]. This observation, together with an ^{11}B chemical shift of 45.4 in THF solution is consistent with a highly polarized $B^{\delta-}$–$Li^{\delta+}$ linkage. Furthermore, the reactivity of **12.2** towards organic electrophiles such as alkyl halides or aldehydes is consistent with a reagent offering strongly nucleophilic/basic properties. As such, **12.2** offers the possibility of a significant step-change in synthetic boron chem-

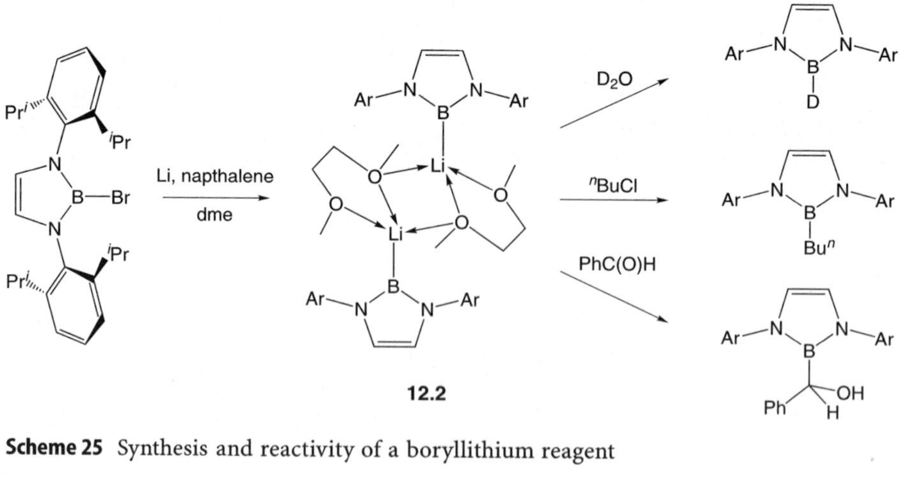

12.2

Scheme 25 Synthesis and reactivity of a boryllithium reagent

istry. For the first time, a nucleophilic source of the BX_2 fragment is available, a development which offers enormous synthetic potential in organic and inorganic boron-containing systems. A list of papers appearing in the calendar year 2007 is appended [243–266, 268].

References

1. Schmid G (1970) Angew Chem Int Ed Engl 9:819
2. Crudden CM, Edwards D (2003) Eur J Org Chem, p 4695
3. Beletskaya I, Pelter A (1997) Tetrahedron 53:4957
4. Burgess K, Ohlmeyer MJ (1991) Chem Rev 91:1179
5. Han L-B, Tanaka M (1999) Dalton Trans, p 395
6. Marder TB, Norman NC (1998) Top Catal 5:63
7. Ishiyama T, Miyaura N (2000) J Organomet Chem 611:392
8. Clegg W, Dai C, Lawlor FJ, Lesley MJG, Marder TB, Nguyen P, Norman NC, Pickett NL, Rice CR, Robbins EG, Scott AJ, Taylor NJ (1997) Royal Society of Chemistry Special Publication—Advances in Boron Chemistry 201:389
9. Hartwig JF (2004) ACS Symposium Series 885:136
10. Ishiyama T, Miyaura N (2003) J Organomet Chem 680:3
11. Irvine GJ, Lesley MJG, Marder TB, Norman NC, Rice CR, Robins EG, Roper WR, Whittell GR, Wright LJ (1998) Chem Rev 98:2685
12. Clark GR, Irvine GJ, Rickard CEF, Roper WR, Williamson A, Wright LJ (2000) Royal Society of Chemistry Special Publication— Contemporary Boron Chemistry 253:379
13. Dickinson AA, Willock DJ, Calder RJ, Aldridge S (2002) Organometallics 21:1146
14. Lam WH, Shimada S, Batsanov AS, Lin Z, Marder TB, Cowan JA, Howard JAK, Mason SA, McIntyre GJ (2003) Organometallics 22:4557
15. Lam KC, Lam WH, Lin Z Marder TB, Norman NC (2004) Inorg Chem 43:2541
16. Wadepohl H (1997) Angew Chem Int Ed Engl 36:2441
17. Braunschweig H (1998) Angew Chem Int Ed Engl 37:1786
18. Hartwig JF, Waltz KM, Muhoro CN, He X, Eisenstein O, Bosque R, Mascaras F (1997) Royal Society of Chemistry Special Publication—Advances in Boron Chemistry 201:373
19. Smith MR III (1999) Prog Inorg Chem 48:505
20. Frenking G, Fröhlich N (2000) Chem Rev 100:717
21. Braunschweig H, Colling M (2001) Coord Chem Rev 223:1
22. Aldridge S, Coombs DL (2004) Coord Chem Rev 248:535
23. Braunschweig H, Kollmann C, Seeler F (2008) Transition Metal Borylene Complexes (in this volume). Springer, Heidelberg
24. Braunschweig H, Rais D, Kollann C (2006) Angew Chem Int Ed 45:5254
25. Muhoro CN, Hartwig JF (1997) Angew Chem Int Ed Engl 36:1510
26. Hartwig JF, Muhoro CN, He X, Eisenstein O, Bosque R, Maseras F (1996) J Am Chem Soc 118:10936
27. Muhuro CN, He X, Hartwig JF (1999) J Am Chem Soc 121:5033
28. Hartwig JF, Muhoro CN (2000) Organometallics 19:30
29. Motry DH, Brazil AG, Smith MR III (1997) J Am Chem Soc 119:2743
30. Baker RJ, Jones C, Murphy DM (2005) Chem Commun, p 1339
31. Lantero DR, Ward DL, Smith III MR (1997) J Am Chem Soc 119:9699
32. Lantero DR, Miller SL, Cho J-Y, Ward DL, Smith III MR (1999) Organometallics 18:235

33. Hartwig JF, De Gala SR (1994) J Am Chem Soc 116:3661
34. Lantero DR, Motry DH, Ward DL, Smith III MR (1994) J Am Chem Soc 116:10811
35. Baker RT, Ovenall DW, Calabrese JC, Westcott SA, Taylor NJ, Willaims ID, Marder TB (1990) J Am Chem Soc 112:9399
36. Waltz KM, Hartwig JF (1997) Science 277:211
37. Waltz KM, Hartwig JF (2000) J Am Chem Soc 122:11358
38. Webster CE, Fan Y, Hall MB, Kunz D, Hartwig JF (2003) J Am Chem Soc 125:858
39. Lam WH, Lin Z (2003) Organometallics 22:473
40. Braunschweig H, Koster M, Wang R (1999) Inorg Chem 38:415
41. Braunschweig H, Koster M (2002) Z Naturforsch B 57:483
42. Braunschweig H, Koster M, Klinkhammer KW (1999) Angew Chem Int Ed 38:2229
43. Braunschweig H, Klinkhammer KW, Koster M, Radacki K (2003) Chem Eur J 9:1303
44. Braunschweig H, Kollann C, Koster M, Englert U, Müller M (1999) Eur J Inorg Chem 2277
45. Braunschweig H, Ganter B, Koster M, Wagner T (1996) Chem Ber 129:1099
46. Kawano Y, Yasue T, Shimoi M (1999) J Am Chem Soc 121:11744
47. Nakazawa H, Ohba M, Itazaki M (2006) Organometallics 25:2903
48. Wadepohl H, Arnold U, Pritzkow H (1997) Angew Chem Int Ed Engl 36:974
49. Wadepohl H, Arnold U, Kohl U, Pritzkow U, Wolf A (2000) J Chem Soc, Dalton Trans, p 3554
50. Rablen PR, Hartwig JF (1994) J Am Chem Soc 116:4121
51. Rablen PR, Hartwig JF (1996) J Am Chem Soc 118:4648
52. Hartwig JF, He X (1996) Angew Chem Int Ed Engl 35:315
53. Hartwig JF, He X (1996) Organometallics 15:5350
54. Braunschweig H, Bera H, Götz D, Radacki K (2006) Z Naturforsch B 61:29
55. Waltz KM, He X, Mohoro C, Hartwig JF (1995) J Am Chem Soc 117:11357
56. Chen H, Hartwig JF (1999) Angew Chem Int Ed 38:3391
57. Waltz KM, Muhoro CN, Hartwig JF (1999) Organometallics 18:3383
58. Aldridge S, Calder RJ, Baghurst RE, Light ME, Hursthouse MB (2002) J Organomet Chem 649:9
59. Schlecht S, Hartwig JF (2000) J Am Chem Soc 122:9435
60. Braunschweig H, Radacki K, Seeler F, Whittell GR (2006) Organometallics 25:4605
61. Braunschweig H, Colling M, Kollann C, Englert U (2002) J Chem Soc, Dalton Trans, p 2289
62. Coombs DL, Aldridge S, Jones C (2002) J Chem Soc, Dalton Trans, p 3851
63. Al-Fawaz A, Aldridge S, Coombs DL, Dickinson AA, Willock DJ, Ooi L-L, Light ME, Coles SJ, Hursthouse MB (2004) Dalton Trans, p 4030
64. Kakizawa T, Kawano Y, Shimoi M (2001) Organometallics 20:3211
65. Yasue T, Kawano Y, Shimoi M (2003) Angew Chem Int Ed 42:1727
66. Hartwig JF, Huber S (1993) J Am Chem Soc 115:4908
67. Schilling BER, Hoffmann R, Lichtenberger D (1979) J Am Chem Soc 101:585
68. Aldridge S, Calder RJ, Dickinson AA, Willock DJ, Steed JW (2000) Chem Commun, p 1377
69. Aldridge S, Calder RJ, Rossin A, Dickinson AA, Willock DJ, Jones C, Evans DJ, Steed JW, Light ME, Coles SJ, Hursthouse MB (2002) J Chem Soc, Dalton Trans, p 2020
70. Rossin A, Aldridge S, Ooi L-L (2005) Appl Organomet Chem 19:181
71. Clark GR, Irvine GJ, Roper WR, Wright LJ (2003) J Organomet Chem 680:81
72. Rickard CEF, Roper WR, Williamson A, Wright LJ (2004) J Organomet Chem 689:1609
73. He X, Hartwig JF (1996) Organometallics 15:400
74. Braunschweig H, Radacki K, Seeler F, Whittell GR (2004) Organometallics 23:4178

75. Kays DL, Rossin A, Day JK, Ooi L-L, Aldridge S (2006) Dalton Trans, p 399
76. Braunschweig H, Radacki K, Rais D, Whittell GR (2005) Angew Chem Int Ed 44:1192
77. Braunschweig H, Whittell GR (2005) Chem Eur J 11:6128
78. Braunschweig H, Radacki K, Scheschkewitz D, Whittell GR (2005) Angew Chem Int Ed 45:1658
79. Braunschweig H, Kollann C, Englert U (1998) Eur J Inorg Chem, p 465
80. Braunschweig H, Kollann C, Klinkhammer KW (1999) Eur J Inorg Chem 1523
81. Kays DL, Day JK, Ooi L-L, Aldridge S (2005) Angew Chem Int Ed 44:7457
82. Coombs DL, Aldridge S, Rossin A, Jones C, Willock DJ (2004) Organometallics 23:2911
83. Aldridge S, Jones C, Gans-Eichler T, Stasch A, Kays DL, Coombs ND, Willock DJ (2006) Angew Chem Int Ed 45:6118
84. Vidovic D, Findlater M, Reeske G, Cowley AH (2006) Chem Commun, p 3786
85. Kays DL, Day JK, Aldridge S, Harrington RW, Clegg W (2006) Angew Chem Int Ed 45:3513
86. Braunschweig H, Koster M (1999) J Organomet Chem 588:231
87. Braunschweig H, Kollann C, Müller M (1998) Eur J Inorg Chem, p 291
88. Aldridge S, Coombs DL, Jones C (2002) Chem Commun, p 856
89. Coombs DL, Aldridge S, Jones C (2003) Appl Organomet Chem 17:356
90. Coombs DL, Aldridge S, Coles SJ, Hursthouse MB (2003) Organometallics 22:4213
91. Coombs DL, Aldridge S, Jones C, Willock DJ (2003) J Am Chem Soc 125:6356
92. Braunschweig H, Radacki K, Rais D, Seeler F (2004) Organometallics 23:5545
93. Aldridge S, Bresner C (2003) Coord Chem Rev 244:71
94. Braunschweig H, Radacki K, Rais D, Seeler F, Uttinger K (2005) J Am Chem Soc 127:1386
95. Braunschweig H, Kraft M, Schwarz S, Seeler F, Stelwag S (2006) Inorg Chem 45:5275
96. Aldridge S, Al-Fawaz A, Calder RJ, Dickinson AA, Willock DJ, Light ME, Hursthouse MB (2001) Chem Commun, p 1846
97. Aldridge S, Kays DL, Al-Fawaz A, Horton PN, Hursthouse MB, Harrington RW, Clegg W (2006) Chem Commun, p 2578
98. Yasue T, Kawano Y, Shimoi M (2000) Chem Lett, p 58
99. Hill AF, Owen GR, White AJP, Williams DJ (1999) Angew Chem Int Ed 38:2759
100. Montiel-Parma V, Lumbierres M, Donnadieu B, Sabo-Etienne S, Chaudret B (2002) J Am Chem Soc 124:5624
101. Lachaize S, Essalah K, Montiel-Palma V, Vendier L, Chaudret B, Barthelat J-C, Sabo-Etienne S (2005) Organometallics 24:2935
102. Kawano Y, Hashiva M, Shimoi M (2006) Organometallics 25:4420
103. Irvine GJ, Roper WR, Wright LJ (1997) Organometallics 16:2291
104. Clark GR, Irvine GJ, Roper WR, Wright LJ (1997) Organometallics 16:5499
105. Rickard CEF, Roper WR, Williamson A, Wright LJ (2000) Organometallics 19:4344
106. Emsley J (1991) The Elements, 2nd edn. Oxford University Press, Oxford
107. Rickard CEF, Roper WR, Williamson A, Wright LJ (1998) Organometallics 17:4869
108. Giju KT, Bickelhaupt FM, Frenking G (2000) Inorg Chem 39:4776
109. Rickard CEF, Roper WR, Williamson A, Wright LJ (1999) Angew Chem Int Ed 38:1110
110. Irvine GJ, Rickard CEF, Roper WR, Williamson A, Wright LJ (2000) Angew Chem Int Ed 39:948
111. Rickard CEF, Roper WR, Wright LJ, Williamson A (2002) Organometallics 21:1714
112. Rickard CEF, Roper WR, Williamson A, Wright LJ (2002) Organometallics 21:4862
113. Westcott SA, Marder TB, Baker RT, Harlow RL, Calabrese JC, Lam KC, Lin Z (2004) Polyhedron 23:2665

114. Westcott SA, Taylor NJ, Marder TB, Baker RT, Jones NJ, Calabrese JC (1991) Chem Commun, p 304
115. Baker RT, Calabrese JC, Westcott SA, Nguyen P, Marder TB (1993) J Am Chem Soc 115:4367
116. Westcott SA, Marder TB, Baker RT, Calabrese JC (1993) Can J Chem 71:930
117. Knorr JR, Merola JS (1990) Organometallics 9:3008
118. Widauer C, Grützmacher H, Ziegler T (2000) Organometallics 19:2097
119. Dorigo AE, Schleyer PvR (1995) Angew Chem Int Ed Engl 34:115
120. Musaev DG, Mebel AM, Morokuma K (1994) J Am Chem Soc 116:10693
121. Musaev DG, Matsubara T, Mabal AM, Koga N, Morokuma K (1995) Pure Appl Chem 34:607
122. Yamamoto Y, Fujikawa R, Umemoto T, Miyaura N (2004) Tetrahedron 60:10695
123. Toshimichi O, Yamamoto Y, Miyaura N (2000) J Am Chem Soc 122:4990
124. Dai C, Robins EG, Scott AJ, Clegg W, Yufit DS, Howard JAK, Marder TB (1998) Chem Commun 1983
125. Nguyen P, Coapes RB, Woodwrard AD, Taylor NJ, Burke JM, Howard JAK, Marder TB (2002) J Organomet Chem 652:77
126. Iverson CN, Smith MR III (1999) J Am Chem Soc 121:7696
127. Smimada S, Batsanov AS, Howard JAK, Marder TB (2001) Angew Chem Int Ed 40:2168
128. Hartwig JF, Cook KS, Hapke M, Incarvito CD, Fan Y, Webster CE, Hall MB (2005) J Am Chem Soc 127:2538
129. Kawamura K, Hartwig JF (2001) J Am Chem Soc 123:8422
130. Bae C, Hartwig JF, Chung H, Harris NK, Switek KA, Hillmyer MA (2005) Angew Chem Int Ed 44:6410
131. Lawrence JD, Takahashi M, Bae C, Hartwig JF (2004) J Am Chem Soc 126:15334
132. Kondo Y, Garcia-Cuadrado D, Hartwig JF, Boaen NK, Wagner NL, Hillmyer MA (2002) J Am Chem Soc 124:1164
133. Chen H, Schlecht S, Semple TC, Hartwig JF (2000) Science 287:1995
134. Lam WH, Lam KC, Lin Z, Shimada S, Perutz RN, Marder TB (2004) Dalton Trans, p 1556
135. Wan X, Wang X, Luo Y, Takami S, Kubo M, Miyamoto A (2002) Organometallics 21:3703
136. Ishiyama T, Takagi J, Ishida K, Miyaura N, Anastasi N, Hartwig JF (2002) J Am Chem Soc 124:390
137. Nguyen P, Blom HP, Westcott SA, Taylor NJ, Marder TB (1993) J Am Chem Soc 115:9329
138. Boller TM, Murphy JM, Hapke M, Ishiyama T, Miyaura N, Hartwig JF (2005) J Am Chem Soc 127:14263
139. Ishiyama T, Nobuta Y, Hartwig JF, Miyaura N (2003) Chem Commun, p 2924
140. Ishiyama T, Takagi J, Hartwig JF, Miyaura N (2002) Angew Chem Int Ed 41:3056
141. Takagi J, Sato K, Hartwig JF, Ishiyama T, Miyaura N (2002) Tetrahedron Lett 43:5649
142. Chotana GA, Rak MA, Smith MR III (2005) J Am Chem Soc 127:10539
143. Maleczka RA Jr, Shi F, Holmes D, Smith MR III (2003) J Am Chem Soc 125:7792
144. Tse MK, Cho J-Y, Smith MR III (2001) Org Lett 3:2831
145. Cho J-Y, Iverson CN, Smith MR III (2000) J Am Chem Soc 122:12868
146. Cho J-Y, Tse MK, Holmes D, Maleczka RE Jr, Smith MR III (2002) Science 295:305
147. Mkhalid IAI, Coventry DN, Albesa-Jove D, Batsanov AS, Howard JAK, Perutz RN, Marder TB (2006) Angew Chem Int Ed 45:489
148. Coventry DN, Batsanov AS, Goeta AN, Howard JAK, Marder TB, Perutz RN (2005) Chem Commun, p 2172
149. Coapes RB, Souza FES, Thomas RL, Hall JJ, Marder TB (2003) Chem Commun, p 614

150. Hata H, Shinokubo H, Osuka A (2005) J Am Chem Soc 127:8264
151. Tamura H, Yamazaki H, Sato H, Sakaki S (2003) J Am Chem Soc 125:16114
152. Musaev D, Morokuma K (1996) J Phys Chem 100:6509
153. Cundari T, Zhao Y (2003) Inorg Chim Acta 345:70
154. Sivignon S, Fleurat-Lessard P, Onno J-M, Volatron F (2002) Inorg Chem 41:6656
155. Ishiyama T, Miyaura N (2006) Pure Appl Chem 78:1369
156. Ishiyama T, Takagi J, Nobuta Y, Miyaura N, Shi Y, Weix DJ, Ellman JA (2005) Org Synth 82:126
157. Ishiyama T, Takagi J, Yonekawa Y, Hartwig JF, Miyaura N (2003) Adv Synth Catal 345:1103
158. Holmes D, Chotana GA, Maleczka RE Jr, Smith MR III (2006) Org Lett 8:1407
159. Feng S, Smith MR III, Maleczka RE Jr (2006) Org Lett 8:1411
160. Dai C, Stringer G, Corrigan JF, Taylor NJ, Marder TB, Norman NC (1996) J Organomet Chem 513:273
161. Adams CJ, Baber RA, Batsanov AS, Bramham G, Charmant JPH, Haddow MF, Howard JAK, Lam WH, Lin Z, Marder TB, Norman NC, Orpen AG (2006) Dalton Trans, p 1370
162. Cook KS, Incarvito CD, Webster CE, Fan Y, Hall MB, Hartwig JF (2004) Angew Chem Int Ed 43:5474
163. Crossley IR, Foreman MRS, Hill AF, White AJP, Williams DJ (2005) Chem Commun, p 221
164. Crossley IR, Hill AF, Humphrey ER, Willis AC (2005) Organometallics 24:4083
165. Crossley IR, Hill AF, Willis AC (2005) Organometallics 24:1062
166. Bontemps S, Gornitzka H, Bouhadir G, Miqeu K, Bourissou D (2006) Angew Chem Int Ed 45:1611
167. Mihalcik DJ, White JL, Tanski JM, Zakharov LN, Yap GA, Incarvito CD, Rheingold AL, Rabinovitch D (2004) Dalton Trans, p 1626
168. Shimoi M, Ikubo I, Kawano Y, Kato K, Ogino H (1998) J Am Chem Soc 120:4222
169. Basil JD, Aradi AA, Bhattacharyya, Rath NP, Eigenbrot C, Fehlner TP (1990) Inorg Chem 29:1260
170. Jun C, Halet J, Rheingold TP, Fehlner TP (1995) Inorg Chem 24:2101
171. Elliot DJ, Levy CJ, Puddephatt RJ, Holah DG, Hughes AN, Magnuson VR, Moser IM (1990) Inorg Chem 29:5014
172. Clegg W, Lawlor FJ, Marder TB, Nguyen P, Norman NC, Orpen AG, Quayle MJ, Rice CR, Robins EG, Scott AJ, Souza FES, Stringer G, Whittell GR (1998) Dalton Trans, p 301
173. Lu Z, Jun C-H, de Gala SR, Sigalas M, Eisenstein O, Crabtree RH (1993) Chem Commun 1877
174. Lu Z, Jun C-H, de Gala SR, Sigalas M, Eisenstein O, Crabtree RH (1995) Organometallics 14:1168
175. Zhu J, Lin Z, Marder TB (2005) Inorg Chem 44:9384
176. Dai C, Stringer G, Marder TB, Scott AJ, Clegg W, Norman NC (1997) Inorg Chem 36:272
177. Souza FES, Nguyen P, Marder TB, Scott AJ, Clegg W (2005) Inorg Chim Acta 358:1501
178. Rossi AR, Hoffmann R (1975) Inorg Chem 14:365
179. Marder TB, Norman NC, Rice CR, Robins EG (1997) Chem Commun, p 53
180. Riehl J-F, Jean Y, Eisenstein O, Pélissier M (1992) Organometallics 11:729
181. Callaghan PL, Fernández-Pacheco R, Jasim N, Lachaize S, Marder TB, Perutz RN, Rivalta E, Sabo-Etienne S (2004) Chem Commun, p 242

182. Curtis D, Lesley MJG, Norman NC, Orpen AG, Starbuck J (1999) Dalton Trans, p 1687
183. Dai C, Stringer G, Marder TB, Baker RT, Scott AJ, Clegg W, Norman NC (1996) 74:2026
184. Churchill MR, Hackbarth JJ (1975) Inorg Chem 14:2047
185. Lu N, Norman NC, Orpen AG, Quayle MJ, Timms PL, Whittell GR (2000) Dalton Trans, p 4032
186. Peterson TH, Golden JT, Bergman R (1999) Organometallics 18:2005
187. Clegg W, Johann TRF, Marder TB, Norman NC, Orpen AG, Peakman TM, Quayle MJ, Rice CR, Scott AJ (1998) Dalton Trans, p 1431
188. Lesley G, Nguyen P, Taylor NJ, Marder TB, Scott AJ, Clegg W, Norman NC (1996) Organometallics 15:5137
189. Iverson CN, Smith MR III (1995) J Am Chem Soc 117:4403
190. Ishiyama T, Matsuda N, Murata M, Ozawa F, Suzuki A, Miyaura N (1996) Organometallics 15:713
191. Iverson CN, Smith MR III (1997) Organometallics 16:2757
192. Iverson CN, Smith MR III (1996) Organometallics 15:5155
193. Bell NJ, Cox AJ, Cameron NR, Evans JSO, Marder TB, Duin MA, Elsevier CJ, Baucherel X, Tulloch AAD, Tooze RP (2004) Chem Commun, p 1854
194. Lawson YG, Lesley MJG, Marder TB, Norman NC, Rice CR (1997) Chem Commun, p 2051
195. Marder TB, Norman NC, Rice CR (1998) Tetrahedron Lett 39:155
196. Ishiyama T, Kitano T, Miyaura N (1998) Tetrahedron Lett 39:2357
197. Ishiyama T, Yamamoto M, Miyaura N (1997) Chem Commun, p 689
198. Anderson KM, Lesley MJG, Norman NC, Orpen AG, Starbuck J (1999) New J Chem 23:1053
199. Sakaki S, Kikuno T (1997) Inorg Chem 36:226
200. Musaev DG, Morokuma K (1997) Organometallics 16:1355
201. Cui Q, Musaev DG, Morokuma K (1997) Organometallics 16:1355
202. Sakaki S, Kai S, Sugimoto M (1999) Organometallics 18:4825
203. Sakaki S, Biswas B, Musashi Y, Sugimoto M (2000) J Organomet Chem 611:288
204. Cui Q, Musaev DG, Morokuma K (1998) Organometallics 17:742
205. Cui Q, Musaev DG, Morokuma K (1998) Organometallics 17:1383
206. Suginome M, Matsuda T, Ito Y (1998) Organometallics 17:5233
207. Suginome M, Matsuda T, Yoshimoto T, Ito Y (2002) Organometallics 21:1537
208. Yamamoto A, Suginome M (2005) J Am Chem Soc 127:15706
209. Aldridge S, Calder RJ, Coles SJ, Hursthouse MB (2003) J Chem Cryst 33:809
210. Onozawa S, Tanaka M (2001) Organometallics 20:2956
211. Onozawa S, Hatanaka Y, Sakakura T, Shimada S, Tanaka M (1996) Organometallics 15:5450
212. Onozawa S, Hatanaka Y, Choi N, Tanaka M (1997) Organometallics 16:5389
213. Onozawa S, Hatanaka Y, Tanaka M (1998) Tetrahedron Lett 39:9043
214. Braunschweig H, Radacki K, Rais D, Uttinger K (2006) Angew Chem Int Ed 45:162
215. Foreman MRS, Hill AF, White AJP, Williams DJ (2004) Organometallics 23:913
216. Braunschweig H, Rais D, Uttinger K (2005) Angew Chem Int Ed 44:3763
217. Habereder T, Nöth H (2003) Appl Organomet Chem 17:525
218. Clegg W, Lawlor FJ, Lesley G, Marder TB, Norman NC, Orpen AG, Quayle MJ, Rice CR, Scott AJ, Souza FES (1998) J Organomet Chem 550:183
219. Braunschweig H, Radacki K, Rais D, Scheschkewitz D (2005) Angew Chem Int Ed 44:5651

220. Sagawa T, Asano Y, Ozawa F (2002) Organometallics 21:5879
221. Bender R, Braunstein P, Jud J-M, Dusausoy Y (1984) Inorg Chem 23:4489
222. Aldridge S, Coombs DL, Jones C (2003) Acta Crystallogr E 59:m584
223. Rajaram J, Pearson RG, Ibers JA (1974) J Am Chem Soc 96:2103
224. Wicht DK, Kourkine IV, Kovacik I, Glueck DS, Concolino TE, Yap GPA, Incarvito CD, Rheingold AL (1999) Organometallics 18:5381
225. Wouters JMA, Vrieze K, Elsevier CJ, Zoutberg MC, Goubitz K (1994) Organometallics 13:1510
226. Alcock NW, Bartlett PN, Eastwick-Field VM, Pike GA, Pringle PG (1991) J Mater Chem 1:569
227. Kerr A, Marder TB, Norman NC, orpen AG, Quayle MJ, Rice CR, Timms PL, Whittell GR (1998) Chem Commun, p 319
228. Braunschweig H, Lutz M, Radacki K (2005) Angew Chem Int Ed 44:5647
229. Scherer OJ, Jungmann H, Hussong K (1983) J Organomet Chem 247:C1
230. Dralle K, Jaffa NL, le Roex T, Moss JR, Travis S, Watermeyer ND, Sivaramakrishna A (2005) Chem Commun, p 3865
231. Ito H, Kawakami C, Sawamura M (2005) J Am Chem Soc 127:16034
232. Laitar S, Müller O, Sadighi JP (2005) J Am Chem Soc 127:17196
233. Laitar S, Tsui E, Sadighi JP (2006) Organometallics 25:2405
234. Takahashi K, Ishiyama T, Miyaura N (2001) J Organomet Chem 625:47
235. Ito H, Yamanaka H, Tateiwa J, Hosomi A (2000) Tetrahedron Lett 41:6821
236. Baker RT, Nguyen P, Marder TB, Westcott SA (1995) Angew Chem Int Ed Engl 34:1336
237. Laitar DS, Tsui EY, Sadighi JD (2006) J Am Chem Soc 128:11036
238. Mankad NP, Gray TG, Laitar DS, Sadighi JP (2004) Organometallics 23:1191
239. Murphy JM, Lawrence JD, Kawamura K, Incarvito C, Hartwig JF (2006) Ruthenium-catalyzed regiospecific borylation of methyl C – H bonds. J Am Chem Soc 128:13684
240. Segawa Y, Yamashita M, Nozaki K (2006) Science 314:113
241. Marder TB (2006) Science 314:69
242. Baker RJ, Jones C (2005) Coord Chem Rev 249:1857
243. Zhao P, Incarvito CD, Hartwig JF (2007) Directly observed transmetalation from boron to rhodium. β-Aryl elimination from Rh(I) arylboronates and diarylborinates. J Am Chem Soc 129:1876
244. Murphy JM, Tzschucke CC, Hartwig JF (2007) One-pot synthesis of arylboronic acids and aryl trifluoroborates by Ir-catalyzed borylation of arenes. Org Lett 9:757
245. Tzschucke CC, Murphy JM, Hartwig JF (2007) Arenes to anilines and aryl ethers by sequential iridium-catalyzed borylation and copper-catalyzed coupling. Org Lett 9:761
246. Zhao H, Lin Z, Marder TB (2006) Density functional theory studies on the mechanism of the reduction of CO_2 to CO catalyzed by copper(I) boryl complexes. J Am Chem Soc 128:15637
247. Campian MV, Harris JL, Jasim N, Perutz RN, Marder TB, Whitwood AC (2006) Comparisons of photoinduced oxidative addition of B – H, B – B, and Si – H bonds at Rhodium(η^5-cyclopentadienyl)phosphine centers. Organometallics 25:5093
248. Lindup RJ, Marder TB, Perutz RN, Whitwood AC (2007) Sequential C – F activation and borylation of fluoropyridines via intermediate Rh(I) fluoropyridyl complexes: A multinuclear NMR investigation. Chem Commun, p 3664
249. Lam KC, Lin Z, Marder TB (2007) DFT studies of β-boryl elimination processes: Potential role in catalyzed borylation reactions of alkenes. Organometallics 26:3149

250. Dang L, Zhao H, Lin Z, Marder TB (2007) DFT studies of alkene insertions into Cu – B bonds in copper(I) boryl complexes. Organometallics 26:2824
251. Charmant JPH, Fan C, Norman NC, Pringle PG (2007) Synthesis and reactivity of dichloroboryl complexes of platinum(II). Dalton Trans, p 114
252. Segawa Y, Yamashita M, Nozaki K (2007) Boryl anion attacks transition-metal chlorides to form boryl complexes: Syntheses, spectroscopic, and structural studies on group 11 borylmetal complexes. Angew Chem Int Ed 46:6710
253. Yamashita M, Suzuki Y, Segawa Y, Nozaki K (2007) Synthesis, structure of borylmagnesium, and its reaction with benzaldehyde to form benzoylborane. J Am Chem Soc 129:9570
254. Paul S, Chotana GA, Holmes D, Reichle RC, Maleczka RE Jr, Smith MR III (2006) Ir-catalyzed functionalization of 2-substituted indoles at the 7-position: Nitrogen-directed aromatic borylation. J Am Chem Soc 128:15552
255. Adhikari D, Huffman JC, Mindiola DJ (2007) Structural elucidation of a nickel boryl complex a recyclable borylation Ni(II) reagent of bromobenzene. Chem Commun, p 4489
256. Pandey KK (2007) Structure and coordinate bonding nature of the manganese σ-borane complexes. J Organomet Chem 692:1997
257. Pandey KK (2007) Structure and coordinate bonding nature of the rhenium σ-borane complexes. Theochem 807:61
258. Caballero A, Sabo-Etienne S (2007) Ruthenium-catalyzed hydroboration and dehydrogenative borylation of linear and cyclic alkenes with pinacolborane. Organometallics 26:1191
259. Alcaraz G, Clot E, Helmstedt U, Vendier L, Sabo-Etienne S (2007) Mesitylborane as a bis(σ B – H) ligand: An unprecedented bonding mode to a metal center. J Am Chem Soc 129:8704
260. Braunschweig H (2007) Boryl anions: lithiumboryl—a synthon for a nucleophilic boryl anion. Angew Chem Int Ed 46:1946
261. Braunschweig H, Lutz M, Radacki K, Schaumlöffel A, Seeler F, Unkelbach C (2006) Facile syntheses of trovacene, the formation of [n]boratrovacenophanes ($n = 1, 2$) and reactivity towards $[Pt(PEt_3)_4]$. Organometallics 25:4433
262. Braunschweig H, Radacki K, Seeler F, Whittell GR (2006) Synthesis and reactivity of dihaloboryl complexes. Organometallics 25:4605
263. Braunschweig H, Kupfer T, Lutz M, Radacki K, Seeler F, Sigritz R (2006) Metal mediated diboration of alkynes with [2]borametallocenophanes under stoichiometric, homogeneous, and heterogeneous conditions. Angew Chem Int Ed 45:8048
264. Braunschweig H, Frenking G, Radacki K, Seeler F, Fernández I (2007) Synthesis and electronic structure of a ferroborirene. Angew Chem Int Ed 46:5215
265. Braunschweig H, Brenner P, Müller A, Radacki K, Rais D, Uttinger K (2007) Experimental Studies on the trans-Influence of Boryl Ligands in Square-planar Platinum(II) Complexes. Chem Eur J 13:7171
266. Braunschweig H, Radacki K, Uttinger K (2007) Syntheses of mono- and dinuclear diiodboryl complexes of platinum. Inorg Chem 46:8796
267. Braunschweig H, Radacki K, Rais D, Schneider A, Seeler F (2007) Reactivity of a platinum iminoboryl complex toward lewis and brønsted acids. J Am Chem Soc 129:10350
268. Pierce GA, Coombs ND, Willock DJ, Day JK, Stasch A, Aldridge S (2007) Insertion reactions of dicyclohexylcarbodiimide with amino-boranes, -boryls and -borylenes. Dalton Trans, p 4405

Struct Bond (2008) 130: 123–148
DOI 10.1007/430_2007_074

Published online: 23 November 2007

Transition Metal σ-Borane Complexes

Zhenyang Lin

Department of Chemistry, The Hong Kong University of Science and Technology,
Clear Water Bay, Kowloon, Hong Kong, China
chzlin@ust.hk

Abstract A variety of structural types of transition metal complexes containing σ-borane ligands are reviewed. Their structure and bonding are discussed. Compared with other types of σ-complexes, σ-borane complexes display quite different structural and bonding properties because of the electron unsaturated boron center in the σ-borane ligands in the precoordination state. The availability of an sp^3-hybridized orbital at the boron center allows stronger back-donation interaction without breaking the coordinated B-H bond. The role played by σ-borane complexes in hydroboration and borylation reactions has also been reviewed.

Keywords Borylation · Hydroboration · σ-Borane complexes · Structure and bonding

1 Introduction

Transition metal complexes containing η^2-coordination of an H-X (X = H, C, B, or Si) bond to a metal center have attracted much attention due to their fundamental importance in coordination chemistry and their role in catalysis [1–5]. In the past two decades we witnessed the progress made in the chemistry of these σ-complexes. σ-Dihydrogen [1, 2, 6] and σ-silane [7–9] complexes have been the most extensively studied among the different types of σ-complexes. σ-Alkane complexes have been studied mainly in solution [10, 11]. The best models for σ-alkane complexes are agostic complexes

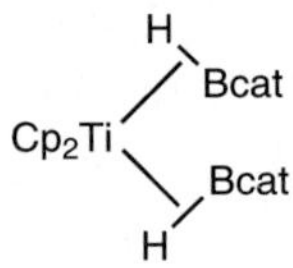

1

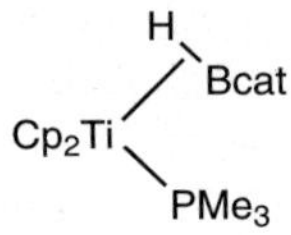

2

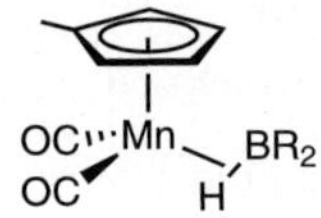

BR_2 = Bcat, Bpin, Cy_2

3

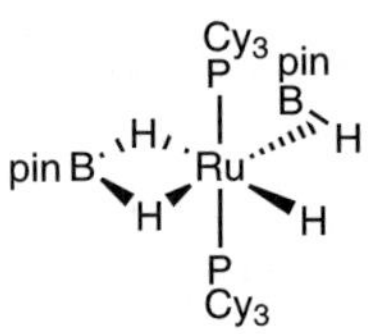

4

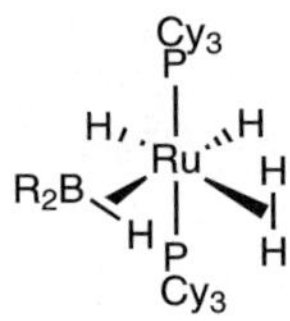

BR_2 = Bcat, Bpin

5

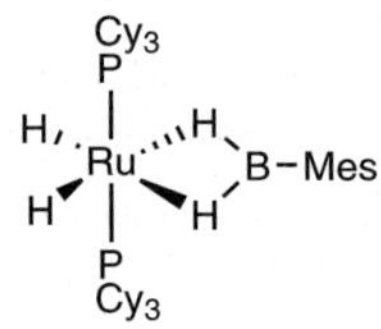

6

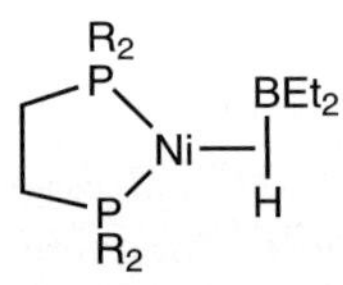

R = Cy, ^{i}Pr, ^{t}Bu

7

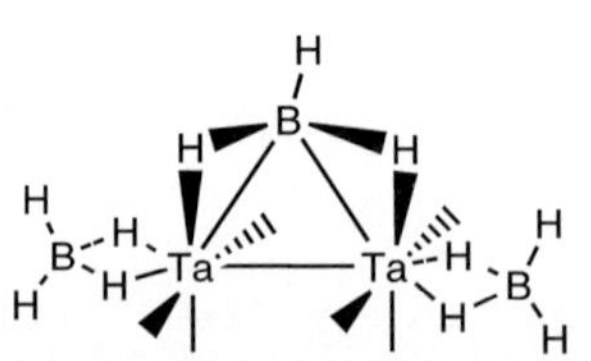

$Ta_2(\mu\text{-}H_2BH)(\mu\text{-dmpm})_3(\eta^2\text{-}BH_4)_2$

8

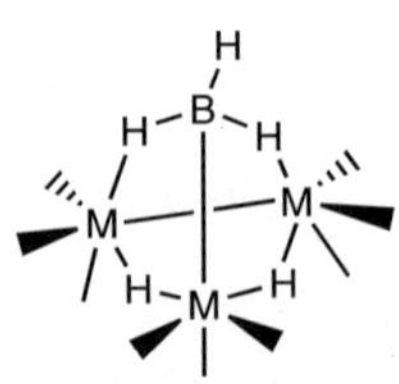

M = Ru, Os

9

having η^2 coordinated H-C bonds via intramolecular C-H bond coordination. The existence of η^2-coordinated H-B bonds in the well-known hydridoborato complexes has been known for some time [12, 13]. In these hydridoborato complexes, hydridoborato ligands carry negative charge and the metal-η^2-H-B interactions are often considered ionic. Despite the importance of the ionic interactions, evidence of electronic factors for the interactions in the M-η^2-H-B linkage is unequivocally clear because neutral borohydride ligands coordinated with metal centers by an M-(η^2-H-B) coordination mode were also known more than twenty years ago [14–16]. In all these σ-complexes mentioned above, a common feature is that the X center of the η^2-coordinated H-X bond in a given σ-ligand achieves the octet (or the full valence shell with two electrons for X = H) in the precoordination state.

Recently, a new and interesting class of transition metal σ-borane complexes containing η^2 coordinated H-B bonds has been discovered [17–25]. Different from the complexes containing hydridoborato or neutral borohydride ligands mentioned above, the boron center in this new class of σ-borane complexes is three-coordinated and highly electron-deficient in the precoordination state. The first well-characterized mononuclear examples of this new class of σ-borane complexes are the titanocene catecholborane complexes **1** and **2** reported by Hartwig et al. in late 2000s [17–19]. More mononuclear examples of σ-borane complexes **3**, which belong to a piano-stool type, were reported in 2000 by Schlecht and Hartwig [20]. Recently, ruthenium complexes (**4**–**6**) containing both hydride and σ-borane ligands were synthesized and characterized by Sabo-Etienne et al. [21–23]. More recently, σ-borane complexes **7** having a nickel d^{10} metal center have also been reported by García and coworkers [24]. Interestingly, we found from our extensive literature search that σ-borane ligands can also be found in multinuclear cluster compounds **8** and **9** [26–30]. Compared with other types of σ-complexes, σ-borane complexes mentioned here display quite different structural and bonding properties because of the electron unsaturated boron center in the σ-borane ligands in the precoordination state. In this chapter, the structure and bonding of this new class of transition metal σ-borane complexes will be discussed and reviewed. Special emphasis will be placed on the different structural and bonding characteristics displayed by this class of complexes in comparison to other σ-complexes.

2
Bonding Nature in σ-Borane Complexes

The current understanding of the nature of the metal-(η^2-H-X) bonding interaction for various types of σ-complexes is based on the traditional Dewar–Chatt–Duncanson model for the well-known π-olefin complexes which emphasizes both ligand-to-metal σ bonding and metal(dπ)-to-ligand(σ^*)

back-donation. In metal σ-silane complexes, the metal(dπ)-to-ligand(σ^*) back-bonding interaction is considered extremely important and plays a dominant role in determining the structural and bonding characteristics [9]. In σ-complexes containing M-(η^2-H_2) or M-(η^2-H-C), the metal(dπ)-to-ligand(σ^*) back-bonding interaction is less significant, although in some cases it determines the relative orientation of the η^2-H_2 ligand with respect to other ligands in the complexes under consideration [31].

Examination of all the structurally characterized σ-borane complexes leads to the discovery of an interesting, common structural feature. Similar to what we see in many transition metal boryl complexes [32], the boron center in these $L_nM(\eta^2\text{-}HBX_2)$ complexes lies in a plane containing the metal center M and the two boron-bonded atoms from the two X groups. On the basis of the Dewar–Chatt–Duncanson model, the metal-η^2-borane bonding interactions can be schematically illustrated in Fig. 1. The ligand-to-metal σ bonding interaction is the donation of the pair of the coordinated H-B σ-bonding electrons to an empty metal d_σ orbital. The metal-to-ligand π interaction involves a concomitant back-bonding interaction of an occupied metal d_π orbital with a boron sp^3-hybridized orbital and the H – B(σ^*) orbital from the η^2-HBX_2 ligand. Several previous theoretical studies [33–35] show that the metal(dπ)-to-ligand(σ^*) back-bonding interaction is less important and the back-bonding interaction of an occupied metal d_π orbital with the "empty" sp^3-hybridized orbital of the three-coordinated boron center dominates the structure and bonding of this class of transition metal σ-borane complexes. As pointed out by Hartwig and his co-workers [20], the most

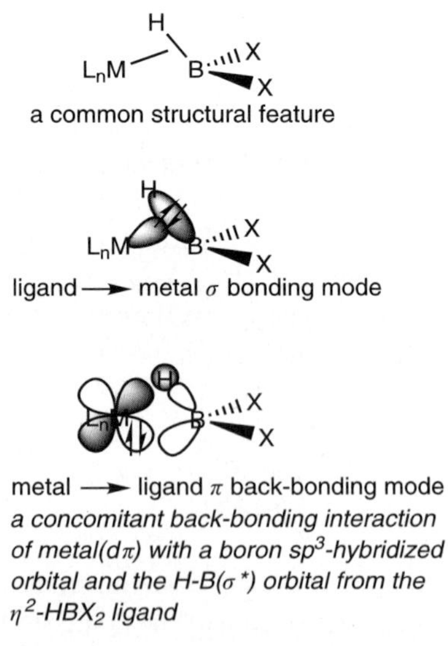

Fig. 1 The metal-η^2-borane bonding interactions based on the Dewar–Chatt–Duncanson Model

important difference between σ-borane complexes and H-X σ-complexes of ligands with electronically-saturated X is the availability of an sp^3-hybridized orbital at boron. This sp^3-hybridized orbital participates in the back-donation interaction with an occupied metal d_π orbital. In alkane, dihydrogen, or silane complexes, back-donation occurs into the H-X σ^*-orbital, which is much higher in energy than the boron sp^3-hybridized orbital.

3 Metallocene σ-Borane Complexes

3.1 $Cp_2Ti(\eta^2$-HBcat$)_2$ and $Cp_2Ti(\eta^2$-HBcat$)(PMe_3)$

In an effort to develop efficient and selective catalysts for alkene hydroborations [36], Hartwig and his co-workers isolated and precisely characterized the new and interesting class of titanocene σ-borane complexes **1** and **2** [17–19]. A surprising structural feature in $Cp_2Ti(\eta^2$-HBcat$)_2$ **1** (HBcat = catecholborane) is that the B···B distance (2.11 Å) is unexpectedly short for a non-bonded interaction but too long for a strong bonding interaction between these atoms. In addition, the B – Ti – B angle (53.8°) is small, suggesting that there is an important difference in the nature of the M-(η^2-H – B) bonding interaction from other types of σ-complexes.

Density functional theory (DFT) calculations [33] on the model complex $Cp_2Ti\{\eta^2$-HB(OH)$_2\}_2$ accurately reproduced the short B···B distance. A Laplacian electron density analysis (Fig. 2) based on the DFT calculations showed that significant electron density concentrations were found along the B···B bonding path, indicating bonding interactions between the two borons. The four concentrations around the Ti center shown in the Laplacian electron density plot have been associated with the electron density contributed from

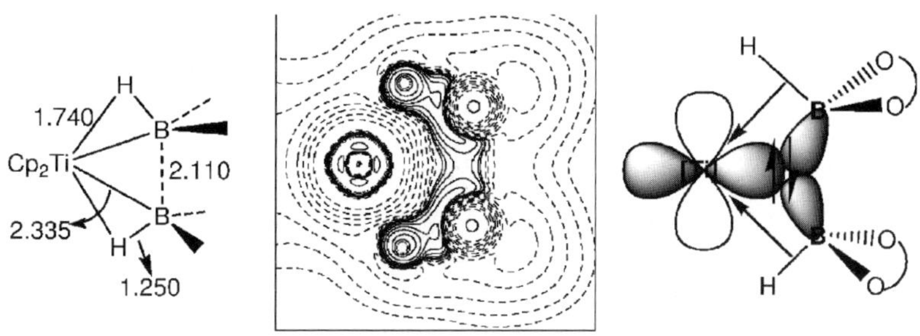

Fig. 2 Selected distances (Å) in $Cp_2Ti(\eta^2$-HBcat$)_2$, a plot of the Laplacian of electron density for $Cp_2Ti\{\eta^2$-HB(OH)$_2\}_2$ on the B – Ti – B plane and a three-center-two-electron bonding molecular orbital describing the bonding interaction of the TiB_2 triangle

a metal d orbital. Figure 3 shows a molecular orbital diagram illustrating the orbital interaction between the Cp_2Ti fragment and the two η^2-HBcat ligands. The bent metallocene Cp_2Ti fragment has three frontier orbitals ($1a_1$, b_2 and $2a_1$, shown in the left column of Fig. 3) available for bonding with the two HBcat ligands. The corresponding symmetry-adapted ligands' orbitals are derived from the linear combinations of the two B-H bonding orbitals as well as the two empty sp^3-hybridized orbitals from the two boron centers, illustrated in the right column of Fig. 3. The orbital interactions between the two sets of orbitals give three bonding and three antibonding orbitals shown in the central column of Fig. 3. Combining the detailed molecular orbital analysis and the results from the Laplacian electron density plot, we proposed a three-

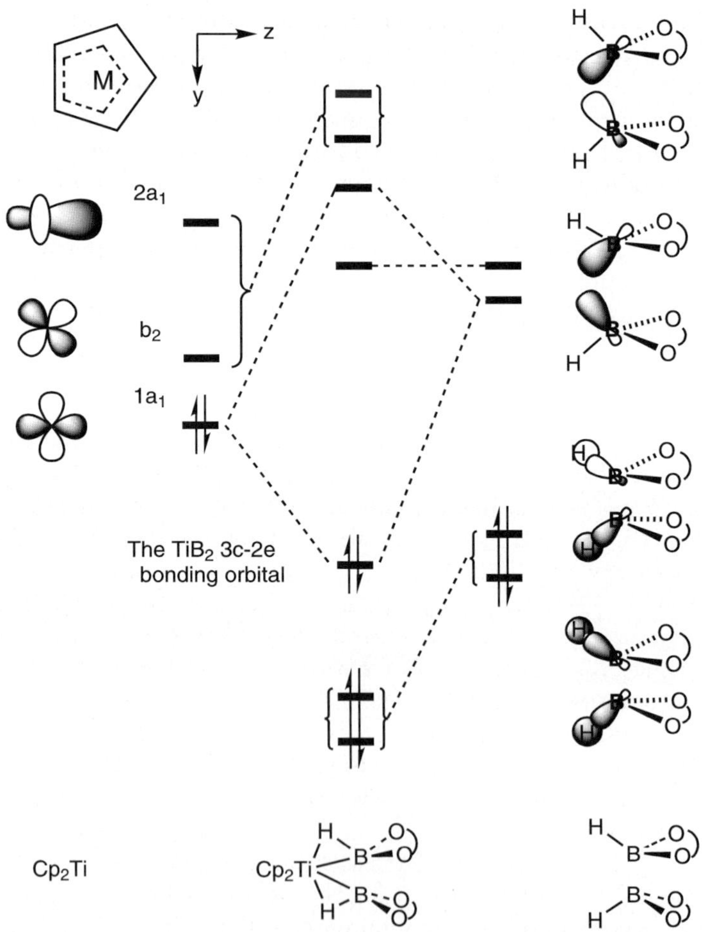

Fig. 3 A schematic orbital interaction diagram for $Cp_2Ti(\eta^2\text{-HBcat})_2$ (Adapted from [33])

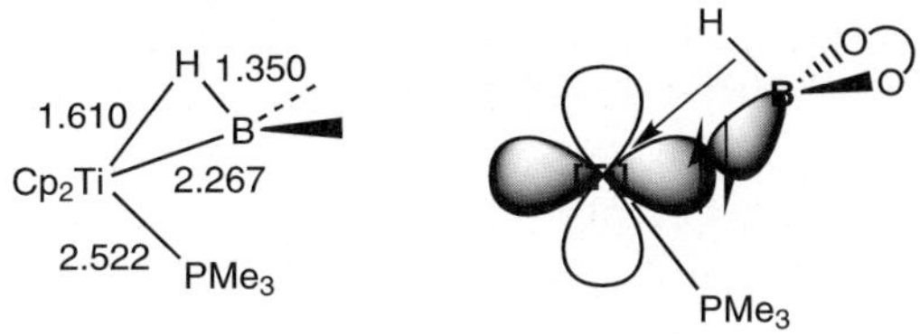

Fig. 4 Selected distances (Å) in $Cp_2Ti(\eta^2\text{-HBcat})(PMe_3)$ and a bonding molecular orbital describing the metal-to-boron back-bonding interaction

center-two-electron bond, shown in Fig. 2, to describe the unique bonding in $Cp_2Ti(\eta^2\text{-HBcat})_2$.

In the $Cp_2Ti(\eta^2\text{-HBcat})(PMe_3)$ complex, a bent Ti-B bond, shown in Fig. 4, has been proposed to account for the Ti-B bonding interaction, which involves the back-bonding interaction of a metal d orbital with an sp^3-hybridized orbital from the HBcat ligand. The nonplanarity around the boron center of the HBcat ligand in the complexes was considered to be a result of maximizing the back-bonding interaction of Ti($d\pi$) with the "empty" sp^3-hybridized orbital of the three-coordinated boron center in the HBcat ligand. Compared with other σ-complexes, the two complexes also have the ligand(σ)-to-metal dative bonds (here ligand(σ) denotes the H-B σ bond of the HBcat ligand) in addition to the unusual bonding characteristics described above.

3.2
$Cp_2Ti(\eta^2\text{-HBcat})(L)$ (L = Acetylene, Olefin or Silane)

Experiments [19] showed that PhC ≡ CPh and $PhSiH_3$ can displace one of the two η^2 catecholborane ligands in $Cp_2Ti(\eta^2\text{-HBcat})_2$ **1**, leading to the formation of the $Cp_2Ti(\eta^2\text{-HBcat})$(acetylene) and $Cp_2Ti(\eta^2\text{-HBcat})$(silane) adducts, respectively. On the basis of the bonding analysis given above, the donation and back-donation interactions between Cp_2Ti and the η^2-HBcat and PMe_3 ligands in $Cp_2Ti(\eta^2\text{-HBcat})(PMe_3)$ **2** can be schematically illustrated in **10**. The situation in $Cp_2Ti(\eta^2\text{-HBcat})$(acetylene) can be quite different. $Cp_2Ti(\eta^2\text{-HBcat})$(acetylene) can have a similar bonding picture, shown in **11** and **12**. Because an acetylene ligand is capable of accepting electrons into its π^* orbital(s), rearrangement of the bonding electron pairs to the situation as shown in **13**, giving a molecular structure like **14**, is possible. A previous density functional theory study [34] demonstrated that $Cp_2Ti(\eta^2\text{-HBcat})$(acetylene) indeed adopts a molecular structure somewhere in between **11** and **14**. Figure 5 shows the structure calculated for the model complex $Cp_2Ti\{\eta^2\text{-HB}(OCH)_2\}$(HC≡CH) together with the Laplacian electron density plot on the plane containing Ti, the η^2-H-B unit, and the two acetylene carbons. From Fig. 5a, we see that the C1-C2 bond (1.330 Å) is

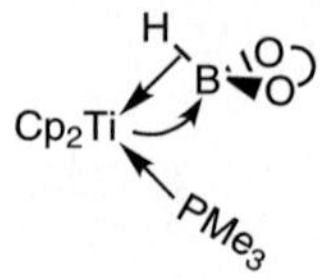

10

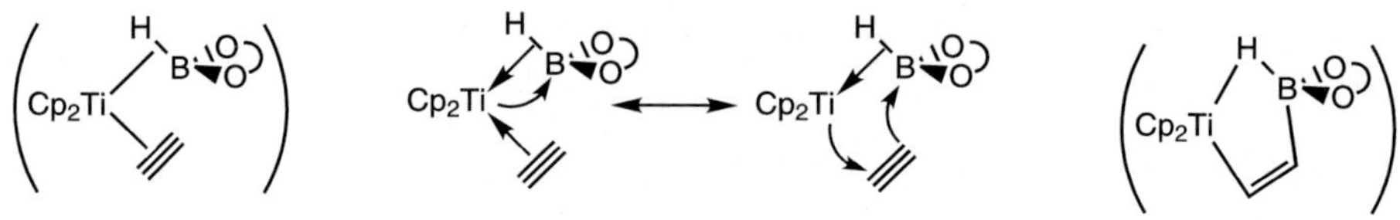

11 12 13 14

longer than the triple bond (1.211 Å) calculated for $HC\equiv CH$ but slightly shorter than the double bond (1.335 Å) calculated for $H_2C=CH_2$. The B-C2 bond (1.683 Å) is much longer than the B-C single bond (1.541 Å) calculated for $H_2C=CH-B(OCH)_2$ [34]. In addition, the short distance between Ti and C2 (2.316 Å) indicates their strong interaction. Clearly, the calculated structure shown in Fig. 5a can only be described by a resonance hybrid between the two important Lewis structures shown in Fig. 5b. The calculated structure supports Hartwig's explanation of the NMR chemical shifts observed experimentally [19]. The hydride chemical shift in the ^1H NMR spectrum indicates a B-H bonding interaction and also implies an electron configuration somewhere between d^0 and d^2 for the metal center while the ^{11}B NMR resonance is nearly equivalent to that of a free catecholborane.

For $Cp_2Ti(\eta^2\text{-HBcat})$(silane), **15** does not correspond to a local minimum on the potential energy surface (PES) [34]. Instead, a structure corresponding to the hydroborato complex **16** was calculated. Structure **17** corresponds to a local minimum on the PES, but the structure was calculated to be 11.4 kcal/mol higher in energy than **16**. It was reasoned that

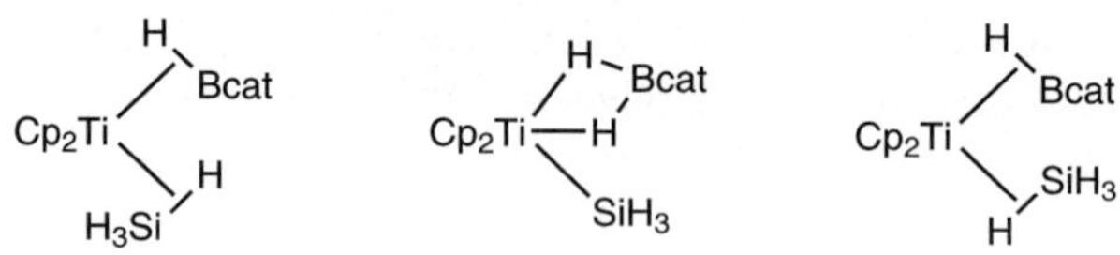

15 16 17

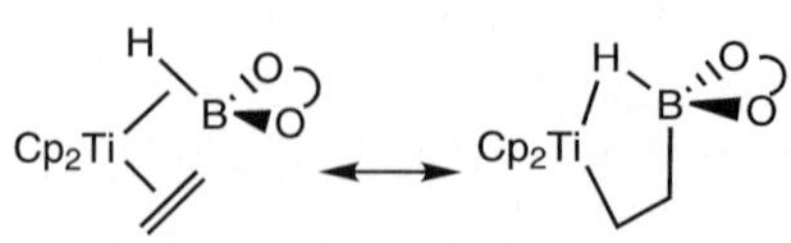

18 19

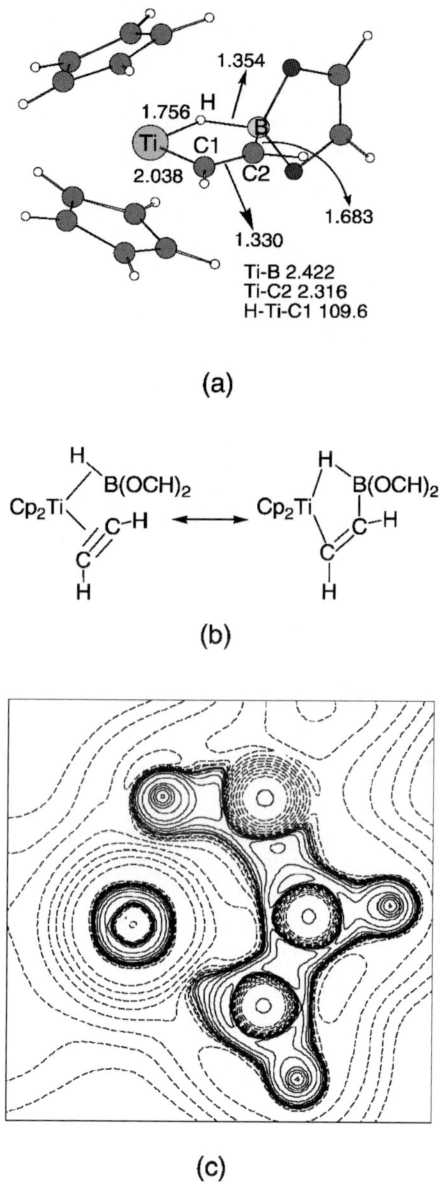

Fig. 5 **a** Calculated structure for $Cp_2Ti(\eta^2\text{-}HB(OCH)_2)(HC{\equiv}CH)$. **b** A resonance hybrid describing the calculated structure. **c** Plot of the Laplacian of electron density on the plane containing the Ti – C1 – C2 – B – H five-membered ring (Adapted from [34])

the high stability of the hydroborato form having strong B-H bonds makes **17** relatively unstable [34]. $Cp_2Ti(\eta^2\text{-}HBcat)$(olefin) are reactive intermediates in the Ti-catalyzed hydroboration of olefins [37, 38]. Theoretical calculations [34] on the model complex $Cp_2Ti\{\eta^2\text{-}HB(OCH)_2\}(H_2C=CH_2)$ lead to a conclusion similar to that made for $Cp_2Ti(\eta^2\text{-}HBcat)$(acetylene). $Cp_2Ti(\eta^2\text{-}HB(OCH)_2)$(olefin) adopts a five-membered ring structure and should be described by a resonance hybrid between the two Lewis structures **18** and **19**. Interestingly, the structural feature of the complexes discussed in this section is governed by how the electron-deficient B center interacts with the ligand L. When L = acetylene or olefin, the complexes still have the character of the σ-borane complexes. When L = silane, the hydroborato form is the dominant character.

3.3 Niobocene and Tantalocene σ-Borane Complexes

In 1994, Smith and his co-workers reported the X-ray crystal structures of endo-Cp_2TaH_2(Bcat) and exo-Cp_2TaH_2(Bcat), which were synthesized through addition of toluene solutions of chlorocatecholborane to Cp_2TaH_2Li [25]. On the basis of the X-ray crystal structures, both of the two regioisomers were considered as a boryl hydride having a d^0 Ta(V) metal center. However, for the exo-isomer, the $^1H\text{-}\{^{11}B\}$ NMR data lend support to a B$\cdots$H interaction [25]. A recent DFT study [35] showed that while the endo-isomer is indeed a boryl hydride, the exo-isomer is a σ-borane complex exo-$Cp_2Ta(H)(\eta^2\text{-}HBcat)$ (see the calculated structure shown in **20**). The DFT calculations predicted that the endo-$Cp_2Ta(H)_2$(Bcat) isomer is more stable than the exo-$Cp_2Ta(H)(\eta^2\text{-}HBcat)$ isomer by 2.2 kcal/mol, consistent with the experimental observation that the exo-isomer is a kinetic product and is gradually converted to the endo-isomer on heating. Ta tends to have a high oxidation state with a formal d^0 electron configuration. It was therefore believed that in the exo-$Cp_2Ta(H)(\eta^2\text{-}HBcat)$ isomer **20** the steric

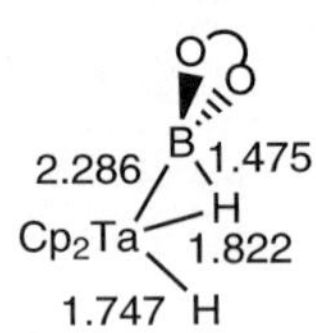

20

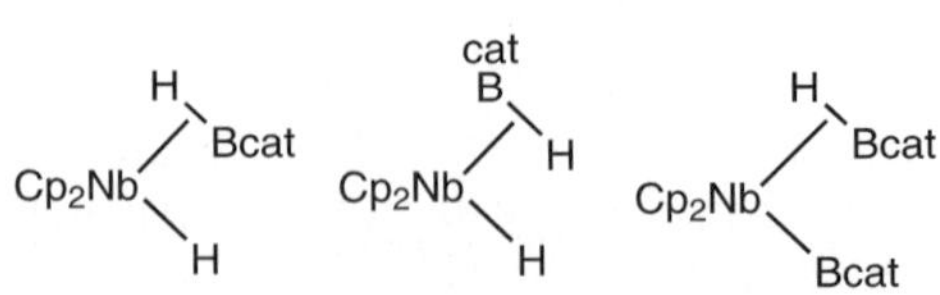

21 **22** **23**

repulsion between the HBcat group and the two Cp rings may prevent an expansion of the ligand spread angle in the wedge, promoting the formation of an η^2-HBcat structure [35]. Thus, the adoption of a σ-complex structure (**20**) by exo-$Cp_2Ta(H)(\eta^2$-$HB(OC_2H_2O))$ is a result of a compromise between the electronic interactions and steric repulsions [35].

As an analogue, $Cp_2NbH_2(Bcat)$, a reactive intermediate observed in the Nb-mediated olefin hydroboration, can be obtained through the reaction of $Cp_2Nb(H)$(olefin) with catecholborane HBcat [39]. Both σ-borane structures **21** and **22** were proposed for the intermediate on the basis of the ^1H-$\{^{11}B\}$ NMR spectra [39]. Theoretical calculations showed that the two structural isomers have similar stability and that structure **21** is more stable by only 0.2 kcal/mol than **22** [35]. The calculations also predicted that the endo isomer **21** can be rapidly equilibrating with the hydridoborate isomer due to the very small barrier connecting them (Scheme 1). The very small barrier connecting the hydridoborate isomer and the σ-borane isomer shown in Scheme 1 explains the observation that $Cp_2Nb(H_2Bcat)$ lies on the continuum between the structural extremes of a boryl complex and a hydridoborate complex [40]. The similar stability of both the endo- and exo-isomers of $Cp_2Nb(H)(\eta^2$-HBcat) and the rapid isomerization between the hydridoborate and endo-isomer structures also explain the easy elimination of HBcat from $Cp_2{}^*Nb(H_2Bcat)$ by various two-electron ligands and the extensive H/D scrambling phenomena between the hydride ligand of $Cp_2Nb(H)(\eta^2$-olefin) and DBcat observed in the olefin hydroboration reactions [39, 41].

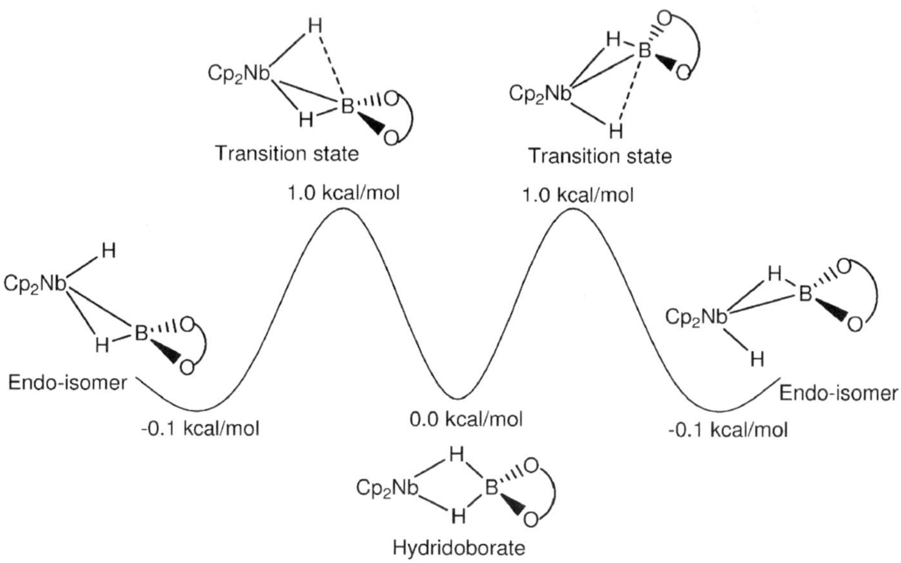

Scheme 1

In the Nb-mediated olefin hydroboration reactions, $Cp_2NbH(Bcat)_2$ was also isolated [39]. Interestingly, the X-ray structure of $Cp_2NbH(Bcat)_2$ shows a crystallographic equivalence of the two Bcat ligands, which was not anticipated because the ^1H and ^{11}B NMR spectra indicate two chemically distinct Bcat units. A boryl hydride $Cp_2Nb(H)(Bcat)_2$ was proposed in which the two Bcat ligands are next to each other and the hydride occupies one of the two exo positions in the metallocene wedge. A borane σ-complex structure was not suggested because borane elimination chemistry (substitution reactions) was not observed [39]. On the contrary, theoretical calculations [35] give a structure corresponding to a σ-borane complex shown in **23**. The inconsistency between the calculated and the proposed structure suggests that there might be other factors affecting the borane substitution reaction by a carbonyl ligand used in the experiment [39].

In the complexes discussed in this section, we can see that the three possible structures, shown in Scheme 2, resulting from the coordination of $HB(OR)_2$ to the L_nMH fragment, differ only slightly in their relative stability. The slight difference in the relative stability reflects a subtle balance of various types of bonding interactions. In endo-$Cp_2TaH_2\{B(OR)_2\}$, the boryl hydride structure **a** is preferred because Ta tends to have a high oxidation state. The orbitals accommodating the lone pair electrons on the oxygen atoms of the OR substituents contribute to the bonding interactions with the "empty" boron orbital in the boryl ligand. In endo-$Cp_2NbH_2\{B(OR)_2\}$, both the σ-borane and hydridoborate structures (**b** and **c**) are possible, in which Nb has a formal oxidation state of + 3 with a d^2 electron configuration. In the σ-borane structure **b**, the back-bonding interaction from Nb($d\pi$) to the "empty" boron sp^3-hybridized orbital stabilizes the electron-deficient boron center. In the hydridoborate structure **c**, the hydrides are responsible for stabilizing the boron center.

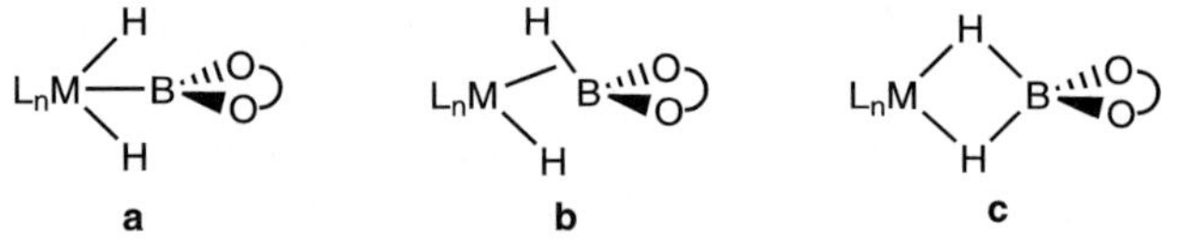

Scheme 2

4 Piano-Stool σ-Borane Complexes

Schlecht and Hartwig reported in 2000 that both photolysis of $MeCpMn(CO)_3$ in the presence of HBcat (catecholborane) or HBpin (pinacolborane) and borylation of $[MeCpMn(CO)_2H]^-$ by ClBcat, ClBpin or $ClBCy_2$ (Cy = cyclohexyl) can give σ-borane complexes of manganese **3** [20]. The synthesis and isolation of these manganese σ-borane complexes show that σ-borane complexes are not restricted to the bent metallocene type containing early transition metals.

Interestingly, the X-ray structural studies [20] show that these manganese σ-borane complexes are isostructural with the previously reported structures of $Cp'Mn(CO)_2(\eta^2\text{-silane})$ (Cp′ = substituted cyclopentadienyl) [7, 8], in which the coordinated H-B bond lies parallel to the plane of the Cp′ ligand. Figure 6 gives selected bond distances relevant to the Mn-(η^2-HB) structural moieties in $MeCpMn(CO)_2(\eta^2\text{-HBcat})$, $MeCpMn(CO)_2(\eta^2\text{-HBpin})$ and $MeCpMn(CO)_2(\eta^2\text{-HBCy}_2)$. In $MeCpMn(CO)_2(\eta^2\text{-HBCy}_2)$, the Mn-B bond distance (2.187 Å) is the longest, and however, the B-H and Mn-H bonds are the shortest among the three σ-borane complexes. These structural data indicate a weaker Mn-B interaction and a more hydride character for the borane hydrogen in the dicyclohexylborane complex. Kinetic studies [20] on the ligand substitution reactions of these manganese σ-borane complexes showed that $HBCy_2$ has a smaller binding energy to the metal fragment compared to the other two alkoxylborane complexes.

From the structural data and the results of the kinetic studies, we can infer that the more electronegative alkoxyl substituents at the boron center enhance the concomitant back-bonding interaction shown in Fig. 1c. This is quite unexpected because one normally believes that the π-donating alkoxyl substituents would make the boron center a poor electron acceptor. In tran-

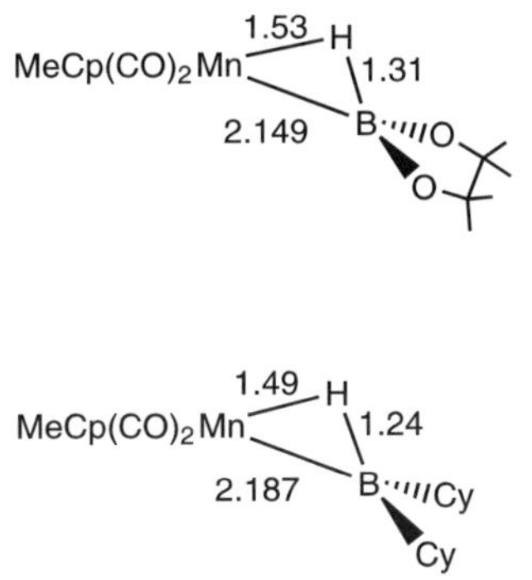

Fig. 6 Selected structural parameters in $MeCpMn(CO)_2(\eta^2\text{-HBcat})$, $MeCpMn(CO)_2(\eta^2\text{-HBpin})$ and $MeCpMn(CO)_2(\eta^2\text{-HBCy}_2)$

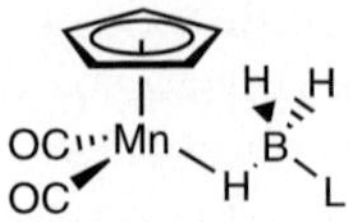

L = NMe3, PMe3

24

sition metal η^2-silane complexes, it has also been found that electronegative substituents at Si also enhance both the metal-silicon and metal-η^2-silane interactions [9, 42].

In addition to the manganese σ-borane complexes, Schlecht and Hartwig also reported the synthesis of $Cp^*Re(CO)_2(\eta^2\text{-HBpin})$ via reaction of cis-$Cp^*Re(CO)_2(Bpin)_2$ with methanol or neopentyl alcohol in benzene [20]. Crystals of this compound suitable for X-ray diffraction were not obtained. It is interesting to note that the group 7 metal fragments $Cp'M(CO)_2$ (M = Mn, Re) are capable of stabilizing various σ-complexes. It is now known that the fragments form σ-complexes with silanes [7, 8], stannanes [43], and three-coordinate boranes [20]. In addition, the fragments also form σ-complexes with borane-Lewis base adducts and alkanes, such as $CpMn(CO)_2(\eta^1\text{-}HBH_2{\cdot}L)$ (L = NMe_3, PMe_3) **24** [44] and *i*-$PrCpRe(CO)_2$(pentane) [10]. In **24**, the Lewis base (NMe_3 or PMe_3) donates its lone pair electrons to the "empty" boron sp^3-hybridized orbital.

5
Cp-free σ-Borane Complexes

In 2002, Sabo-Etienne and her co-workers reported that reaction of $RuH_2(H_2)_2(PCy_3)_2$ with excess HBpin gives the Cp-free σ-borane complex $RuH(H_2Bpin)(\eta^2\text{-HBpin})(PCy_3)_2$ **4** [21]. In 2005, they reported further that the stoichiometric reaction of $RuH_2(H_2)_2(PCy_3)_2$ with HBcat or HBpin produces the σ-borane complexes $RuH_2(\eta^2\text{-}H_2)(\eta^2\text{-HBcat})(PCy_3)_2$ and $RuH_2(\eta^2\text{-}H_2)(\eta^2\text{-HBpin})(PCy_3)_2$ **5**, respectively [22]. In 2007, they obtained and characterized $RuH_2(\eta^2:\eta^2\text{-}H_2BMes)(PCy_3)_2$ **6**, having an unprecedented bonding mode that contains mesitylborane as a bis(η^2-H – B) ligand [23].

These complexes are very interesting because many possible isomeric structural forms, including σ-borane, hydridoborate and boryl hydride, are possible. Sabo-Etienne and her co-workers carried out DFT calculations on two possible isomers for the model complex $RuH_4(Bpin')_2(PMe_3)_2$ (Bpin′ = $B(OCH_2)_2$) and four possible isomers for each of the model complexes $RuH_5(Bcat)(PMe_3)_2$ and $RuH_5(Bpin')(PMe_3)_2$ [22]. Figure 7 shows the relative energies. For $RuH_4(Bpin')_2(PMe_3)_2$, the σ-borane structural isomer **a1** is more stable than the hydridoborate structural isomer **a2** by only 3.8 kcal/mol.

The four structural isomers for $RuH_5(Bcat)(PMe_3)_2$ or $RuH_5(Bpin')(PMe_3)_2$ do not differ much in the stability, again reflecting the subtle differences among the various types of bonding interactions discussed above. It is interesting to note from structures **b1** and **b2** shown in Fig. 7 that a simple rotation of the η^2-H_2 ligand by 90° switches a σ-borane structure (**b1**) to a hydridoborate structure (**b2**). It was argued [22] that when the η^2-H_2 ligand lies on the equatorial plane (the plane perpendicular to the P – Ru – P axis) it competes for back-bonding interaction with the occupied metal d orbital on the equatorial plane, switching the structure from that of σ-borane to hydridoborate.

When 9-borabicyclo[3.3.1]nonane (9-BBN), instead of HBcat or HBpin, was used as the boron reagent, only the hydridoborate structural form, similar to structure **b2** shown in Fig. 7, was obtained [22]. 9-BBN is a much stronger Lewis acid than HBcat or HBpin. Apparently, only the boron-hydride interaction can stabilize the boron center.

Very recently, García and coworkers found that a low valent late-transition metal such as Ni(0) is able to stabilize a three-coordinate borane [24]. They reported that reactions of $[(P\text{-}P)NiH]_2$ (P-P = dcype, dippe, dtbpe) with

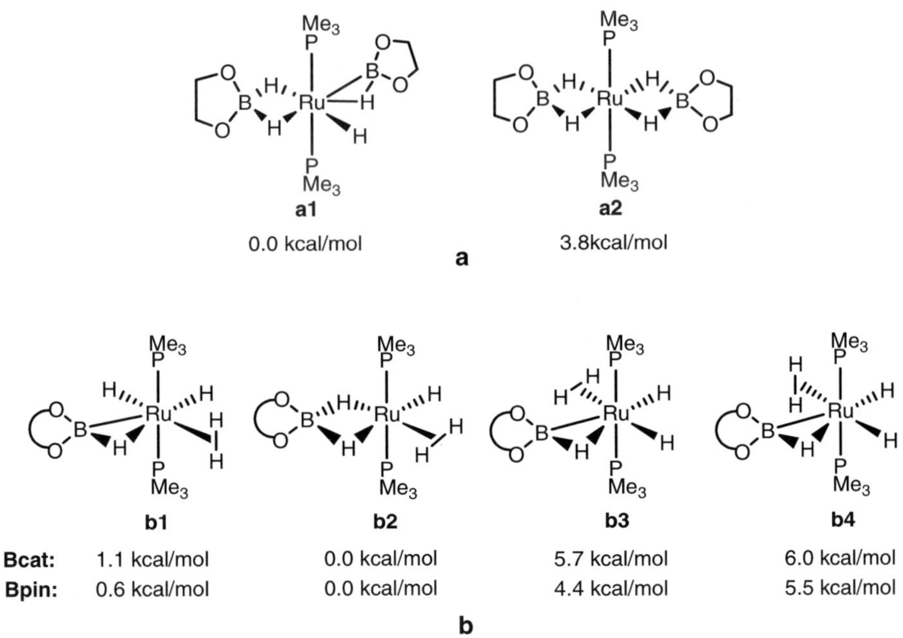

Fig. 7 **a** Two structural isomers of $RuH_4(Bpin)_2$ together with their relative energies. **b** Four structural isomers of $RuH_5(BR_2)(PMe_3)_2$ (BR_2 = Bcat and Bpin) together with their relative energies. In the DFT calculations carried out by Sabo-Etienne and her coworkers, four hydrogen atoms were used to replace the four methyl groups of the Bpin ligand

a mixture of BEt_3 and $LiHBEt_3$ afford the σ-borane nickel(0) complexes (P-P)Ni(η^2-$HBEt_2$) **7**. The X-ray crystal structure of (dcype)Ni(η^2-$HBEt_2$) was reported. The Ni-B, B-H, and Ni-H bond distances are 2.172 Å, 1.226 Å, and 1.477 Å, respectively. Although a detailed bonding analysis of these Ni(0) σ-borane complexes is not yet available, we expect that a significant back-bonding interaction exists between an occupied Ni(d) orbital and the "empty" boron sp^3-hybridized orbital from the η^2-$HBEt_2$ ligand. Since the Et substituents at boron of the η^2-$HBEt_2$ ligand have no lone pairs of electrons, the boron-hydride interaction is necessary to stabilize the boron center, providing a structure having an η^2-$HBEt_2$ coordination other than that of a boryl hydride.

6
Multinuclear Cluster Compounds Containing σ-Borane Ligands

As mentioned in the introduction, σ-borane ligands can also be found in multinuclear cluster compounds [26–30]. Ta_2(μ-H_2BH)(μ-dmpm)$_3$(η^2-BH_4)$_2$ **8** represents an interesting dinuclear compound containing both tetrahydroborato and σ-borane ligands. The coordination mode of the σ-borane ligand is quite unique in which there are two Ta-(η^2-H-B) units in the molecular structure. In the Ta(μ-H_2BH)Ta unit, the bridging boron atom is symmetrically coordinated to the two Ta atoms with a distance of 2.306 Å, 0.26 Å shorter than the Ta-B distance in either one of the two terminal Ta(η^2-BH_4) units and slightly longer than the Ta-B distances (2.263 Å and 2.295 Å) found in the exo- and endo-Cp_2TaH_2(Bcat) isomers [25].

Formally, **8** can be described as a ditantalum(I,I) compound when the σ-borane ligand, i.e., the bridging μ-BH_3 ligand, is considered to be neutral [26]. According to the 18-electron rule, the dinuclear cluster has a Ta = Ta double bond. On the basis of the local metal frontier orbital approach proposed earlier for metal clusters containing π-donor ligands [45], which considers metal-metal bonding as arising from the interaction of the fragment frontier orbitals derived from the individual local coordination, we can describe the bonding picture as follows. Each metal center has a pentagonal-bipyramidal local coordination geometry when the Ta – Ta σ bond is taken into account. Each metal center has only two d orbitals available for metal–metal π and δ bonding interactions. We can deduce here that four of the eight metal d electrons in the cluster occupy the Ta = Ta ($\sigma + \pi$) bonding molecular orbitals and the remaining four d electrons occupy the Ta-Ta $\delta + \delta^*$ molecular orbitals. We can also envision that the occupied Ta-Ta π bonding orbital acts as a donating orbital for back-bonding interaction with the "empty" boron sp^3-hybridized orbital in the μ-H_2BH bridging ligand, giving rise to a three-center-two-electron bond in the Ta_2B triangle shown in **25**.

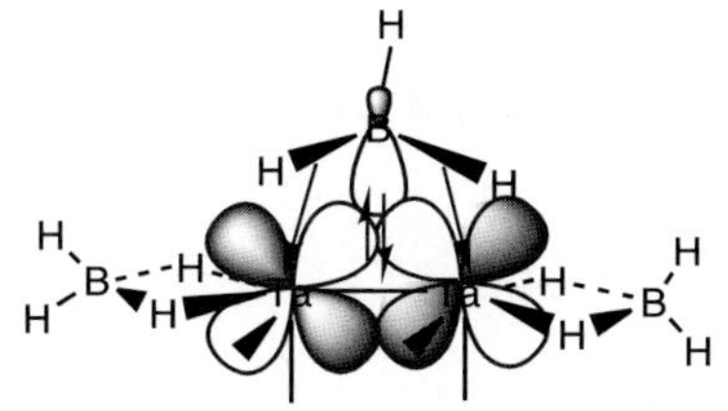

$Ta_2(\mu\text{-}H_2BH)(\mu\text{-dmpm})_3(\eta^2\text{-}BH_4)_2$

25

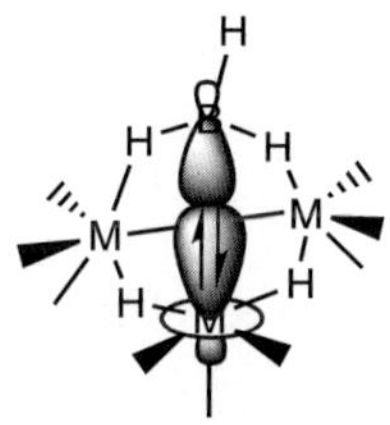

26

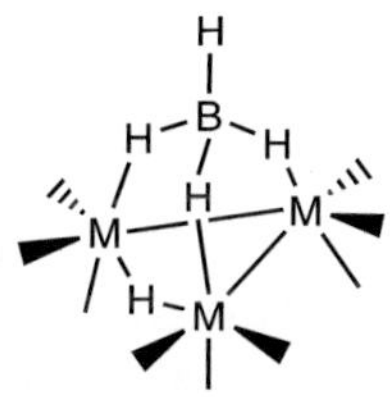

M = Ru, Fe

27

Bridging BH_3 to multiple metal centers is also known in trinuclear clusters of the type $M_3(CO)_6L_3(\mu\text{-}H_2BH)(\mu_2\text{-}H)_2$ (M = Ru, Os; L = CO, phosphine) **9** [27–30]. These clusters satisfy the 18-electron rule and each has an electron count of 48, typical for triangular trinuclear clusters [46]. Examining the local coordination in these clusters, we expect a significant back-bonding interaction between an occupied metal d orbital from the metal center, which is directly bonded to the boron center without a bridging hydride of the $\mu\text{-}BH_3$ ligand, and the "empty" boron sp^3-hybridized orbital from the $\mu\text{-}BH_3$ ligand. This is shown in **26**. A complete series of group 8 trinuclear cluster analogs of composition $M_3(CO)_9H_5B$ (M = Fe, Ru, Os) are now known [27–30]. Interestingly, all these cluster analogs have their structures based on a tetrahedral core M_3B with a terminal BH bond. The specifics of the bridging hydrogen atoms, however, differs from one transition metal to another. When M = Fe a structural isomer (**27**) that is different from **9** is observed, i.e., $Fe_3(CO)_9(\mu_2\text{-}H)(\mu\text{-}H_3BH)$ [27]. The ruthenium analog was found to exist in both isomeric forms, $Fe_3(CO)_9(\mu_2\text{-}H)_2(\mu\text{-}H_2BH)$ (**9**) and $Fe_3(CO)_9(\mu_2\text{-}H)_2(\mu\text{-}H_3BH)$ (**27**) [29]. When M = Os, only the structural type **9** was observed [30]. The relative strength in the M-B back-bonding interaction (**26**) among different group 8 metals may be able to explain the interesting observations. Os is expected to have greater M-B back-bonding interaction because of its more diffuse d orbitals when compared with Fe or Ru. Therefore, Os is able to stabilize a three-coordinate boron center. The Fe d orbitals are rather contracted. Therefore, a different structural form having a hydride to stabilize the boron center is necessary. Ru is somewhere in between Os and Fe in terms of the diffuseness in the d orbitals. Therefore, both structural forms exist.

7
σ-Borane Complexes in Hydroboration and Borylation Reactions

7.1
Hydroboration of Alkenes

In many catalytic processes and transition metal mediated reactions, σ-borane complexes have been shown to be intermediates. The bis(borane) complex $Cp_2Ti(\eta^2\text{-HBcat}')_2$ (HBcat′ = HBcat-4-t-Bu) is a highly active catalyst for the hydroboration of vinylarenes [37]. A mechanism, shown in Scheme 3, has been proposed for the Ti-catalyzed hydroboration on the basis of a detailed mechanistic study [37]. Theoretical calculations provided further support to the proposed reaction mechanism and showed that the reductive elimination step, giving the product molecules, is rate-determining [38]. In the $Cp_2Ti(CO)_2$ catalyzed hydroboration of alkynes [36, 37], the proposed reaction mechanism (Scheme 4) also involves a σ-borane complex similar to **11** and **14**. In the titanium-catalyzed decaborane-olefin hydroborations [47, 48], σ-borane complexes were also considered as intermediates. In the Cp_2MH (M = Nb, Ta) mediated hydroboration reactions of olefins [39, 41], Smith and his coworkers observed several interesting σ-borane complexes, such as **21**–**23** discussed above.

Hydroboration of alkenes catalyzed by the Wilkinson complex has been well studied [49–52]. Although there were several high level theoretical studies [53–56] employing the model reaction of $HB(OR)_2 + H_2C=CH_2 \rightarrow H_3CCH_2B(OR)_2$ ($(OR)_2$ = (OH), OCH = CHO) catalyzed by $[RhCl(PH_3)_2]$, its mechanism remains disputed. For example, associative versus dissociatve with regard to phosphine and H versus boryl migration to coordinate akene mechanisms have been proposed. Despite the ambiguities, it is clear that the first step in the catalytic cycle is the (oxidative) addition of $HB(OR)_2$ to

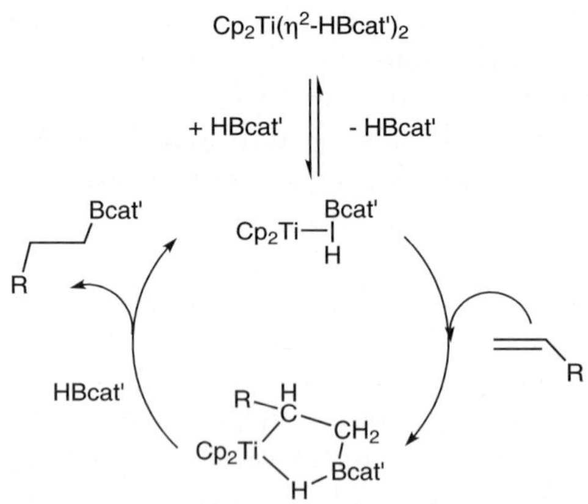

Scheme 3

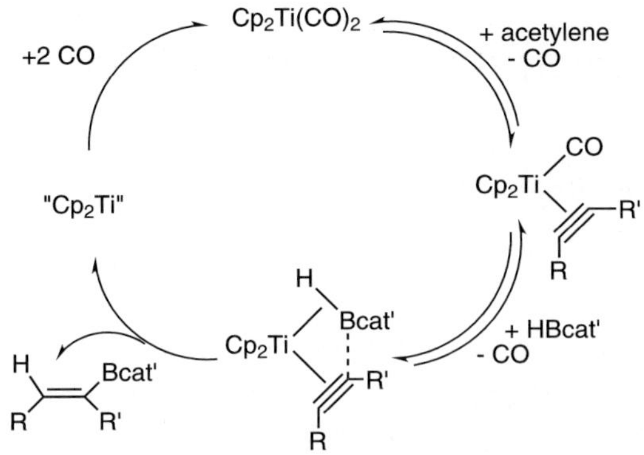

Scheme 4

the Rh center. In the various theoretical studies, the calculations based on the density functional theory predict the intermediate formed from the addition of $HB(OR)_2$ to the Rh center to be a pseudo square-planar 16-electron Rh(I) σ-borane complex $(PH_3)_2Rh(Cl)(\eta^2)$-$HB(OR)_2$ [55]. The calculations based on the second-order Møller Plesset perturbation theory, however, give a 5-coordinate trigonal-bipyramidal, 16-electron Rh(III) hydridoboryl complex [53]. Interestingly, a recent paper [57] reported the results of low temperature single-crystal X-ray, neutron diffraction and computational studies of $[(P^i Pr_3)_2RhHCl(boryl)]$ (boryl = Bcat, Bpin), providing the accurate location of the hydride ligands and indicating that the two complexes are best described as Rh(III) hydrido boryl rather than Rh(I) σ-borane complexes, although there is a modest residual H-B interaction in both complexes.

7.2 Borylation of Alkanes and Arenes

7.2.1 Stoichiometric Borylation Reactions

Recently, Hartwig and co-workers found that the transition metal boryl complexes $Cp^*M(CO)_{n+1}$(boryl) ($n = 1$: M = Fe, Ru; $n = 2$: M = W; boryl = Bcat or Bpin) can efficiently and selectively functionalize alkane C-H bonds after photodissociation of one CO ligand [58–61] and that $Cp^*Re(CO)_3$ can catalyze borylation of alkanes with B_2pin_2 under photochemical conditions [62]. Photodissociation of one CO ligand from $Cp^*M(CO)_{n+1}$(boryl) generates a coordinately unsaturated 16e intermediate, which can be trapped with phosphine. In the absence of phosphine, the intermediate readily cleaves and functionalizes a terminal alkane C-H bond or an arene C-H bond (Scheme 5). The related complexes of alkyl and aryl ligands, $CpM(CO)_{n+1}R$, do not show a similar reactivity [63]. The boryl ligands are clearly important in the reac-

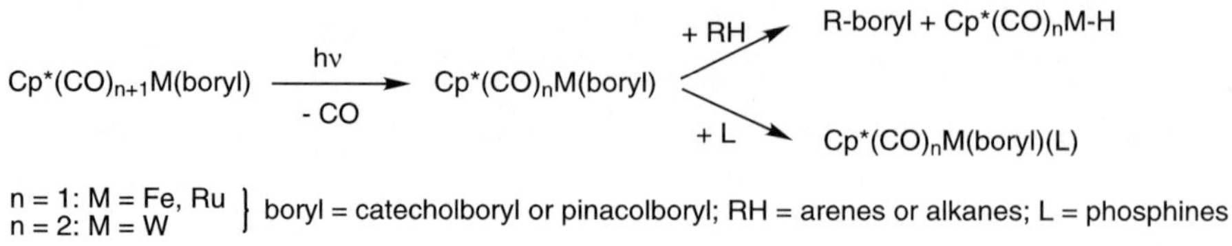

Scheme 5

tions. Theoretical calculations at the B3LYP level of density functional theory allow a reaction mechanism, shown in Scheme 6, to be defined [63]. In the reaction mechanism, σ-borane intermediates play important roles.

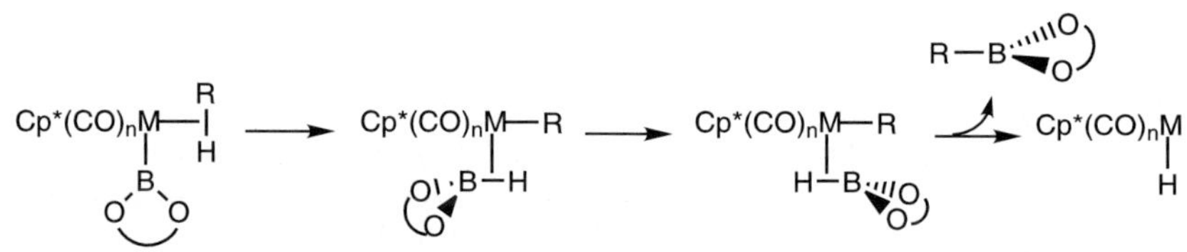

Scheme 6

7.2.2 Catalytic Borylation Reactions

Intensive studies also showed that many transition metal complexes are able to catalyze aromatic C-H borylation of various arenes (Scheme 7), e.g., Cp*Ir(H)(Bpin)(PMe_3) [64, 65], Cp*Rh(η^4-C_6Me_6) [65, 66], (η^5-Ind)Ir(COD) [67], (η^6-mesitylene)Ir$(Bpin)_3$ [67], $[IrX(COD)]_2$/bpy (X = Cl, OH, OMe, OPh) [68–70]. A very recent study by Marder and his coworkers showed that $[Ir(OMe)(COD)]_2$ can also catalyze borylation of C-H bonds in N-containing heterocycles [71]. For the Ir-catalyzed borylation reactions, it is now believed that tris(boryl)iridium(III) complexes [67, 69], 40c, [72] are likely the reactive intermediates and a mechanism involving an Ir(III)-Ir(V) catalytic cycle is operative [67, 69]. A recent theoretical study [73] provided further support for this hypothesis. A mechanism, shown in Scheme 8, was proposed. Interestingly, there are no σ-borane complexes involved in the Ir-catalyzed reactions. The very electropositive boryl and hydride ligands may play important roles in stabilizing the iridium(V) intermediates.

Recently, it was also shown [74] that pentamethylcyclopentadienyl rhodium and iridium complexes with labile dative ligands catalyze the borylation of alkanes at the terminal position under thermal conditions (Scheme 9). It was proposed that Ir(V) and Rh(V) boryl complexes act as intermediates in these reactions. Reaction of Cp*IrH_4 with HBpin gives Cp*Ir(Bpin)$(H)_3$ and Cp*$Ir(Bpin)_2(H)_2$. Both of the two Ir(V) boryl complexes react with octane to give octylboronate ester [75], though it is still unclear regarding the role

$$\mathrm{Ar{-}H} + \mathrm{HBpin} \xrightarrow[\sim 150^{\circ}\mathrm{C}]{\text{catalyst}} \mathrm{ArBpin} + \mathrm{H_2}$$

catalyst: $Cp^*Ir(H)(Bpin)(PMe_3)$, $Cp^*Rh(\eta^4\text{-}C_6Me_6)$, $(\eta^5\text{-Ind})Ir(COD)$ or $(\eta^6\text{-mesitylene})Ir(Bpin)_3$

$$2\ \mathrm{Ar{-}H} + \mathrm{B_2pin_2} \xrightarrow[\text{Room Temperature}]{\text{catalyst}} 2\ \mathrm{ArBpin} + \mathrm{H_2}$$

catalyst: $[IrX(COD)]_2$/bpy (X = Cl, OH, OMe, OPh; bpy = subsituted bipyridine)

Scheme 7

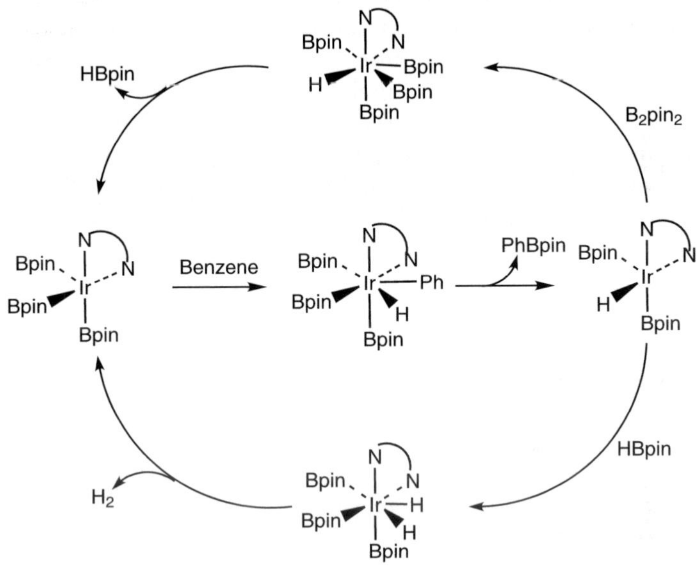

Scheme 8

played by the Ir(V) species. The latest detailed experimental and theoretical studies on the reactivity of the $Cp^*Rh(\eta^4\text{-}C_6Me_6)$ catalyst indicate that the boryl complexes $Cp^*Rh(H)_2(Bpin)_2$ **28** and $Cp^*RhH(Bpin)_3$ **29** are the likely intermediates in the regioselective borylation of alkanes catalyzed by $Cp^*Rh(\eta^4\text{-}C_6Me_6)$ [76]. Both X-ray diffraction studies and DFT calculations suggested that the two complexes should be regarded as elongated σ-borane complexes, having the B-H distances in the range of 1.50–1.70 Å [76]. Experi-

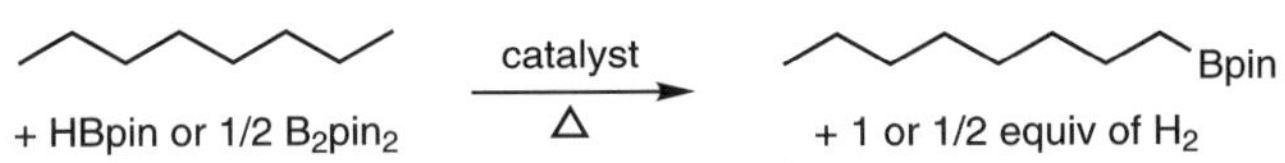

catalyst: $Cp^*Rh(\eta^4\text{-}C_6Me_6)$, $Cp^*Rh(\eta^2\text{-}C_2H_4)_2$, Cp^*IrH_4 or $Cp^*Ir(\eta^2\text{-}C_2H_4)_2$

Scheme 9

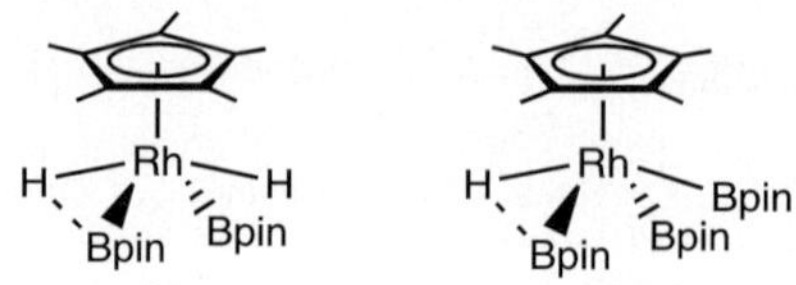

28 **29**

mental studies further showed that the two boryl complexes react with alkanes and arenes to form alkyl- and arylboronate esters at temperatures similar to or below those of the catalytic borylation of alkanes and arenes [76]. A mechanism, shown in Scheme 10, has been proposed for the reactions on the basis of the DFT calculations [76]. Again, σ-borane complexes were considered to be the important intermediates for the reactions. Scheme 10 ends at the Cp*Rh(H)(X) (X = H, Bpin) fragment. Presumably, in the Rh-catalyzed borylation reaction, the fragment can further react with HBpin followed by elimination of H_2 [77], regenerating the reactive species Cp*Rh(Bpin)(X) (X = H, Bpin) to complete a catalytic cycle.

Independently, in 2001 Marder and his coworkers reported a high degree of benzyl selectivity in the borylation reactions of C-H bonds in toluene, *p*-xylene, and mesitylene with HBpin (pin = $OCMe_2CMe_2O$) using the catalyst precursor $[(P^iPr_3)_2Rh(Cl)(N_2)]$ (Scheme 11) [78]. Oxidative addition of HBpin to $[(P^iPr_3)_2Rh(Cl)(N_2)]$ yields *trans*-$[(P^iPr_3)_2Rh(Cl)(H)(Bpin)]$, the structure of which was determined by low temperature single-crystal X-ray

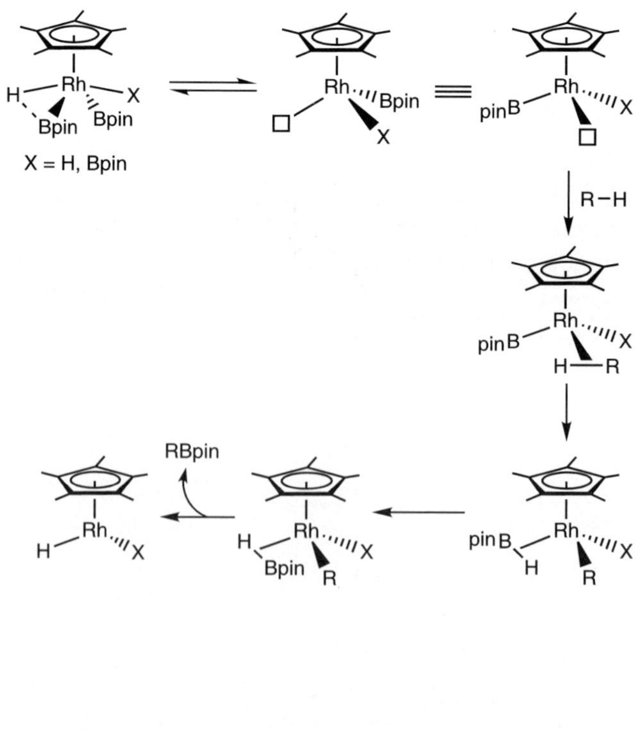

Scheme 10

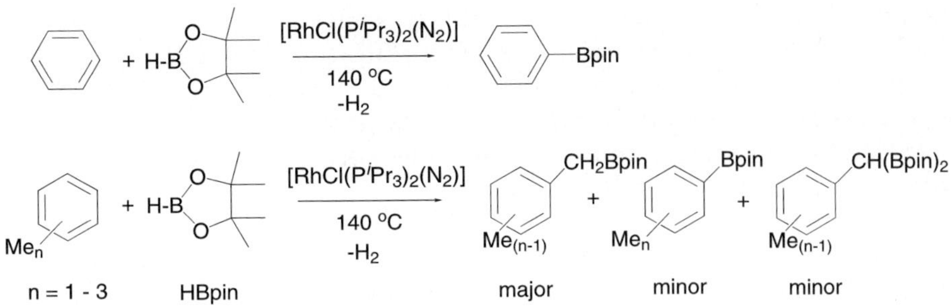

Scheme 11

studies [57, 78]. The isolated *trans*-[(P^{i}Pr$_3$)$_2$Rh(Cl)(H)(Bpin)] complex also catalyzed the reaction of toluene with HBpin to give borylated products. Two possible reaction mechanisms, shown in Scheme 12, for the formation of PhCH$_2$Bpin were proposed [78]. We can see that a σ-borane complex is proposed as an intermediate. Our theoretical calculations [79] suggest that the catalytic cycle in the left-hand side of Scheme 12 is energetically more favorable. The exceedingly strong *trans*-influence of the boryl ligands [80] was

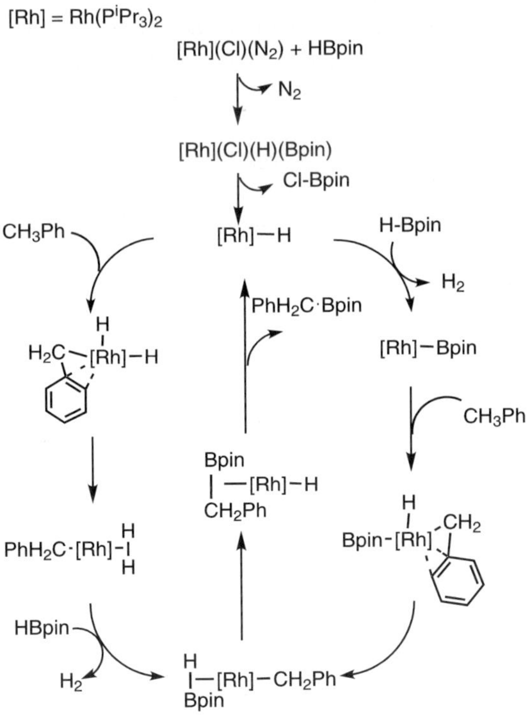

Scheme 12

found to make C-H activation from $[RhL_2(boryl)]$ (L = phosphine) energetically much less favourable than that from $[RhL_2(H)]$. It was found that the formation of η^3-intermediates in the C-H bond borylation pathway is critical to the benzylic over aromatic selectivity of the system, and is a direct consequence of the geometric and electronic preferences of the $[RhL_2]$ moiety in contrast to the Cp*Rh and Ir catalysts.

8 Conclusion

In this chapter, we have reviewed the structure and bonding of transition metal σ-borane complexes. We see that a variety of structural types exist, including bent-metallocene, piano-stool and Cp-free, etc. The metal centers are also varied from early to late transition metals. σ-Borane complexes also display quite interesting bonding properties. Because of the electron unsaturated boron center in the σ-borane ligands in the precoordination state, the metal-to-ligand π interaction involves mainly the back-bonding interaction of an occupied metal d_π orbital with the "empty" sp^3-hybridized orbital of the three-coordinated boron center. The back-bonding interaction with the H-B(σ^*) orbital from the η^2-HBX_2 ligand is less important. Strong metal-to-ligand back donation does not necessarily break the coordinated B-H bond. As a result, σ-borane complexes are unexpectedly common among early transition metals. Because of the "empty" orbital of the three-coordinated boron center, σ-borane complexes play important roles in transition metal catalyzed hydroboration of olefins and borylation of alkanes and arenes.

Acknowledgements I'd like to thank my co-workers for their contribution to the work reported here: Wai Han Lam, Dan Liu, and King Chung Lam. I'd like to acknowledge financial support from the Hong Kong Research Grants Council (Grant No. HKUST 601507).

References

1. Kubas GJ (2001) Metal-dihydrogen and sigma-bond complexes: structure, theory, and reactivity. Kluwer, Academic, New York
2. Kubas GJ (2001) J Organomet Chem 635:37
3. Crabtree RJ (1993) Angew Chem Int Ed Engl 32:789
4. Maseras F, Lledós A, Clot E, Eisenstein O (2000) Chem Rev 100:601
5. Clot E, Eisenstein O (2004) Struct Bond 113:1
6. Heinekey DM, Oldham WJ (1993) Chem Rev 93:913
7. Schubert H (1990) Adv Organomet Chem 30:151
8. Corey JY, Braddock-Wilking J (1999) Chem Rev 99:175
9. Lin Z (2002) Chem Soc Rev 31:239
10. Lawes DJ, Geftakis S, Ball GE (2005) J Am Chem Soc 127:4134
11. Hall C, Perutz RN (1996) Chem Rev 96:3125

12. Marks TJ, Kolb JR (1977) Chem Rev 77:263
13. Xu Z, Lin Z (1996) Coord Chem Rev 156:139
14. Snow SA, Shimoi M, Ostler CD, Thompson BK, Kodama G, Parry RW (1984) Inorg Chem 23:511
15. Yasue T, Kawano Y, Shimoi M (2003) Angew Chem Int Ed 42:1727
16. Merle N, Koicok-Köhn G, Mahon MF, Frost CG, Ruggerio GD, Weller AS, Willis MC (2004) Dalton Trans 2004:3883
17. Hartwig JF, Muhoro CN, He X, Eisenstein O, Bosque R, Maseras F (1996) J Am Chem Soc 118:10936
18. Muhoro CN, Hartwig JF (1997) Angew Chem Int Ed Engl 36:1510
19. Muhoro CN, He X, Hartwig JF (1999) J Am Chem Soc 121:5033
20. Schlecht S, Hartwig JF (2000) J Am Chem Soc 122:9435
21. Montiel-Palma V, Lumbierres M, Donnadieu B, Sabo-Etienne S, Chaudret B (2002) J Am Chem Soc 124:5624
22. Lachaize S, Essalah K, Montiel-Palma V, Vendier L, Chaudret B, Barthelat JC, Sabo-Etienne S (2005) Organometallics 24:2935
23. Alcaraz G, Clot E, Helmstedt U, Vendier L, Sabo-Etienne S (2007) J Am Chem Soc 129:8704
24. Crestani MG, Muñoz-Hernández M, Arévalo A, Acosta-Ramírez A, García JJ (2005) J Am Chem Soc 127:18066
25. Lantero DR, Motry DH, Ward DL, Smith III MR (1994) J Am Chem Soc 116:10811
26. Cotton FA, Murillo CA, Wang X (1998) J Am Chem Soc 120:9594
27. Vites JC, Eigenbrot C, Fehlner TP (1984) J Am Chem Soc 106:4633
28. Chipperfield AK, Housecroft CE (1988) J Organomet Chem 349:C17
29. Housecroft CE, Matthews DM, Edwards AJ, Rheingold AL (1993) J Chem Soc Dalton Trans 1993:2727
30. Chung JH, Boyd EP, Liu J, Shore SG (1997) Inorg Chem 36:4778
31. Lin Z, Hall MB (1994) Coord Chem Rev 135/136:845
32. Irvine GJ, Lesley MJG, Marder TB, Norman NC, Rice CR, Robins EG, Roper WR, Whittell GR, Wright LJ (1998) Chem Rev 98:2685
33. Lam WH, Lin Z (2000) Organometallics 19:2625
34. Liu D, Lam KC, Lin Z (2003) Organometallics 22:2827
35. Liu D, Lam KC, Lin Z (2003) J Organomet Chem 680:148
36. He X, Hartwig JF (1996) J Am Chem Soc 118:1696
37. Hartwig JF, Muhoro CN (2000) Organometallics 19:30
38. Liu D, Lin Z (2002) Organometallics 21:4750
39. Lantero DR, Miller SL, Cho JY, Ward DL, Smith III MR (1999) Organometallic 18:235
40. Hartwig JF, De Gala SR (1994) J Am Chem Soc 116:3661
41. Lantero DR, Ward DL, Smith III MR (1997) J Am Chem Soc 119:9699
42. Choi SH, Feng JW, Lin Z (2000) Organometallics 19:2051
43. Schubert U, Kunz E, Harkers B, Willnecker J, Meyer J (1989) J Am Chem Soc 111:2572
44. Kakizawa T, Kawano Y, Shimoi M (2001) Organometallics 20:3211
45. Lin Z, Fan MF (1997) Struct Bond 87:35
46. Mingos DMP, Wales DJ (1990) Introduction to Cluster Chemistry. Prentice Hall, Englewood Cliffs, New Jersey
47. Pender MJ, Wideman T, Carroll PJ, Sneddon LG (1998) J Am Chem Soc 120:9108
48. Pender MJ, Carroll PJ, Sneddon LG (2001) J Am Chem Soc 123:12222
49. Männing D, Nöth H (1985) Angew Chem Int Ed Engl 24:878
50. Burgess K, Ohlmeyer MJ (1991) Chem Rev 91:1179
51. Evans DA, Fu GC, Anderson BA (1992) J Am Chem Soc 114:6679

52. Burgess K, van der Donk WA, Westcott SA, Marder TB, Baker RT, Calabrese JC (1992) J Am Chem Soc 114:9350
53. Musaev DG, Mebel AM, Morokuma K (1994) J Am Chem Soc 116:10693
54. Dorigo AE, Schleyer PVR (1995) Angew Chem Int Ed Engl 34:115–118
55. Widauer C, Grützmacher H, Ziegler T (2000) Organometallics 19:2097
56. Huang X, Lin Z (2002) In: Maseras F, Lledós A (eds) Computational Modeling of Homogeneous Catalysis. Kluwer Academic Publishers, Amsterdam, p 189
57. Lam WH, Shimada S, Batsanov AS, Lin Z, Marder TB, Cowan JA, Howard JAK, Mason SA, McIntyre GJ (2003) Organometallics 22:4557
58. Waltz KM, He X, Muhoro CN, Hartwig JF (1995) J Am Chem Soc 117:11357
59. Waltz KM, Hartwig JF (1997) Science 277:211
60. Waltz KM, Muhoro CN, Hartwig JF (1999) Organometallics 18:3383
61. Waltz KM, Hartwig JF (2000) J Am Chem Soc 122:11358
62. Chen H, Hartwig JF (1999) Angew Chem Int Ed 38:3391
63. Webster CE, Fan Y, Hall MB, Kunz D, Hartwig JF (2003) J Am Chem Soc 125:858
64. Iverson CN, Smith MR III (1999) J Am Chem Soc 121:7696
65. Cho JY, Iverson CN, Smith MR III (2000) J Am Chem Soc 122:12868
66. Tse MK, Cho JY, Smith MR III (2001) Org Lett 3:2831
67. Cho JY, Tse MK, Holmes D, Maleczka RE Jr, Smith MR III (2002) Science 295:305
68. Ishiyama T, Takagi J, Hartwig JF, Miyaura N (2002) Angew Chem Int Ed 41:3056
69. Ishiyama T, Takagi J, Ishida K, Miyaura N, Anastasi NR, Hartwig JF (2002) J Am Chem Soc 124:390
70. Boller TM, Murphy JM, Hapke M, Ishiyama T, Miyaura N, Hartwig JF (2005) J Am Chem Soc 127:14263
71. Mkhalid IA, Coventry DN, Albesa-Jove D, Batsanov AS, Howard JAK, Perutz RN, Marder TB (2006) Angew Chem Int Ed 45:489
72. Nguyen P, Blom HP, Westcott SA, Taylor NJ, Marder TB (1993) J Am Chem Soc 115:9329
73. Tamura H, Yamazaki H, Sato H, Sakaki S (2003) J Am Chem Soc 125:16114
74. Chen H, Schlecht S, Semple TC, Hartwig JF (2000) Science 287:1995
75. Kawamura K, Hartwig JF (2001) J Am Chem Soc 123:8422
76. Hartwig JF, Cook KS, Hapke M, Incarvito CD, Fan Y, Webster CE, Hall MB (2005) J Am Chem 127:2538
77. Wan X, Wang X, Luo Y, Takami S, Kubo M, Miyamoto A (2003) Organometallics 21:3703
78. Shimada S, Batsanov AS, Howard JAK, Marder TB (2001) Angew Chem Int Ed 40:2168
79. Lam WH, Lam KC, Lin Z, Shimada S, Perutz RN, Marder TB (2004) Dalton Trans 2004:1556
80. Lam KC, Lam WH, Lin Z, Marder TB, Norman NC (2004) Inorg Chem 43:2541

Struct Bond (2008) 130: 149–202
DOI 10.1007/430_2007_076

Published online: 7 December 2007

Coordination Modes and Hydride Exchange Dynamics in Transition Metal Tetrahydroborate Complexes

Maria Besora · Agustí Lledós (✉)

Departament de Química, Universitat Autònoma de Barcelona, 08193 Bellaterra, Spain
agusti@klingon.uab.es

Abstract This contribution reviews the structural and dynamical properties of mononuclear transition metal complexes with tetrahydroborate ligands. Three different coordination modes involving one (η^1), two (η^2) or three (η^3) bridging hydrogen atoms are possible for the BH_4^- ligand. A structural classification of the X-ray characterised complexes is presented. The metal-boron distances and the vibrational frequencies of the coordinated borohydrides, which are the key experimental data usually used to determine the hapticity of tetrahydroborate binding, have been surveyed and trends along the Periodic Table established. Electronic factors governing the coordination mode have been rationalized by means of simple orbital arguments supported by quantitative calculations. In solution, most of the transition metal tetrahydroborate complexes show fluxional behaviour, displaying a single resonance for the four B – H hydrogens in the ^1H NMR spectrum at ambient temperature. This fast intramolecular exchange between bridging and terminal hydrogens has been analysed. Experimental and computational data for these processes have been collected and the exchange mechanisms are discussed. In summary, several examples illustrating the perspectives on the field are presented.

Keywords Tetrahydroborate Complexes · Coordination Modes · Structural Trends · Bonding Analysis · Hydrogen Exchange Dynamics

Abbreviations

(COT)″	η^8-1,4-bis(trimethylsilyl)cyclo-octatetraene
(*t*BuNCH)$_2$	(*N*,*N*′-di-*t*-butylethylenediamine)
{Me(16)aneS4}	3,3,7,7,11,11,15,15-octamethyl-1,5,9,13-tetrathiacyclohexadecane
Ad	Adamantyl (3,5-dimethylphenyl) amide
binap	2,2′-bis(di-*p*-tolylphosphino)-1,1-binaphtyl
bp	methylene-2,2′-bis(4-methyl-6-tert-butylphenoxy)
$C_7H_6C_6H_5P(iPr)_2$	η^7-(2-(di-isopropylphosphino)phenyl)-heptatrienyl
CHtBu	neopentylene-C,H
CNNC	*N*,*N*′-bis(2-isopropyl-6-isopropylenephenyl)pentane-2,4-diiminato
Cp	cyclopentadienyl
Cp*	η^5-pentamethyl-cyclopentadienyl
Cp′	η^5-$C_5H_4SiMe_3$
CpB	1-(di-isopropylamino(η^5-indenylidene)phosphino)-1,2-dicarba-dodecaboran-2-yl(10)
CpBu	(1,3-Di-*t*butyl-cyclopentadienyl)
CpMe	η^5-methyl-cyclopentadienyl
CpMet	η^5-Ethyl-tetramethyl-cyclopentadienyl
Cp-O	η^5,σ1-2-Methoxyethylcyclopentadienyl
CpSi	η^5-trimethylsilyl-cyclopentadienyl
CpSi$_2$	η^5-1,3-trimethylsilyl-cyclopentadienyl
Cy	Cyclohexyl
Cyclam	1,4,8,11-tetra-azacyclotetradecane
diam	1,2-diphenylethylenediamine
dmdphen	2,9-dimethyl-4,7-diphenyl-1,10-phenantroline
dmoe	1,2-dimethoxyethane
dmpe	1,2-bis(dimethylphosphino)ethane
dmphen	2,9-dimethyl-1,10-phenantroline
EtCp$_2$	1,1,2,2-tetramethyl-1,2-ethylene-bis(η^5-cyclopentadienyl)
EtCpBu$_2$	1,1,2,2-tetramethyl-1,2-ethylene-bis(η^5-3-*t*-butyl-cyclopentadienyl)
$HC(CMeNC_6H_4iPr_2)_2$	(2-((2,6-diisopropylphenyl)amino)-4-((2,6-diisopropylphenyl)-imino)pent-2-enyl)
hdmpb	hydrogen tris(3,5-dimethylpyrazolyl)borato anion
In-O	η^5,σ1-2-Methoxyethylindenyl
MeCp$_2$	2,2-bis(η^5-cyclopentadienyl)propane
MetCp$_2$	3,3-bis(η^5-cyclopentadienyl)pentane
mphi	*N*-P-tolyl-*N*′-(*m*-methyl-*o*-phenylene)imidazolidine
NCH	η^2-*N*-(2,6-diisopropylphenyl)-4,6,6-trimethylhept-2,4-dien-2-aminato
N-CO	η^2-3,3,5,5-tetramethyl-1,1,7,7-tetra-isopropyl-1-trihydroborane-4-aza-1,7-diphospha-3,5-disila-octan-1-one-4-yl
NNN	*N*,*N*′,*N*″-tris(trimethylsilyl)diethylenetriamine dianion

N-NPro	*N,N′*-bis(2,6-diisopropylphenyl)pentane-2,4-diiminato)
NPh	2,6-di-isopropylphenylimino
NSi_2	bis(trimethylsilyl)amine
OPh	2,6-di-isopropylphenolato
$PBuPh_2$	butyl-diphenylphosphine
phen	1,10-phenantroline
PMe_3	trimethylphosphine
Pnor	((1*R*,2*R*)-1-(diphenylphosphino)-1-phenylprop-2-ylamine-N,P)
P-N-P	bis(di-isopropylphosphinomethyl(dimethyl)silyl)amino
PSiP	*t*Butyl-tris((dimethylphosphino)methyl)silane
py	pyridine
SiNCNSi	*N,N′*-bis(trimethylsilyl)benzamidinato
*t*Bu	tertiary butyl
terpy	2,2′ : 6,2″-terpyridyl
thf	tetrahydrofuran
TMAC	5,7,12,14-tetramethyl-1,4,8,11-tetra-azacyclotetradecane
tmeda	*N,N,N′,N′*-tetramethylethylenediamine
$TpMe_2$	(hydrogen tris(3, 5-dimethylpyrazolyl)borate-*N,N′,N″*)
tripod	1,1,1-tris((diphenylphosphino)-methyl)-ethane

1 Introduction

The simplest hydroborate anion, BH_4^-, has rich and diverse coordination chemistry. Borohydride covalent complexes of varying stability and reactivity are known for the majority of transition metals. Transition metal tetrahydroborate complexes have found a practical use in synthetic and catalytic procedures. They stabilize hydrogen-rich environments and can be the hydrogen source in metal-hydrogen interactions of significant commercial, synthetic and biochemical applications. Tetrahydroborate complexes are used as selective reducing agents, starting compounds in the synthesis of organometallic derivatives, precursors for the production of borides, hydrides and other inorganic materials, and as catalysts of hydrogenation, isomerisation and polymerisation processes. BH_4^- ligands also present very interesting properties from the perspectives of structure and bonding. BH_4^- is a very effective polyhapto ligand for which up to six coordination modes have been reported. The ligation is invariably through bridging hydrogen atoms. In mononuclear complexes, three different coordination modes involving one (η^1), two (η^2) or three (η^3) bridging hydrogen atoms have been found (Scheme 1). When more than one BH_4^- ligand is present, several coordination modes can coexist. If bimetallic complexes are also considered, in addition to the η^1, η^2 and η^3 modes mentioned above, examples of bridging modes are also known in which BH_4^- is ($\mu\eta^2$,η^2), ($\mu\eta^3$) and $\mu(\eta^4)$.

Despite the existence of complexes with η^1, η^2 and η^3, BH_4^- ligands have been proved by various structural methods. Most of the transition-metal com-

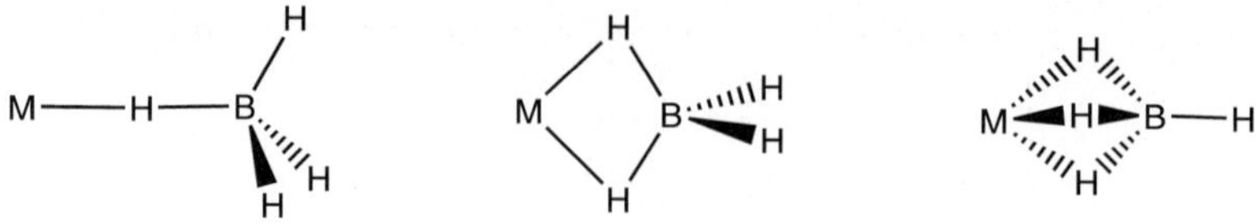

η^1 η^2 η^3

Scheme 1 Coordination modes of tetrahydroborate in covalent mononuclear complexes: monodentate (η^1), bidentate (η^2) and tridentate (η^3)

plexes present a fluxional behaviour in solution, displaying a single resonance for the four B – H hydrogens in the ^{1}H NMR spectrum at ambient temperature. Bridging and terminal hydrogens of the tetrahydroborate ligands are able to undergo fast intramolecular exchange (on the NMR time scale).

Excellent reviews spanning different periods of the development of borohydride chemistry have been published. In 1977, a thorough review covering the chemistry of covalent transition metals including lanthanide and actinide tetrahydroborate complexes was published by Marks and Kolb [1]. The borohydride compounds of f-elements were reviewed by Ephritikhine in 1997 [2]. More recently, profuse experimental data on the structures and properties of tetrahydroborate complexes, including main group metals and f-elements, have been collected by Makhaev [3]. A general theoretical analysis of the structure and bonding relationship for a variety of transition metal borohydride complexes based on orbital interaction arguments was reported by Lin [4].

This contribution focuses on the structural and dynamic properties of only mononuclear transition metal complexes with BH_4^- ligands. The first section is dedicated to the analysis and systematization of the large amount of experimental data reported. In the first section, the borohydride complexes structurally characterised or theoretically studied is classified according to the coordination geometry of the transition metal fragment to which the BH_4^- ligand is bonded. Then, a critical discussion of the two most usual experimental measurements used to determine the borohydride coordination mode, the metal–boron distance and the stretching vibrations of the B – H bonds, is presented. A theoretical analysis of the coordination modes follows. After outlining the bonding scheme deduced from a qualitative orbital analysis to account for the factors determining the structural properties of the BH_4^- ligands in complexes, selected examples of quantitative calculations giving an energetic ordering of the structures depending on the coordination mode are discussed. This section begins with monotetrahydroborate complexes and transitions to discuss then the more complex compounds with several BH_4^- ligands. The last part is devoted to the dynamical properties of tetrahydroborate complexes. Experimental data, theoretical calculations and mechanism proposal for the H-bridging/H-terminal exchange are examined.

The final section presents a summary of this contribution and current perspectives in the field.

2
Analysis of the Bonding Modes from Experimental Data

2.1
Structural Classification of Mononuclear Transition Metal Borohydrides

A large variety of structures can be found for organometallic complexes containing tetrahydroborate ligands. In this section, we will classify the X-ray characterised mononuclear transition metal tetrahydroborate complexes according to the coordination geometry of the metal. In the following discussion, each BH_4^- ligand is considered to contribute one to the coordination number for the use of the coordination polyhedra definition and bridging hydrogens are not counted. Calculated complexes will also be included. Compounds will be labelled with a first number indicating the coordination mode of the BH_4 followed by second number indicating the coordination number of the metal, considering each BH_4 group is occupying only one coordination site. For instance, **2.4** refers to a η^2-BH_4 in a tetracoordinate complex. Theoretically studied complexes will be labelled with the same convention, though starting with **T**.

2.1.1
One Tetrahydroborate Ligand

The η^1-BH_4^- coordination is much less common than the other two coordination modes. Most of the η^1 complexes are constituted by late transition metals and a large variety of phosphino ligands. Only tetrahedral (**1.4/T1.4**) and octahedral (**1.6**) structures have been crystallographically characterised (see Scheme 2). Complexes adopting tetrahedral geometries are d^{10} and contain three phosphorus atoms coordinated to the metal centre. Although some examples of pentacoordinated complexes are present in the literature, these

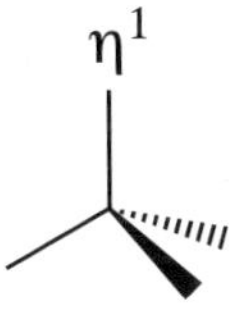

1.4
T1.4

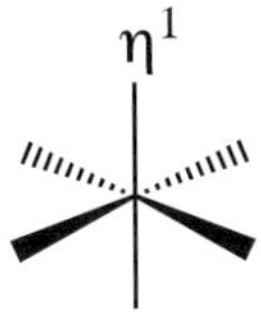

1.6

1.4
$(Ph_2MeP)_3Ag(\eta^1\text{-}BH_4)$ [18,19]
$(Ph_2MeP)_3Cu(\eta^1\text{-}BH_4)$ [20,21]
$Cu(\eta^1\text{-}BH_4)$\{tripod\} [22,23]
T1.4
$[Cu(BH_4)(PH_3)_3]$ [11]

1.6
$(dmpe)_2HFe(\eta^1\text{-}BH_4)$ [24-26]
(binap)RuH(diam)(η^1-BH_4) [27]
(Pnor)2RuH(η^1-BH_4) [28]
RuH(η^1-BH_4){Me(16)aneS4} [29]
$(dmpe)_2Mo(NO)$ (η^1-BH_4) [30]
$(NO)(PMe_3)_3(CO)Mo(\eta^1\text{-}(BH_4)$[31]
$(NO)(PMe_3)_3(CO)W(\eta^1\text{-}BH_4)$ [32]

Scheme 2 X-ray characterised or calculated complexes with one η^1-BH_4

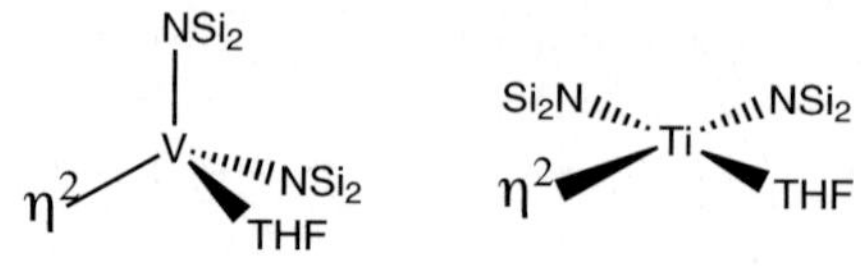

Scheme 3 Coordination geometries for **a** $(NSi2)_2V(\eta^2-BH_4)(thf)$ [42] and **b** $(NSi2)_2Ti(\eta^2-BH_4)(thf)$ (NSi2 = bis(trimethylsilyl)amine) [52]

structures do not correspond to covalent η^1 complexes but to ionic $M^+ BH_4^-$ attachments. In these ionic complexes, the BH_4^- is not acting as a ligand but plays the role of a counteranion. Previous reviews have summarized examples of these species [3, 4], though they will not be considered in this chapter dealing with covalent coordinated tetrahydroborate ligands.

For complexes with a η^2 coordination of the BH_4^- ligand, a large number of structures are known with a large variety of coordination numbers and geometries: trigonal-planar (TP), tetrahedral (TET), square-planar (SQP), trigonal-bipyramidal (TBP), square-pyramidal (SQPY) octahedral (OCT) and distorted octahedral (DOCT) (see Scheme 4). Nevertheless, some structures are more common than others. Only one example of both the SQPY [5] and DOCT [6] and two examples of OCT [7, 8] have been reported. Linear structures $M\text{-}\eta^2-BH_4$ (**T2.1**) and $LM\text{-}(\eta^2-BH_4^-)$ (**T2.2**) have only been computationally studied [9, 10] and are not models of any experimental compound. Several computational studies have also been performed in order to study η^2 complexes with higher coordination numbers [11–17].

Many TP complexes with a η^2-BH_4 group can be found. Between them, two populated classes are those with two phosphino ligands (or a bidentate phoshine) (**2.3/T2.3**) and those with two η^5-cyclopentadienyl ligands (**2.Cp2a/C2.Cp2**). A profuse series of complexes containing two Cp or Cp-derivate ligands, the η^2-BH_4 ligand and a Group 5 or Group 6 transition metal center exist. A large number of niobium complexes are known, however several complexes of Ti, V and Sc have also been crystallised. Most of the non-containing Cp ligand complexes have late transition metals (**2.3/T2.3**) and a large number of them respond to the formula $CuL_2(\eta^2-BH_4)$. These complexes can be described as tetrahedral if the η^2-BH_4 group is considered to occupy two coordination positions.

Both TET (**2.4a/T2.4**) and SQP (**2.4b**) tetracoordinated complexes with a η^2-BH_4 ligand have been characterised. The favoured geometry depends on the ligands, but also on the metal centre. For instance, in the couple $(NSi2)_2V(\eta^2-BH_4)$ (thf) and $(NSi2)_2Ti(\eta^2-BH_4)$ (thf), the two complexes have the same ligands (tetrahydrofuran and bis(trimethylsilyl)amine) and differ only on the metal centre, and the d^2-vanadium complex prefers the tetrahedral geometry while the square-planar geometry is adopted by the d^1-titanium (Scheme 3). In Sect. 2.2.6, more details about $(NSi2)_2V(\eta^2-BH_4)$

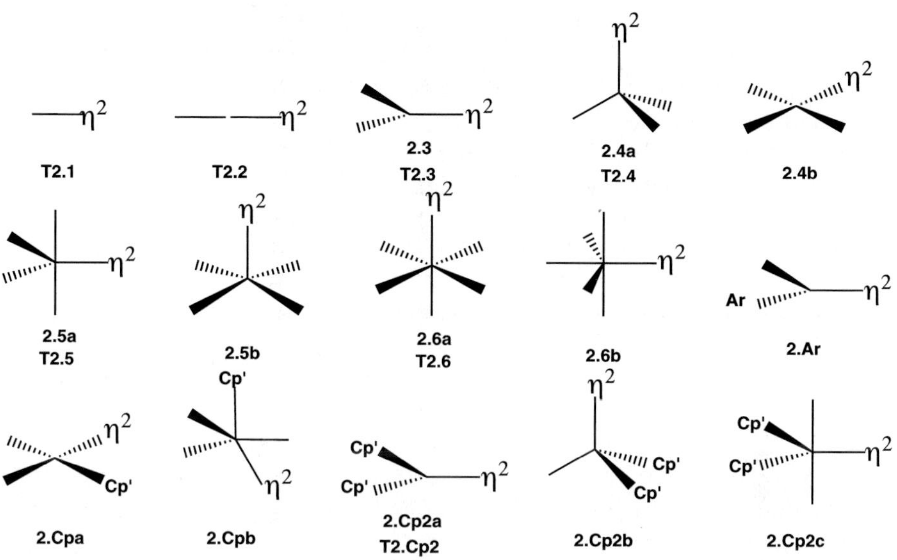

T2.1
$Au(BH_4)$ [9]
T2.2
$[(CH_3)Zn(BH_4)]$ [10]

2.3
$(Ph_3P)_2Cu(\eta^2\text{-}BH_4)$ [1,33]
(dmphen)$Cu(\eta^2\text{-}BH_4)$ [34]
(dmdphen)$Cu(\eta^2\text{-}BH_4)$ [35]
$(PBuPh_2)_2Cu(\eta^2\text{-}BH_4)$ [36,37]
$(Ph_3P)_2Cu(\eta^2\text{-}BH_4)$ [33,38,39]
$(SiNCNSi)_2Ti(\eta^2\text{-}BH_4)$ [40]
$HC(CMeNC_6H_4iPr_2)_2Zn(\eta^2\text{-}BH_4)$ [41]
T2.3
$Cu(BH_4)(PH_3)_2$ [11]
$[Cu(BH_4)(PH_3)_2]$ [12]

2.4a
(tripod)$Co(\eta^2\text{-}BH_4)$ [22]
$(NSi_2)_2V(\eta^2\text{-}BH_4)$(thf) [42]
Cu(2-BH_4)(phen) $[P(OEt)_3]$ [43]
(hdmpb)$Cd(\eta^2\text{-}BH_4)$ [44]
(tmeda)$ZnCl(\eta^2\text{-}BH_4)$ [45]
(phen)$(Ph_3P)Cu(\eta^2\text{-}BH_4)$ [35]
$(NSi_2)_2V(\eta^2\text{-}BH_4)$(py) [46]
T2.4
[(tripod)$Ni(\eta^2\text{-}BH_4)$] [13]

2.4b
(terpy)$Co(\eta^2\text{-}BH_4)$ [47]
$(PCy_3)_2HNi(\eta^2\text{-}BH_4)$ [48,49]

$(PCy_3)_2HCo(\eta^2\text{-}BH_4)$ [50,51]
$(NSi_2)_2Ti(\eta^2\text{-}BH_4)$(thf) [52]
[NNN]$Rh(\eta^2\text{-}BH_4)$ [53]

2.5a
$[(CO)_4Cr(\eta^2\text{-}BH_4)]^-$ [54]
$[(CO)_4Mo(\eta^2\text{-}BH_4)]^-$ [55]
$(Ph_3P)_2$(mphi)$Ru(\eta^2\text{-}BH_4)$ [56]
(tripod)$HFe(\eta^2\text{-}BH_4)$ [57,58]
$(Me_3P)_3HRu(\eta^2\text{-}BH_4)$ [59]
$(PMe_3)_2W(NO)(CO)(\eta^2\text{-}BH_4)$ [60]
$(PMe_3)_3Re(CO)$ ($\eta^2\text{-}BH_4$) [61]
T2.5
$[Cr(BH_4)(CO)_4]^-$ [16,17]
$[Mo(BH_4)(CO)_4]^-$ [16,17]
$[Mn(BH_4)(CO)_4]$ [12]

2.5b
(CNNC)Ti($\eta^2\text{-}BH_4$) [5]

2.6a
$((C_5H_9)_3P)_2H_3Os(\eta^2\text{-}BH_4)$ [8]
$(dmpe)_2HCr(\eta^2\text{-}BH_4)$ [7]
T2.6
$OsH_3(BH_4)(PiPr_3)_2]$ [14,15]

2.6b
$(PMe_3)_4HMo(\eta^2\text{-}BH_4)$[6]

2.Ar
$C_7H_6C_6H_5P(iPr)_2Mo(\eta^2\text{-}BH_4)$ [62,63]

2.Cpa
Cp(dmpe)$V(\eta^2\text{-}BH_4)$ [64,65]
$Cp^*Ti(\eta^2\text{-}BH_4)$(PSiP) [66]
$CpZr(\eta^2\text{-}BH_4)(N^-CO)$ [67]

2.Cpb
$CpZr(\eta^2\text{-}BH_4)(P^-N^-P)$ [67]

2.Cp2a
$Cp_2Nb(\eta^2\text{-}BH_4)$ [68-70]
$(MeCp_2)Nb(\eta^2\text{-}BH_4)$ [71]
$(MetCp_2)Nb(\eta^2\text{-}BH_4)$ [72]
$(EtCp_2)Nb($ 2-$BH_4)$ [72]
$(EtCpBu_2)Nb(\eta^2\text{-}BH_4)$ [72]
$Me_2Si(C_5H_4)_2Nb(\eta^2\text{-}BH_4)$ [72]
$Cp^*{}_2Nb(\eta^2\text{-}BH_4)$ [73]
$Cp_2Ti(\eta^2\text{-}BH_4)$ [68,74-76]
$(CpBu)_2Ti(\eta^2\text{-}BH_4)$ [75]
$(CpSi_2)_2Sc($ $\eta^2\text{-}BH_4)$ [77]
$(EtCp_2)V(\eta^2\text{-}BH_4)$ [68,72]
T2.Cp2
$[Nb((CpSi)_2(\eta\text{-}H_2BH_2)]$ [78]

2.Cp2b
$CpMet)_2Y(\eta^2\text{-}BH_4)$(thf) [79]
$[(CpMe)_2Zr(\eta^2\text{-}BH_4)(thf)]^+BPh_4^-$[80]

2.Cp2c
$Y(Cp\text{-}O)_2(\eta^2\text{-}BH_4)$ [81]
$Y(In\text{-}O)_2(\eta^2\text{-}BH_4)$ [82]

Scheme 4 X-ray characterised or calculated complexes with one $\eta^2 - BH_4$

(thf) will be given. The effect of the ligands can be appreciated by comparing (tripod)Co(η^2-BH_4) (TET) and (terpy)Co(η^2-BH_4) (SQP). For these Co (I) complexes, we should expect a square-planar geometry which is the geometry adopted when the ligands do not constrain the system. However, with the tripod ligand, the square-planar geometry is not possible due to the constraints imposed by the tridentate P-ligand and a tetrahedral arrangement is adopted. Tetracoordinated η^2-BH_4 complexes with cyclopentadienyl or its derivatives and thf ligands have been crystallographically characterised (**2.Cp2b**, Scheme 4).

Almost all the pentacoordinated complexes adopt the usual TBP geometry (**2.5a**, Scheme 4), which is in fact a formal octahedron if we consider the η^2-BH_4 occupying two coordination sites. A variety of ligands can be found in the complexes crystallised with this geometry: carbonyl, phosphino, hydrides, ... and also several metal centres from Groups 5 to 7. Some of these complexes have been studied computationally (**T2.5**). Yttrium pentacoordinated complexes with both cyclopentadienly and indenyl derivatives have been reported (**2.Cp2c**, Scheme 4). Only one example of a square-pyramidal structure can be found (**2.5b**). It corresponds to the (CNNC)Ti(η^2-BH_4) complex which can not adopt a TBP structure due to the ligand geometry.

Examples of hexacoordinated η^2-BH_4 complexes are known with octahedral (**2.6a/T2.6**) and distorted octahedral molecular geometries (**2.6b**). All of these structures present at least one hydride and two phosphino ligands. The metal centres are d^4-Group 6 or Group 8 transition metals. The $OsH_3(BH_4)(P^iPr_3)_2$ complex exhibits interesting dynamical features which have been computationally studied [14, 15] and will be discussed with more detail in Sect. 4.

Complexes with one η^3-BH_4 ligand basically adopt two geometries, TP (**3.3/T3.3**) and TET (**3.4/T3.4**) (Scheme 5). Linear $M-\eta^3-BH_4^-$ (**T3.1**) and $LM-\eta^3-BH_4$ (**T3.2**) complexes have been only theoretically studied. It is

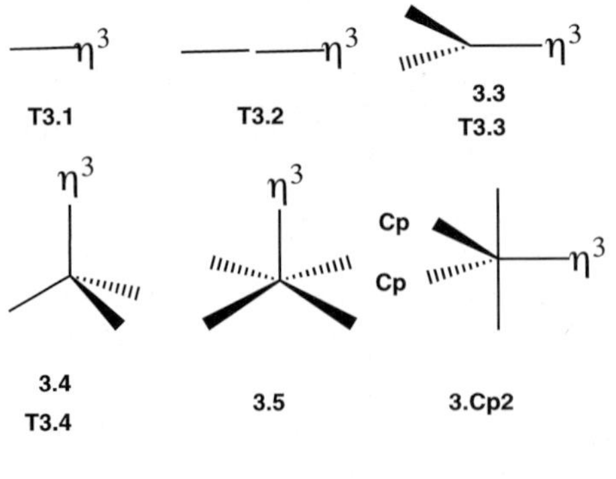

T3.1
$Cu(BH_4)$ [9]
$Ag(BH_4)$ [9]

T3.2
$Cu(BH_4)(PH_3)$ [11]

3.3
$YCl_2(\eta^3-BH_4)$ [1]
$(CH_3O)_2Y(\eta^3-BH_4)$ [1]

T3.3
$TiH_2(BH_4)$ [84,85]

3.4
(tmeda)TiCl(η^3-BH_4) [86]
$[OPh]_3Ti(\eta^3-BH_4)$ [87]
$(tBuNCH)_2TiCl(\eta^3-BH_4)$ [86]
$Zr(\eta^3-BH_4)(Ad)_3$ [88]
(NCH)(NPh)Ti(η^3-BH_4) [5]
(N-NPro)(CHtBu)Ti(η^3-BH_4) [5]
$TpMe_2Ni(\eta^3-BH_4)$ [89]

3.5
$[Ti(CO)_4(\eta^3-BH_4)]^-$[83]

T3.5
$[Ti(BH_4)(CO)_4]^-$ [90]

3.Cp2
$(In-O)_2La(\eta^3-BH_4)$[82]

Scheme 5 X-ray characterised or calculated complexes with one η^3-BH_4

worth mentioning the two structures which do not belong to these geometries: the SQPY complex $[Ti(CO)_4(\eta^3-BH_4)]^-$ **(3.5)** [83] and a TBP lanthanide complex with an aromatic ligand $(In-O)_2La(\eta^3-BH_4)$ **(3.Cp2)**. The distorted geometry of the titanium complex is probably due to the fact that it corresponds to a fragment of a chain. The second complex most likely adopts trigonal bipyramid geometry due to the particularities of lanthanides and the two indenyl-derivated ligands which could constrain the geometry.

2.1.2
More than One Tetrahydroborate Ligand

Complexes with more than one BH_4^- ligand are less usual, and the set of experimental structures found is considerably lower than that for monotetrahydroborate complexes. First, we will discuss structures with the same coordination mode for all the BH_4^- ligands and then the "mixed" structures, for which several BH_4^- coordination modes coexist in the same compound.

Only one kind of structure with two η^1-BH_4^- ligands is known (Scheme 6). Two octahedral complexes with two η^1 ligands in *trans* have been reported. However, one of the both, the chromium complex, is better described as ionic due to the long Cr – B distance it shows (3.253, 3.309 Å). The other one, $(dmpe)_2V(\eta^1-BH_4)_2$, is a high-spin d^3 complex.

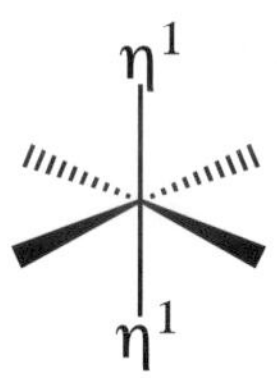

1.6.2

1.6.2
$(dmpe)_2V(\eta^1\text{-}BH_4)_2$ [91,92]
$(py)_4Cr(\eta^1\text{-}BH_4)_2$ [93], ionic

Scheme 6 X-ray characterised or calculated complexes with more than one η^1-BH_4

Several examples of compounds with two or three η^2-BH_4 ligands coordinated to the metal centre have been described, adopting different kind of structures (Scheme 7). Tetrahedral complexes with two cyclopentadienyl and two BH_4^- ligands (**2.Cp2.2**) have been characterised. Only one tetracoordinated complex without Cp or Cp-derivate ligands has been crystallised: $(tmeda)Cr(\eta^2-BH_4)_2$. The two tetrahydroborates keep the η^2 coordination, after the addition of a pyridine ligand, leading to the pentacoordinated complex $(tmeda)(py)Cr(\eta^2-BH_4)_2$. Some examples of pentacoordinated complexes with the η^2-BH_4 ligands in equatorial positions (**2.5.2**) and octahedral complexes with the BH_4^- ligands in *trans* (**2.6.2**) are known.

Complexes with three η^2-BH_4 ligands present trigonal-planar, tetrahedral and trigonal-bipyramidal geometries.

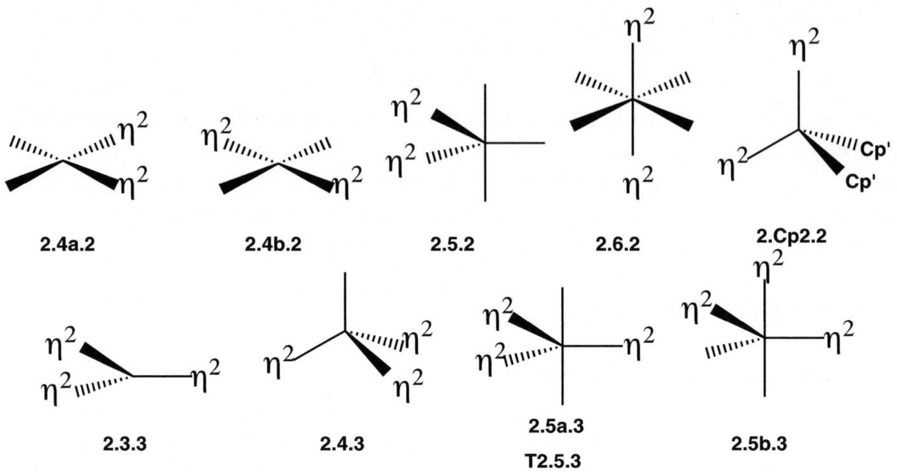

2.4a.2
(tmeda)Cr(η^2-BH_4)$_2$ [93,94]

2.6.2
(dmpe)$_2$Ti(h$_2$–^{B}H$_4$)$_2$ [91,96]

2.4.3
[Mn(η^2-BH_4)$_3$(thf)]$^-$[PPh_4]$^+$ [103,104]

2.5.2
(thf)$_3$Mn(η^2-BH_4)$_2$ [95]
(NSi$_2$)Ti(η^2-BH_4)$_2$(py)$_2$ [52]
[OPh]Ti(η^2-BH_4)$_2$(PMe_3)$_2$ [87]
(tmeda)(py)Cr(η^2-BH_4)$_2$ [93]

2.Cp2.2
(CpMe)$_2$Hf (η^2-BH_4)$_2$ [97]
Cp_2Zr(η^2-BH_4)$_2$ [98,99]
(EtCp$_2$)Zr(η^2-BH_4)$_2$ [100]

2.3.3
(Ph_4P)[Zn(η^2-BH_4)$_3$] [101,102]

2.5a.3
(PMe_3)$_2$V(η^2-BH_4)$_3$ [92,105]

T$_2$.5.3
[V(BH_4)$_3$(PH_3)$_2$] [106]

2.5b.3
(dmoe)Ti(η^2-BH_4)$_3$ [107,108]

Scheme 7 X-ray characterised or calculated complexes with more than one η^2-BH_4

Complexes with more than one η^3-BH_4 ligand are found only with Group 3 and Group 4 transition metals (Scheme 8). Most of them have no other ligands than η^3-BH_4 and they can accommodate until four η^3-BH_4^-. These compounds raise important questions about the nature of the M-BH_4 bonding since they are in evident contradiction with the 18-electron rule. Although most of them are tetrahedral, examples of trigonal-planar, trigonal bipyramid, octahedron, and pentagonal-bipyramidal are known.

A large amount of computational work has been developed in complexes with more than one η^3-BH_4 ligand (**T3.3.3** and **T3.4.4**).

Several complexes combining η^2 and η^3-BH_4 ligands in the coordination sphere have been reported with TP, TET, TBP and OCT geometries (Scheme 9). It is worth comparing the tetrahydrobrate coordination modes found in a series of M(BH_4)$_3$L$_2$ (M = Sc, Ti, V) complexes. Despite the three compounds exhibiting TBP structures, vanadium shows the three BH_4^- coordinated η^2, scandium prefers two η^3 and one η^2, and titanium prefers two η^2 and one η^3-BH_4^- ligands. These series have been theoretically studied in order to discuss the importance of electronic effects as a main factor

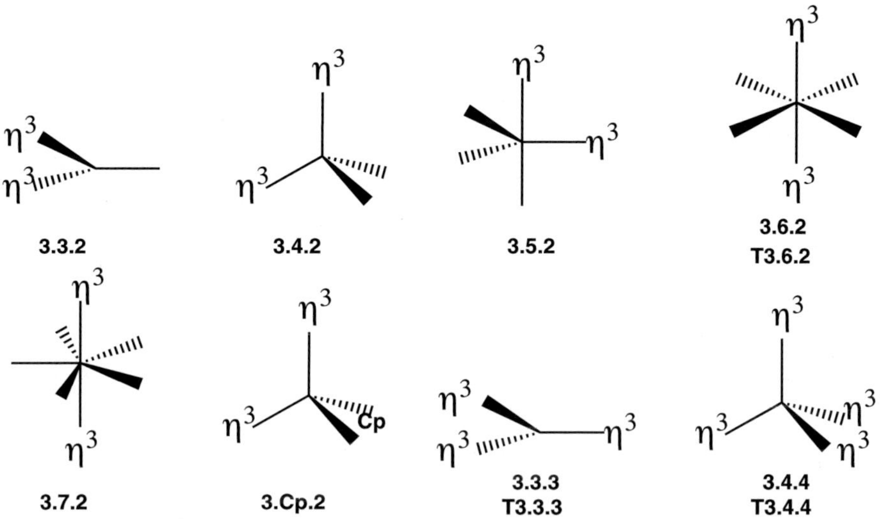

T3.3.2
YCl(η^3-BH_4)$_2$ [1]

3.4.2
(bp)Ti(η^3-BH_4)$_2$ [109]
(tBuNCH)$_2$Zr(η^3-BH_4)$_2$ [86]

3.5.2
(NNN)Zr(η^3-BH_4)$_2$ [110]

3.6.2
[(thf)$_4$Y(η^3-BH_4)$_2$]$^+$ [111,112]

T3.6.2
[Y(thf)$_4$(BH_4)$_2$]$^+$ [113]

3.7.2
La(η^3-BH_4)$_2$(thf)$_5$]$^+$ [114]

3.Cp.2
(Cp-B)(η^3-BH_4)$_2$Y [115]

3.3.3
Ti(η^3-BH_4)$_3$ [116]

T3.3.3
Y(BH_4)$_3$ [1]
Ti(BH_4)$_3$ [117]
Ti(BH_4)$_3$ [118]

3.4.4
Y(η^3-BH_4)$_4^-$ [111,112]
Hf(η^3-BH_4)$_4$ [119-121]
Zr(η^3-BH_4)$_4$ [122,123]

T3.4.4
Hf(BH_4)$_4$ [117,124,125]
Zr(BH_4)$_4$ [124,126,1]

Scheme 8 X-ray characterised or calculated complexes with more than one $\eta^3 - BH_4$

governing coordination modes of tetrahydroborate ligands [128]. The main conclusions from these studies will be presented in Sect. 3.

2.2 Transition Metal–boron Distances

Due to the difficulties involved in locating hydrogen atoms using the usual techniques of crystal structure resolution (X-ray crystallography), the coordination mode of the BH_4^- ligand can not be univocally determined by X-ray diffraction analysis. A much better elucidation can be achieved from the metal–boron distance. The $M - B$ distance does not have the uncertainties associated with the hydrogen localization and has been proven to be a good indicator of the coordination mode of the BH_4^-. As it can be observed by simple

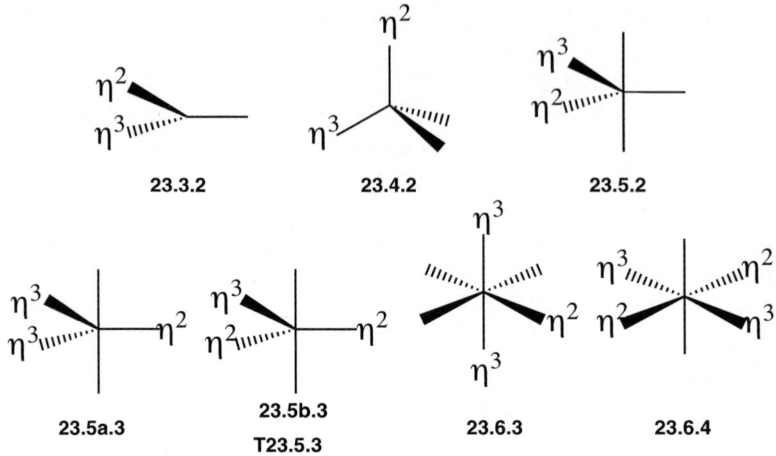

23.3.2
$Zr(COT)''(\eta^2\text{-}BH_4)(\eta^3\text{-}BH_4)$ [129]

23.4.2
$(tBuNCH)_2Ti(\eta^2\text{-}BH_4)(\eta^3\text{-}BH_4)$ [86]

23.5.2
$(bp)Zr(thf)(\eta^2\text{-}BH_4)(\eta^3\text{-}BH_4)$ [109]

23.5a.3
$(thf)_2Sc(\eta^2\text{-}BH_4)(\eta^3\text{-}BH_4)_2$ [112,130]

23.5b.3
$(Me_2PhP)_2Ti(\eta^2\text{-}BH_4)_2(\eta^3\text{-}BH_4)$ [131]
$(Et_3P)_2Ti(\eta^2\text{-}BH_4)_2(\eta^3\text{-}BH_4)$ [131]

T23.5.3
$[Ti(BH_4)_3(PH_3)_2]$ [128]

23.6.3
$(thf)_3Y(\eta^3\text{-}BH_4)_2(\eta^2\text{-}BH_4)$ [132]

23.6.4
$[La(\eta^2\text{-}BH_4)_2(\eta^3\text{-}BH_4)_2(thf)_2]$ [114]

Scheme 9 X-ray characterised or calculated complexes with both η^2 and $\eta^3 BH_4$ ligands

inspection of the $M\text{-}BH_4$ geometries associated to the different coordination modes of the tetrahydroborate in covalent complexes (Scheme 1), the η^1 coordination mode has the longer M – B distances, the η^2 has the intermediate ones, and the η^3 has the shorter distances. Naturally, these distances do not only depend on the coordination mode of the ligand, but also on the transition metal complex where it is included. The M – B distance for each coordination mode strongly depends on the transition metal centre of the complex. That is obviously due to the strong differences on the transition metal ionic radii. Metals at the right of the periodic table show larger distances than those of the left, and the first-row transition metals reveal shorter distances than those of the third row. Good correlations between the metal ionic radius and M – B distance was demonstrated for bidentate and tridentate coordinations [133]. As we will discuss later on, the M – B distance is also influenced for the set of ligands different to the tetrahydroborate coordinated to the metal. This aspect is made evident by comparing complexes with the same metal and the same coordination of the BH_4^- but with different ancillary ligands (see Sect. 2.2.5).

In addition to the η^1, η^2 and η^3 covalent transition metal tetrahydroborates, another type of complexes exist where the interaction between M^+ and BH_4^- is mainly electrostatic. These compounds which are quite common are defined as ionic tetrahydroborate complexes. They are characterised by the non-coordination of the tetrahydroborate and consequently by a long metal-boron distance. Several compounds of this class have been wrongly classified as monodentate covalent complexes.

We have analysed the experimental transition metal–boron distances for covalent transition metal complexes with BH_4^- ligands. The distances have mainly been obtained from the Cambridge Data Base [134], however, the most recent have been collected from the corresponding articles. In the following subsections, plots of the experimental M – B distances as a function of the metals are presented (Figs. 1–10). A separate discussion for each transition metal row is given and within each row, η^1, η^2 and η^3, coordination modes are analysed individually. A distinction is also done depending on the number of tetrahydroborate ligands of the complexes: in grey, complexes with only one BH_4^-, in black, complexes with two BH_4^-, in white with black lines, complexes with three or more BH_4^-, and black with white lines indicate complexes which combine two coordination modes. Squared bands show the average of the M – B distances while non-squared show the individual distances located. This classification allows for the comparison between the different complexes, and general trends about the evolution of the M – B distances along the periodic table arise from this analysis. Representation of the M – B distances vs. atomic number has also allowed the localization of some complexes likely with the coordination mode wrongly assigned.

2.2.1 First-row Transition Metals

A large number of covalent tetrahydroborate complexes with first-row transition metals has been described. The most common coordination mode of the tetrahydroborate ligand is η^2, being known as at least one example of $\eta^2 - BH_4$ for all the transition metals of this row. Although less abundant, several η^1 and η^3 complexes have been characterised experimentally, though only for a limited number of metals. In particular, most of the monodentate complexes are Cu compounds and most of the tridentate are Ti compounds.

Before discussing M – B distances in monodentate complexes, it must be emphasised that in all the real η^1-BH_4 compounds, the tetrahydroborate is far from the ideal η^1-BH_4 coordination defined by a linear arrangement of the $M - H_b - B$ atoms (Scheme 10). This angle is considerably lower than 180°. What is distinctive from all the η^1 complexes is that although two terminal hydrogens are pivoting to the metal, they are still far away enough to preclude any M – H strong interaction.

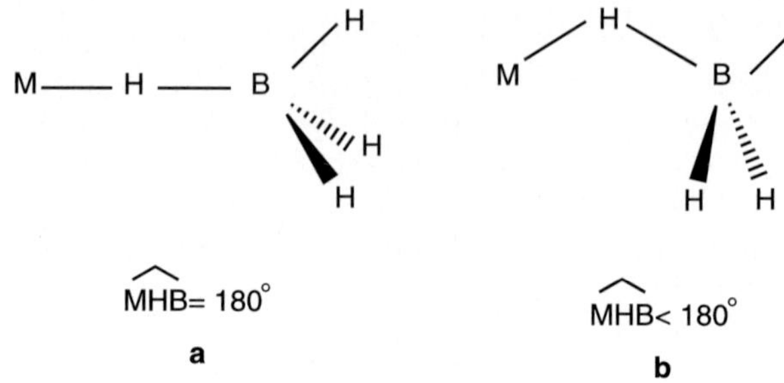

Scheme 10 Coordination in ideal (**a**) and real (**b**) η^1-BH_4 complexes

In Fig. 1, the M – B distances for first-row η^1-BH_4^- complexes are presented. The expected tendency of longer metal–boron distances at the left of the periodic table is found, although vanadium and iron complexes show very close M – B distances. The difference between the longer (V) and the shorter (Cu) distances is 0.4 Å. It is worth mentioning that there is a difference of more than 0.2 Å between the set of Cu-(η^1-BH_4) complexes, stressing the influence of the ligand on the M – B distance. The shortest Cu – B distance (2.441 Å) is found in Cu(η^1-BH_4){MeC(CH_2PPh_2)$_3$} and the longest one (2.646 Å) in Cu(η^1-BH_4)($PMePh_2$)$_3$. Both complexes are four-coordinated and tetrahedral (**1.4**, Scheme 2) with the Cu coordinated to three phosphorus and the BH_4^-. It seems that the more constrained ligand {MeC(CH_2PPh_2)$_3$}

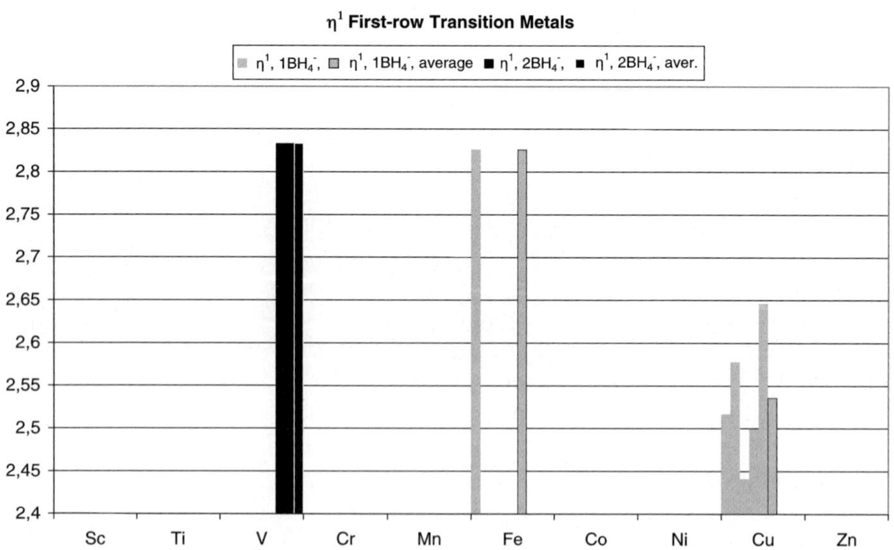

Fig. 1 M – B distances for η^1-BH_4 first-row transition metal complexes. In *grey*, complexes with one $\eta^1-BH_4^-$ and in black, complexes with two $\eta^2-BH_4^-$. *Squared bands* show the average values of a group data

allows a better interaction between the metal and BH_4^- (lower M – B distance) than the three $PMePh_2$, due to a lower P – Cu – P angle in the constrained ligand. However, $Cu(\eta^1-BH_4)(PMePh_2)_3$ has been largely studied and Cu – B distances of 2.499, 2.517, 2.578 and 2.646 Å have been obtained for the same complex in different studies [20, 21]. The value of 2.517 Å is the only one obtained by neutron diffraction. These notable variations in the M – B distances are related to the decreasing of the $M-H_b-B$ angle in η^1 structures at a very low energy cost. Closing of this angle entails the decreasing of the M – B distance.

Figure 2 is the representation which contains more data since η^2-BH_4 ligands are known for all the first-row transition metals. Titanium, copper and chromium are the metals for which more examples of η^2 coordination can be found. For copper, all the examples correspond to complexes with only one BH_4^- ligand. While for titanium, complexes with more than one tetrahydroborate have also been reported.

Looking at the figure, the longer distances are observed at the left and the shorter distances are at right, as expected. By comparing the average M – B distances obtained for the different metals, a remarkable difference of 0.4 Å is found between the longer (scandium: 2.528 Å) and the shorter distances (zinc: 2.120 Å). Vanadium data does not precisely follow the general tendency of longer distances on the left, and has a shorter metal–boron mean distance than the metals on its right (Cr, Mn). Four examples of vanadium complexes with one $\eta^2-BH_4^-$ ligand are known and three of them present short distances around 2.3 Å or less (lower distances correspond to complexes containing Cp ligands). Analogue to these vanadium complexes, several titanium complexes with Cp ligands have been characterised, even though they do no present shortened distances. An easy explanation for these short distances in vanadium complexes is not evident. Complexes at the right of the periodic table present similar mean values of M – B distances, deviations being below 0.1 Å.

Differences found in complexes with the same metal but different number of tetrahydroborate ligands can be significant. For instance, for titanium, the mean value of the Ti – B distance in complexes with two BH_4^- ligands is about 0.1 Å longer than the average obtained for complexes with more than two BH_4^- ligands. The same tendency is observed for manganese and zinc (in this case, comparing complexes with one and more than two BH_4^- ligands). As already found in η^1 complexes, the copper η^2 complexes present a broad range of Cu – B distances. Very low distances of 2.079 Å (in $Cu(\eta^2-BH_4)$(dmphen)) and 2.112 Å (in $Cu(\eta^2-BH_4)$(dmdphen)) coexist with long Cu – B distances as 2.344 Å (in $Cu(\eta^2-BH_4)(phen)(P(OEt)_3))$ and 2.320 Å (in $Cu(\eta^2-BH_4)(phen)(PPh_3)$). The $Cu(\eta^2-BH_4)(phen)(PPh_3)$ complex has been crystallised and resolved by Green and co-workers in different studies and conditions [34]. The crystallographic data obtained produce different Cu – B distances: 2.274, 2.290, 2.295 and 2.320 Å. The 2.290 Å distance belongs to the unique neutron diffraction analysis.

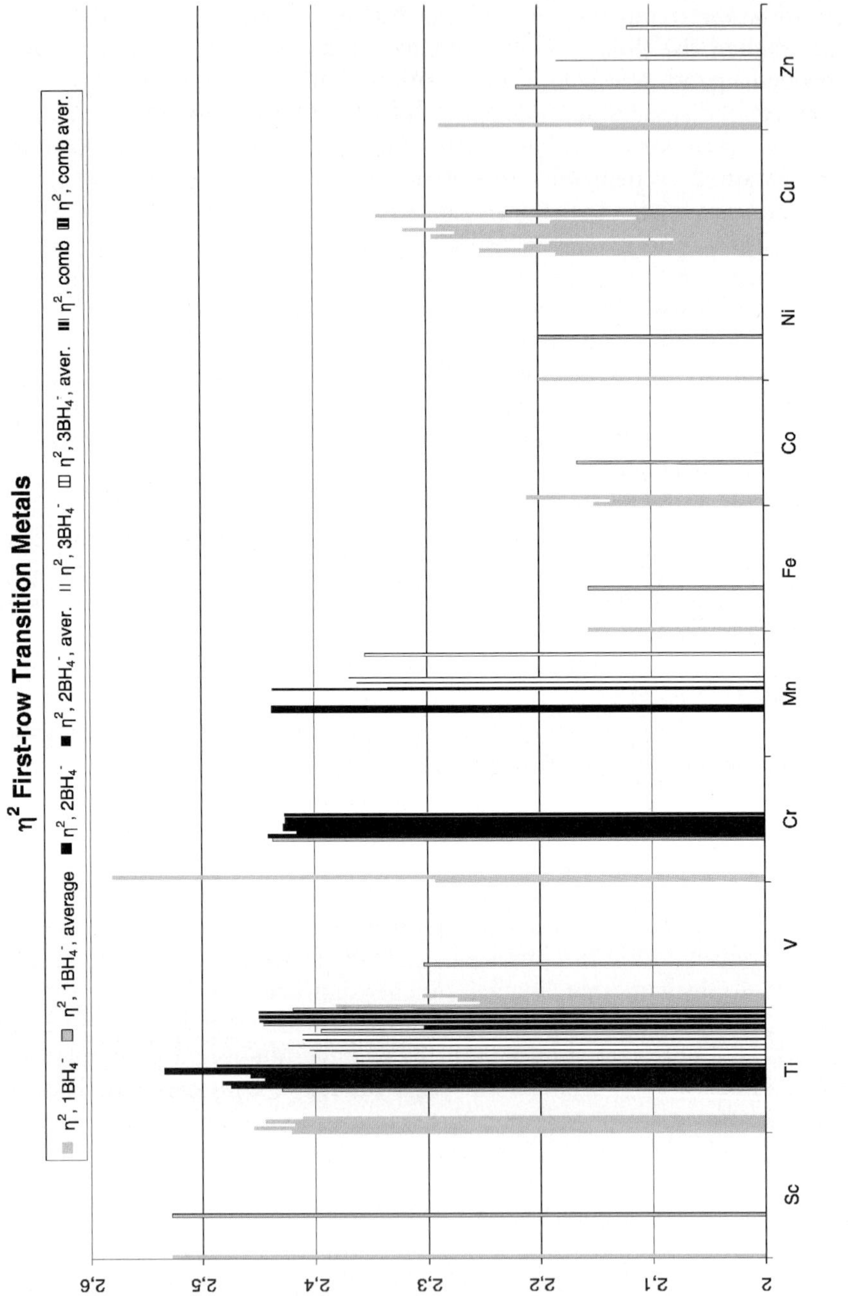

Fig. 2 M – B distances for $\eta^2 - BH_4$ first-row transition metal complexes. In *grey*, complexes with one $\eta^2 - BH_4^-$, in *black*, complexes with two $\eta^2 - BH_4^-$, in *white with black lines*, complexes with more than two $\eta^2 - BH_4^-$ and in *black with white lines*, complexes combining η^2 and η^3 ligands. *Squared bands* show the average values of data group

Between the first-row transition metals, only titanium and nickel complexes with BH_4^- ligands coordinated in the a η^3 coordination mode have been characterised (Fig. 3). For titanium, both complexes with one and more than one η^3-BH_4 can be found, whilst for nickel, only one compound with one η^3-BH_4 is known. The longest Ti – B distances are found in complexes with more than two BH_4^-, in contrast with observations done in the case of

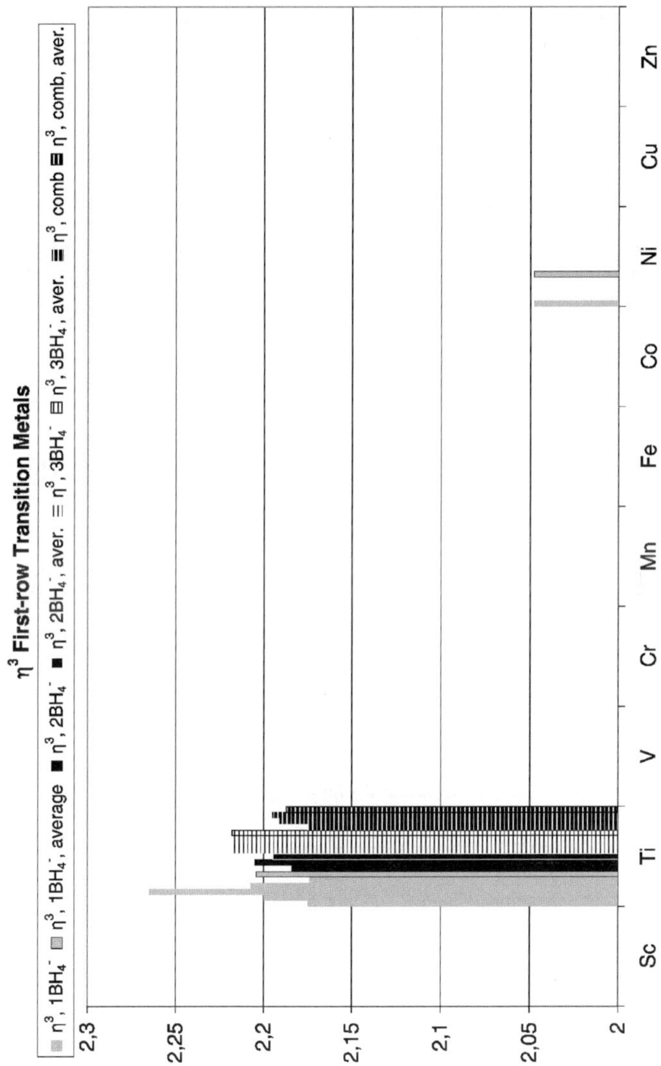

Fig. 3 M – B distances for $\eta^3-BH_4^-$ first-row transition metal complexes. In *grey*, complexes with one $\eta^3-BH_4^-$, in *black*, complexes with two $\eta^3-BH_4^-$, in *white with black lines*, complexes with more than two $\eta^3-BH_4^-$ and in *black with white lines*, complexes combining η^2 and η^3 ligands. *Squared bands* show the average values of a data group

η^2 coordination mode, where complexes with more BH_4^- ligands have shorter distances. However, differences are slight, the greater discrepancy (less than 0.1 Å) being found between two Ti η^3 complexes with one BH_4^-.

2.2.2 Second-row Transition Metals

Presence of BH_4^- ligands in complexes of the second-row transition metals row is not as usual as in complexes of the first row. As for the first row metals, the most frequent coordination mode is the η^2 one.

The η^1 coordination mode has been crystallographically observed for three metals of this row: molybdenum, ruthenium and silver, although only one silver complex has been characterised (Fig. 4). The ruthenium mean distance is larger than the molybdenum one, not following the general tendency of larger distances on the left. However, the difference is small. In the ruthenium data, the difference between higher and lower distances is about 0.14 Å, the longest distance (2.947 Å) corresponding to the Ru – B distance of one of the two molecules found in the crystal cell of the $[(binap)(diam)RuH(\eta^1-BH_4)](THF)(C_6H_{14})$ [27]. The other molecule has a distance of 2.849 Å. The shorter Ru – B distance (2.810 Å) has been found in $(Pnor)_2RuH(\eta^1-BH_4)$ [28].

η^2-BH_4 complexes are found for all the second-row transition metals, with the exception of technetium, palladium and silver. The most populated classes correspond to the early metals of the row: yttrium, zirconium and niobium. The general tendency of larger distances at left and shorter at right is followed with the exception of niobium and cadmium. The niobium mean distance (and individual Nb – B distances) is notably shorter than the

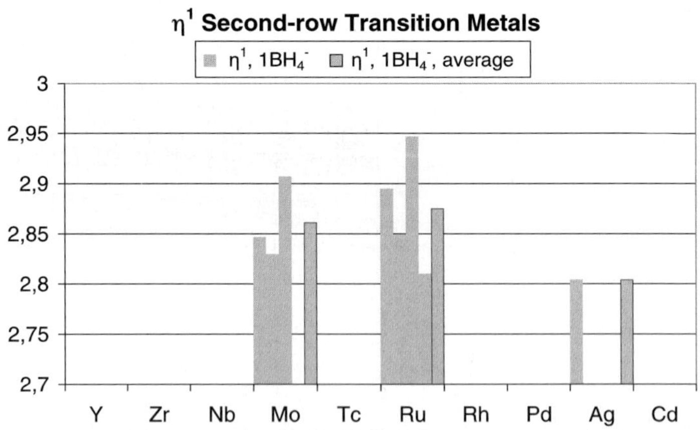

Fig. 4 M – B distances for η^1-BH_4 second-row transition metal complexes. Only complexes with one $\eta^1-BH_4^-$ have been reported. *Squared bands* show the average values of a data group

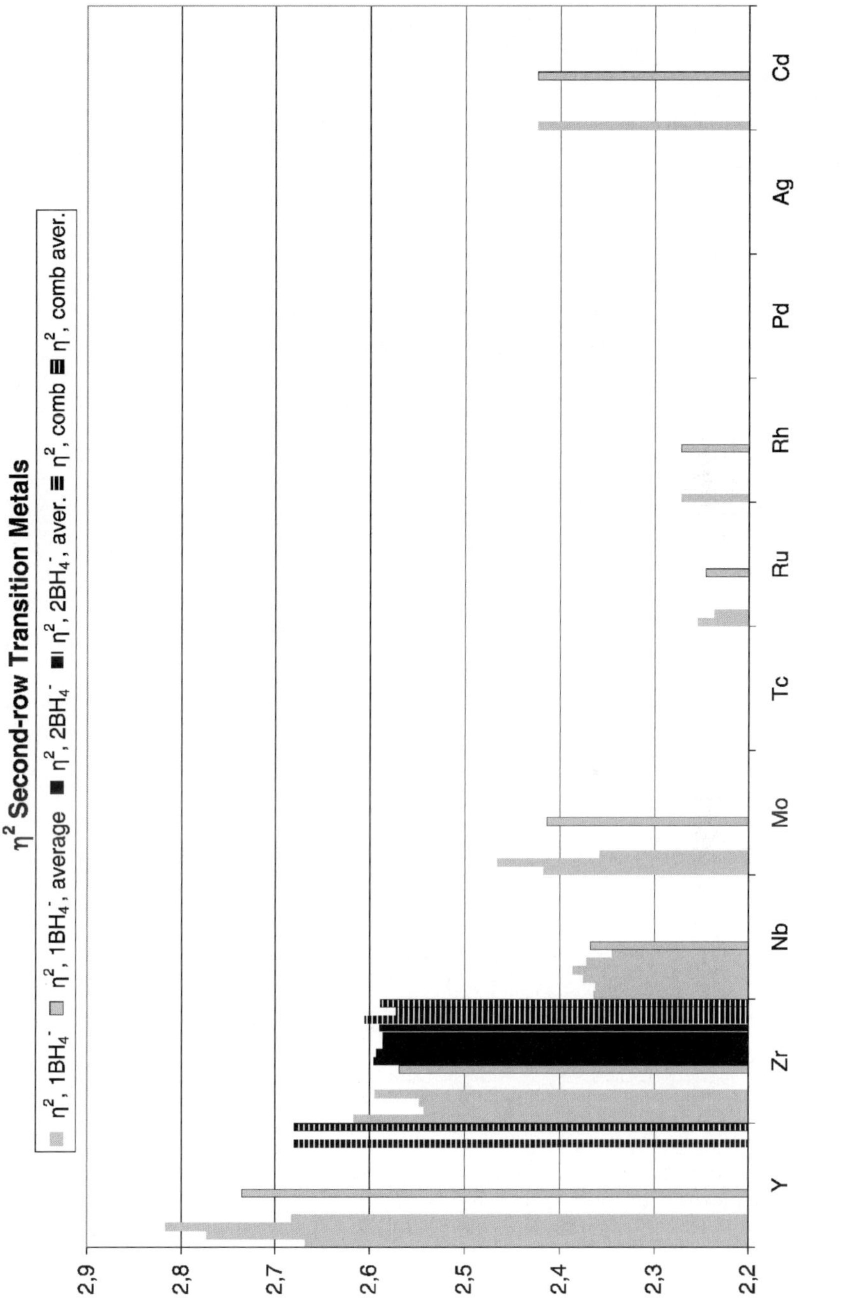

Fig. 5 M – B distances for $\eta^2 - BH_4^-$ second-row transition metal complexes. In *grey*, complexes with one $\eta^2 - BH_4^-$, in *black*, complexes with two $\eta^2 - BH_4^-$ and in *black with white lines*, complexes combining η^2 and η^3 ligands. *Squared bands* show the average values of a data group

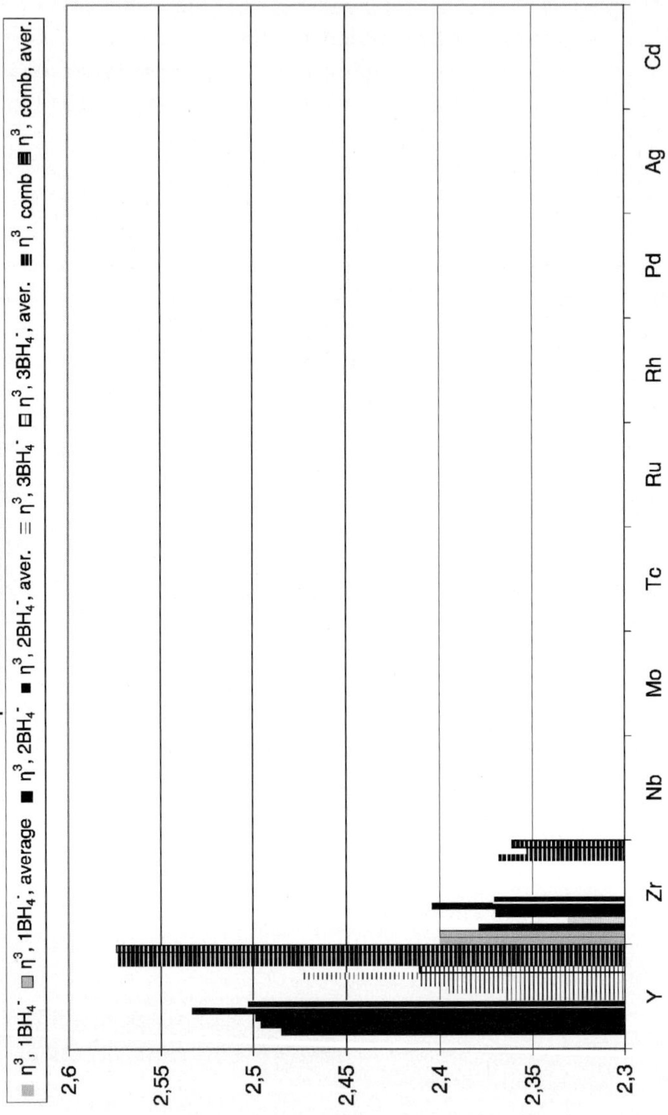

Fig. 6 M – B distances for $\eta^3 - BH_4^-$ second-row transition metal complexes. In *grey*, complexes with one $\eta^3 - BH_4^-$, in *black*, complexes with two $\eta^3 - BH_4^-$, in *white with black lines*, complexes with more than two $\eta^3 - BH_4^-$ and in *black with white lines*, complexes combining η^2 and η^3 ligands. *Squared bands* show the average values of a data group

zirconium one and also slightly shorter than the molybdenum one. This is probably due to the fact that all the niobium data collected corresponds to the same kind of complexes which have two cyclopentadienyl ligands (**2.Cp2a** in Scheme 3). These complexes present shorter metal–boron distances than related compounds without Cp ligands. Only one cadmium complex with a $\eta^2-BH_4^-$ ligand has been studied crystallographically. The Cd — B distance is longer than those of ruthenium and rhodium compounds.

Only yttrium and zirconium complexes with η^3 coordination of BH_4^- can be found between the second-row transition metals. The corresponding data is presented in Fig. 6. Notable differences exist between the complexes formed by the two metals. While averages of the different kind of zirconium complexes present close distances, averages of yttrium complexes show important variations. In complexes containing more than two BH_4^- ligands, the mean value of the Y – B distance is 2.411 Å, while for complexes combining η^2 and η^3 coordination modes, it is 2.574 Å. In $(thf)_3Y(BH_4)_3$, the Y – B distances are significantly longer (2.680 Å) than the other two Y – B distances (2.574 Å), which is in agreement with a $(thf)_3Y(\eta^2-BH_4)(\eta^3-BH_4)_2$ structure.

2.2.3
Third-row Transition Metals

Tetrahydroborate complexes with third-row transition metals are much less abundant than for first or second row metals. Only complexes with lanthanum, hafnium, tungsten, rhenium and osmium have been characterised by diffraction techniques. For each metal, only a few compounds are known, making conclusions less reliable regarding the M – B distances.

η^1-BH_4 ligands have only been found in two tungsten compounds (Fig. 7). The two distances obtained differ in less than 0.1 Å.

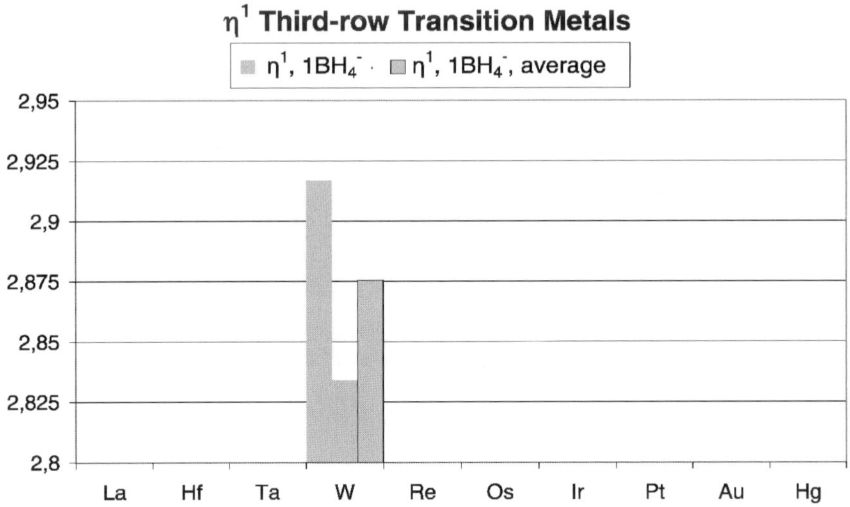

Fig. 7 M – B distances for $\eta^1-BH_4^-$ third-row transition metal complexes. Only complexes with one $\eta^1-BH_4^-$ have been reported. *Squared bands* show the average values of a data group

As for the first and second row metals, η^2 coordination is the most common coordination mode found. It is known for the five metals, although only one complex for each metal is known. The distances are presented in Fig. 8.

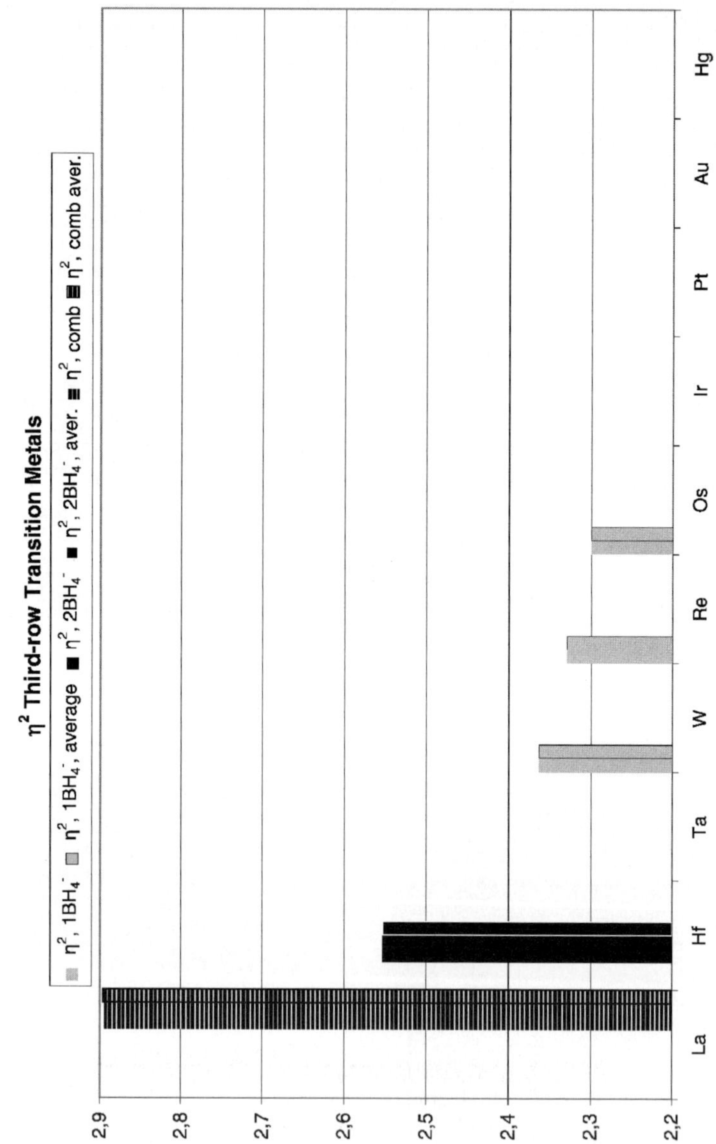

Fig. 8 M – B distances for $\eta^2 - BH_4^-$ third-row transition metal complexes. In *grey*, complexes with one $\eta^2 - BH_4^-$, in *black*, complexes with two $\eta^2 - BH_4^-$ and in *black with white lines*, complexes combining η^2 and η^3 ligands. *Squared bands* show the average values of a data group

The general tendency of longer distances on the left and shorter on the right is maintained.

It can be observed in Fig. 9 that the η^3 coordination mode is quite usual for lanthanum. The only example know with a different metal is a hafnium

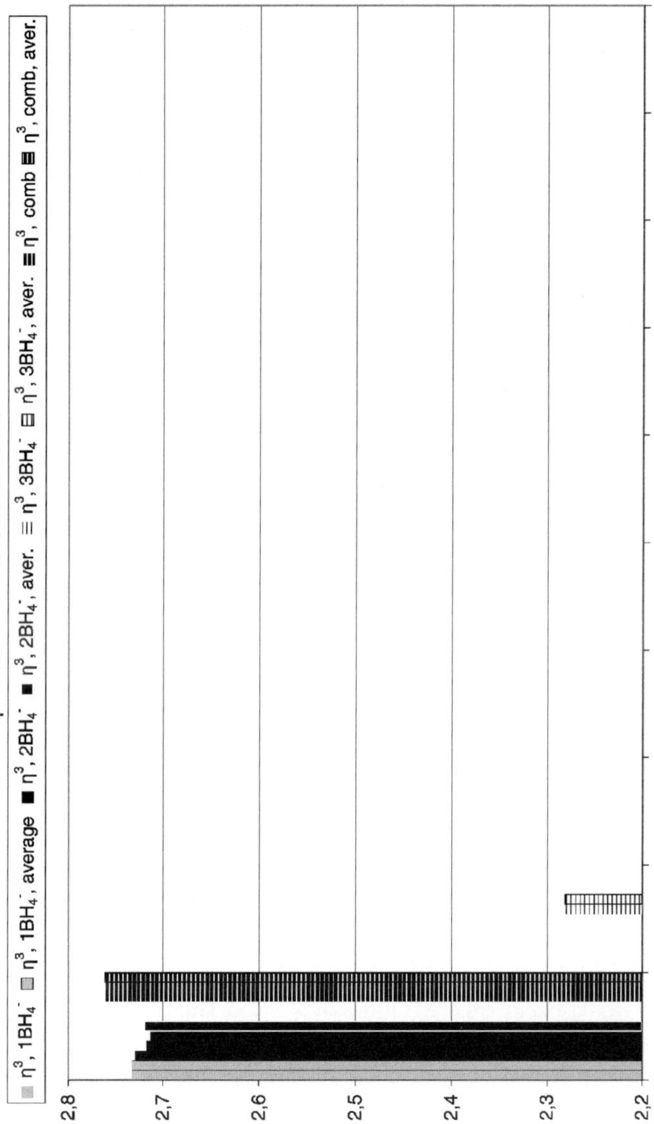

Fig. 9 M – B distances for $\eta^3-BH_4^-$ third-row transition metal complexes. In *grey*, complexes with one $\eta^3-BH_4^-$, in *black*, complexes with two $\eta^3-BH_4^-$, in *white with black lines*, complexes with more than two $\eta^3-BH_4^-$ and in *black with white lines*, complexes combining η^2 and η^3 ligands. *Squared bands* show the average values of a data group

complex. The Hf – B distance is notably shorter than the La – B ones. In fact, the "particular" distance is not that of hafnium, but those of lanthanum complexes. Lanthanum presents very long distances for a η^3 coordination (more than 2.7 Å). That is probably due to the fact that the lanthanum should be considered as a Lanthanide and not a transition metal. Electrostatic components could be dominant in the La – BH_4 interaction.

2.2.4
General Trends

In Fig. 10, the average values of the metal–boron distances obtained from crystallographic data are depicted for each transition metal and each coordination mode. The results are given following the periodic order. In white, the results for the first transition metal row are presented, in grey, results for the second, and in black, results for the third. There are three bars for each transition metal: on the left, the average of η^1 distances (plain), in the middle, the η^2 (squared), and at the right, the η^3 mean values (vertical bordered).

The assumption that for a particular transition metal the M – B distances follow the order: $\eta^1 > \eta^2 > \eta^3$ can be clearly appreciated from Fig. 10. By comparing metals from the same group, third-row transition metals have M – B distances longer than second-row metals and second-row longer than first-row, according to the radius variation. Also, according to the radius variation, the M – B distances usually decrease along a period, the longest ones being found for complexes with metals left of the period.

It can also be observed in Fig. 10 that the η^2 coordination mode is the most common. It is the only one that can be found for the entire series of first-row transition metals, and for almost all the second-row metals. Moreover, η^2 coordination is known for all the third-row metals which coordinate BH_4^-. The η^3 coordination mode is quite usual for elements at the left of the periodic table, however it is not found for elements at the right. The number of well characterised η^1-BH_4 complexes is still scarce. The η^1-tetrahydroborate complexes are found with metals of groups 5–6, 8 and 10.

2.2.5
Ligand Effects in Transition Metal–boron Distances

In the previous subsections, we have discussed the dependence of the M – B distances on the coordination mode and the metal. For a particular metal and coordination mode, we have already noted marked differences in the M – B distances depending on the ancillary ligands that accompany the tetrahydroborate on the coordination sphere. In this subsection, we will further analyze the ligand effects taking two metals, one at the left of the periodic table (titanium) and the other one at the right (copper), for which the large amount of structural data allow the comparison to be done. Tables with the available

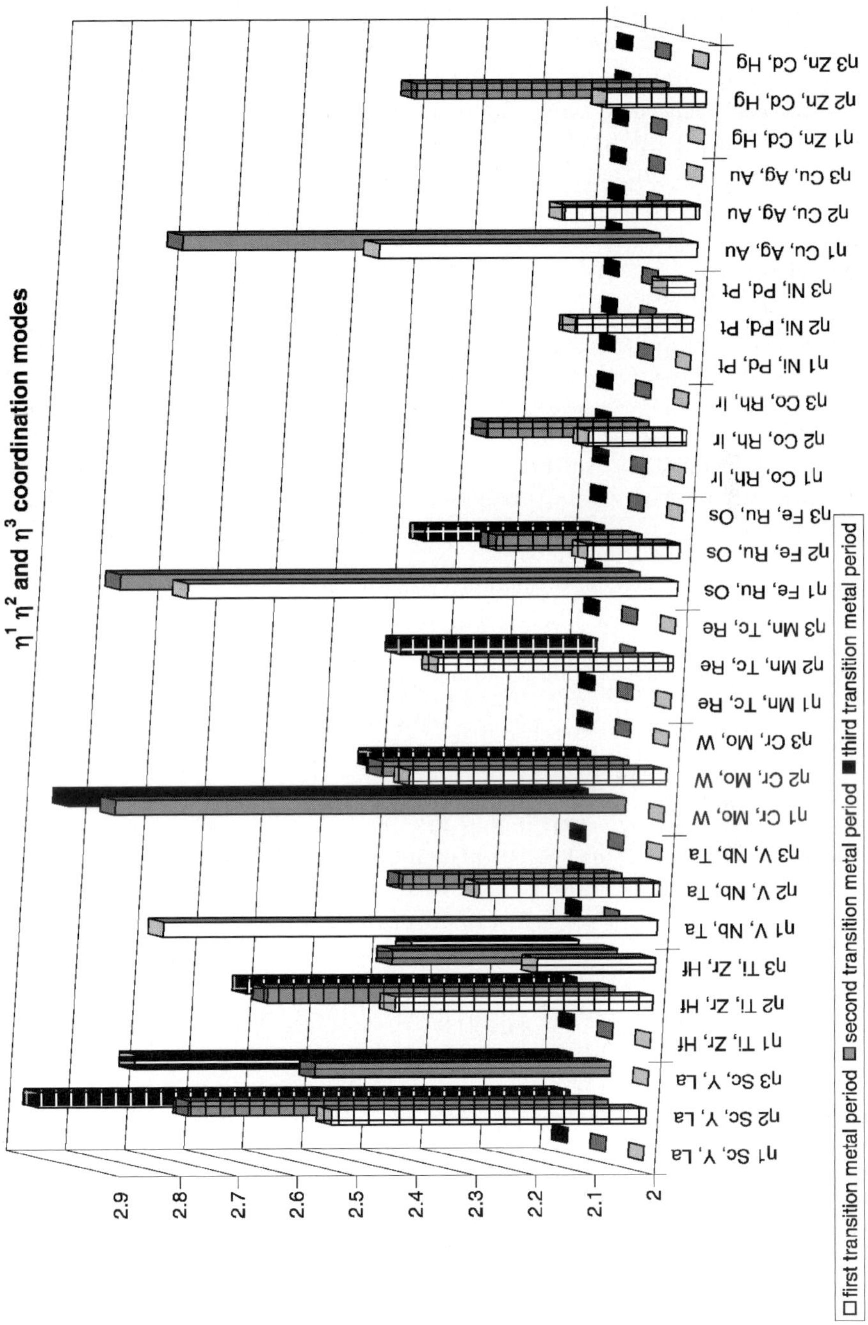

Fig. 10 Average of the η^1 (*plain columns*), η^2 (*squared columns*), and η^3 (*vertical bordered columns*) M – B distances for the complexes of the first (*white*), second (*grey*) and third (*black*) transition metal rows

Table 1 M – B distances (Å) in titanium η³ tetrahydroborate complexes

Ti–B distance (Å)	Ti-η^3 BH$_4^-$ coordination mode and ancillary ligands					Refs.
2.174	η^3-BH$_4$	CO	CO	CO	CO	[83]
2.175	η^3-BH$_4$	η^2-BH$_4$		(^{t}BuNCH)$_2$		[86]
2.175	η^3-BH$_4$	Cl		(^{t}BuNCH)$_2$		[86]
2.184	η^3-BH$_4$	η^3-BH$_4$		bp		[109]
2.191	η^3-BH$_4$	η^2-BH$_4$	η^2-BH$_4$	PEt$_3$	PEt$_3$	[131]
2.195	η^3-BH$_4$	η^2-BH$_4$	η^2-BH$_4$	PMe$_2$Ph	PMe$_2$Ph	[131]
2.2	η^3-BH$_4$		N-Npro		CH^{t}Bu	[5]
2.205	η^3-BH$_4$	η^3-BH$_4$		bp		[109]
2.207	η^3-BH$_4$	Oph	Oph	Oph		[87]
2.218	η^3-BH$_4$	η^3-BH$_4$	η^3-BH$_4$			[116]
2.265	η^3-BH$_4$		NCH	NPh		[5]

Table 2 M – B distances (Å) in titanium η² tetrahydroborate complexes

Ti–B distance (Å)	Ti-η^2 BH$_4^-$ coordination mode and ancillary ligands						Refs.
2.304	η^2-BH$_4$	η^3-BH$_4$		(^{t}BuNCH)$_2$			[86]
2.367	η^2-BH$_4$	Cp	Cp				[68]
2.401 (a)	η^2-BH$_4$	η^2-BH$_4$	η^2-BH$_4$		dmoe		[108]
2.405 (a)	η^2-BH$_4$	η^2-BH$_4$	η^2-BH$_4$		dmoe		[108]
2.409 (b)	η^2-BH$_4$	η^2-BH$_4$	η^2-BH$_4$		dmoe		[141]
2.411	η^2-BH$_4$	CpBu	CpBu				[75]
2.411 (b)	η^2-BH$_4$	η^2-BH$_4$	η^2-BH$_4$		dmoe		[141]
2.411 (b)	η^2-BH$_4$	η^2-BH$_4$	η^2-BH$_4$		dmoe		[141]
2.418	η^2-BH$_4$			CNNC			[5]
2.421	η^2-BH$_4$		SiNCNSi		SiNCNSi		[40]
2.424 (a)	η^2-BH$_4$	η^2-BH$_4$	η^2-BH$_4$		dmoe		[108]
2.444	η^2-BH$_4$	Cp*		PSiP			[66]
2.445	η^2-BH$_4$	η^2-BH$_4$	PMe$_3$	PMe$_3$	Oph		[87]
2.446	η^3-BH$_4$	η^2-BH$_4$	η^2-BH$_4$	PEt$_3$	PEt$_3$		[131]
2.449	η^3-BH$_4$	η^2-BH$_4$	η^2-BH$_4$	PEt$_3$	PEt$_3$		[131]
2.45	η^3-BH$_4$	η^2-BH$_4$	η^2-BH$_4$	PMe$_2$Ph	PMe$_2$Ph		[131]
2.45	η^3-BH$_4$	η^2-BH$_4$	η^2-BH$_4$	PMe$_2$Ph	PMe$_2$Ph		[131]
2.454	η^2-BH$_4$	N(SiMe$_3$)$_2$	N(SiMe$_3$)$_2$	(thf)			[52]
2.457	η^2-BH$_4$	η^2-BH$_4$	PMe$_3$	PMe$_3$	OPh		[87]
2.475	η^2-BH$_4$	η^2-BH$_4$	N(SiMe$_3$)$_2$	py	py		[52]
2.482	η^2-BH$_4$	η^2-BH$_4$	N(SiMe$_3$)$_2$	py	py		[52]
2.534	η^2-BH$_4$	η^2-BH$_4$		dmpe		dmpe	[91]
2.534	η^2-BH$_4$	η^2-BH$_4$		dmpe		dmpe	[91]

metal–boron distances for Ti (Tables 1 and 2) and Cu (Table 3) are presented in increasing order of the M – B distance. The ligands are added to the tables according to the number of coordination sites they occupy. A ligand occupying one coordination site will occupy a cell in the table, while a ligand occupying two coordination sites will occupy two cells. In the complexes with more than one BH_4^- ligands, two M – B distances are given. Titanium tetrahydroborate complexes present η^2 and η^3 coordination modes, whereas copper compounds show η^1 and η^2 coordination modes. Some examples of ionic copper compounds are also included in Table 3 for comparative purposes. For each metal, the expected order (Ti – B in η^3 < Ti – B in η^2 and Cu – B in η^2 < Cu – B in η^1) is found. However, significant differences in the M – B distances are obtained for complexes with the same coordination mode. The η^2 Ti – B distances span a range of 0.13 Å (from 2.17 to 2.30 Å) and a slightly broad range is found for the η^3 ones (0.16 Å, from 2.37 to 2.53 Å). Larger variations in the M – B distance are even found in the series of Cu complexes (0.26 Å in η^2 and 0.20 in η^1). Longer distances are found in the more sterically encumbered complexes, pointing out the influence of steric factors in the M – B distances.

Table 3 M – B distances in copper tetrahydroborate complexes

Cu–B distance (Å)	Cu BH_4^- coordination mode and ancillary ligands				Refs.
2.079	η^2-BH_4		dmphen		[34]
2.112	η^2-BH_4		dmdphen		[35]
2.184 (b)	η^2-BH_4	PPh_3	PPh_3		[1, 33]
2.189	η^2-BH_4	PPh_3	PPh_3	(C_5H_5N)	[142]
2.190	η^2-BH_4	$PBuPh_2$	$PBuPh_2$		[36, 37]
2.212 (c)	η^2-BH_4	PPh_3	PPh_3		[38]
2.252 (d)	η^2-BH_4	PPh_3	PPh_3		[39]
2.274 (a)	η^2-BH_4	PPh_3		phen	[35]
2.290 (b)	η^2-BH_4	PPh_3		phen	[35]
2.295 (c)	η^2-BH_4	PPh_3		phen	[143]
2.320 (d)	η^2-BH_4	PPh_3		phen	[35]
2.344	η^2-BH_4	$P(OEt)_3$		phen	[43]
2.441	η^1-BH_4		tripod		[22, 23]
2.499 (a)	η^1-BH_4	$PMePh_2$	$PMePh_2$	$PMePh_2$	[20]
2.517 (b)	η^1-BH_4	$PMePh_2$	$PMePh_2$	$PMePh_2$	[20]
2.578 (c)	η^1-BH_4	$PMePh_2$	$PMePh_2$	$PMePh_2$	[20]
2.646 (d)	η^1-BH_4	$PMePh_2$	$PMePh_2$	$PMePh_2$	[144]
3.093	BH_4^-		Cyclam		[145]
3.18	BH_4^-		TMAC		[145]
7.197	BH_4^-		TMAC		[145]

Usually the addition of a new ligand entails a decrease in the denticity of the tetrahydroborate ligand, as expected from qualitative orbital analysis (Sect. 3.1). In this way, copper complexes with ligands occupying four coordination sites do not coordinate BH_4^- and are described as ionic tetrahydroborate complexes. Analysis of several $Cu(\eta^2-BH_4)L_2$ examples demonstrate the importance of ligand effects both in M – B distances and on choice of the coordination mode. Both complexes with L = PR_3 or L_2 = dmphen and dmdphen (phenantroline derived ligands) have a bidentate tetrahydroborate, though with slightly longer distances for the phosphine complexes (Table 3). After coordination of another PR_3 ligand, both complexes evolve in a new complex with a longer Cu – B distance. However, phenantroline derived complexes conserve the η^2 coordination mode, while the biphosphino complexes evolve to a η^1 coordination mode of the BH_4^-.

2.2.6 Wrongly Assigned Complexes

In absence of a precise location of the hydrogens, the difficulties in univocally assigning the coordination mode of the tetrahydroborate, especially in complexes with more than one BH_4^-, can be illustrated with several examples of wrongly assigned complexes which are found in the literature. Some of them will be discussed in this section, exemplifying the usefulness and limitations of M – B distances for assigning tetrahydroborate coordination modes.

$[Mn(BH_4)_3(thf)]^-[PPh_4]^+$ was initially described by Lobkovskii et al. [103] as having three $\eta^1-(BH_4)_3$. However, the η^1 complex should be 13e-complex following the electron counting described in Sect. 3. Moreover, the M – B distances are too short for a η^1 coordination (2.368, 2.335 and 2.362 Å). The M – B distances in this compound are in the range of those presented on Fig. 2 for a $\eta^2-BH_4^-$ coordination mode, and the complex is better described as $[Mn(\eta^2-BH_4)_3(thf)]^-$. This $\eta^2-BH_4^-$ coordination was already reported in the Makhaev review [3].

The $(^tBuNCH)_2Ti(BH_4)_2$ was described in the review of Xu and Lin as presenting two η^3-BH_4^- ligands coordinated to the titanium [4]. However, the X-ray diffraction results indicate that both η^3- and η^2-BH_4 are found in this complex, although in the ^1H and ^{11}B NMR results, no difference between these two ligands is apparent. The Ti – B distances are 2.175 and 2.304 Å, which support the hypothesis of a η^2-BH_4 and a η^3-BH_4 coordination. The 2.175 Å distance is too short to be considered as a Ti – B η^2-BH_4 distance and it is in agreement with the other Ti – B η^3-BH_4 distances. It should be noted that Herrmann et al. reported this complex as having one η^3- and one $\eta^2-BH_4^-$ ligand [86]. The same complex with zirconium $(^tBuNCH)_2Zr(\eta^3-BH_4)_2$ is actually a bis-(η^3-BH_4) complex with two similar and long Zr – B distances of 2.380 and 2.331 Å.

Another interesting complex which has been an object of study is the d^1 titanium complex $[(Me_3P)_2Ti(BH_4)_3]$ synthesized by Girolami et al. in 1986 [135]. The authors initially concluded that two of the three BH_4^- were coordinated to titanium through formation of a Ti – B bond and a α B – H agostic interaction with the metal centre, and the third BH_4^- was as a "normal" $\eta^2 - BH_4$ ligand. Subsequent theoretical studies on the model of this system $(H_3P)_2Ti(BH_4)_3$ by Volatron et al. [128] showed that the most stable structure does not contain any Ti – B bond nor an agostic interaction. The authors of the computational work put forward a $(H_3P)_2Ti(\eta^2 - BH_4)_2(\eta^3 - BH_4)$ structure. This theoretical proposal has been recently confirmed by a thorough reinvestigation of the nature of the BH_4 binding in these complexes which demonstrated that the crystallographic structure of $(Et_3P)_2Ti(\eta^2 - BH_4)_2(\eta^3 - BH_4)$ and $(Me_2PhP)_2Ti(\eta^2 - BH_4)_2(\eta^3 - BH_4)$ correspond to a complex with two BH_4^- coordinated as $\eta^2 - BH_4$ and one as $\eta^3 - BH_4$ [131]. The bonding in this complex is discussed in Sect. 3.1.

The vanadium complex $(NSi2)_2V(\eta^2 - BH_4)(thf)$ presents a V – B distance (2.382 Å) too long for a $\eta^2 - BH_4$ coordination [42]. Moreover, it should be a 14e-complex. Related vanadium complexes as $Cp(Me_2P(CH_2)_2PMe_2)V(\eta^2 - BH_4)$ and $Me_4C_2(C_5H_4)_2V$ $(\eta^2 - BH_4)$ have shorter V – B distances of 2.254 and 2.274 Å, respectively [64, 65, 68, 72]. The $(NSi2)_2V(\eta^2 - BH_4)(py)$ also has a shorter V – B distance (2.305 Å) [46]. Reinvestigation of the structure is suggested.

The Yttrium complex $[(thf)_4Y(BH_4)_2]^+$ has been studied experimentally [111, 112] and computationally [113]. Experimental results (X-ray diffraction) can not discriminate between η^2 and η^3 coordination modes [111, 112]. The Y – B distances (2.533 and 2.496 Å) are located between the averages of $\eta^3 - BH_4$ and $\eta^2 - BH_4$. Computational work performed by Lin and coworkers [113] showed that the most stable isomer was $[(thf)_4Y(\eta^3 - BH_4)_2]^+$. This study illustrates the importance of the electrostatic effects on the M – BH_4 interaction and will be discussed in Sect. 3.2.

The analysis of the experimental M – B distances has also allowed the location of some examples of ionic complexes described in the literature as $ML_n(\eta^1 - BH_4)$. The long Cu – B distances preclude BH_4^- coordination. These examples are the copper complexes $[Cu(Cyclam)]^+[BH_4]^-$, $[Cu(TMAC)]^+[BH_4]^-$, (TMAC = $CyclamMe_4$) (see Table 3) and the chromium complex $[(py)_4Cr(\eta^1 - BH_4)]^+[BH_4]^-$.

2.3 Stretching Vibrations of the B–H Bonds

Due to the difficulties concerning crystallization and the non-precise localization of hydrogen atoms using X-ray diffraction, the examination of IR frequencies has been very useful for the assignment of the coordination mode of the BH_4^- ligands. However, this method may be imprecise since

$B-H$ stretching vibrations cannot be uniquely related to one coordination mode.

When a BH_4^- is coordinated, it suffers several deformations. On one side, its local symmetry is lowered from the T_d symmetry in the free anion to the C_{3v} in the η^1 and η^3 coordination modes and C_{2v} in the η^2 one. Moreover the $B-H_{bridging}$ bonds lengthen and the $B-H_{terminal}$ ones shorten. In this way, as a consequence of the coordination, both the number of bands and their positions are modified respect the vibrational spectrum of the free anion. The tetrahedral BH_4^- ($B-H = 1.26$ Å) shows two stretching frequencies at 2264 and 2240 cm^{-1}. When coordinated, two types of stretching frequencies appear depending on the terminal or bridging nature of the hydrogen. The $B-H_b$ bonds are weakened and the $B-H_t$ strengthened with respect to the free anion (for instance, in diborane: $B-H_b = 1.34$ Å and $B-H_t = 1.20$ Å). These changes in the $B-H$ bonds are also reflected in the $B-H$ stretching frequencies, and high-frequency terminal $\nu(B-H_t)$ are well-separated from the low-frequency bridging $\nu(B-H_b)$.

According to the literature, in mononuclear MBH_4 systems, $B-H$ stretching frequencies are localized within 1900–2600 cm^{-1}. The main characteristics of the IR frequencies of the three coordination modes are:

- η^1-BH_4 complexes exhibit two bands within the region 2300–2450 cm^{-1} assigned to the $B-H_t$ stretching and a strong absorption at 2000 cm^{-1} corresponding to the $B-H_b$ stretching;
- η^2-BH_4 coordination is characterised by the presence of second $B-H_b$ stretching between 1650–2150 cm^{-1}. $B-H_t$ stretching frequencies are identified by a doublet between 2400–2600 cm^{-1};
- η^3-BH_4 coordination are identified by a $B-H_t$ stretching singlet between 2450–2600 cm^{-1} and a $B-H_b$ doublet between 2100–2200 cm^{-1}.

A large amount of vibrational data can be found spanning a great diversity of complexes in all the coordination modes. In Fig. 11 we present only frequencies of first-row transition metal complexes which have also been characterised by X-ray diffraction. In Fig. 11, complexes are presented from the left to the right of the periodic table, and classified by coordination mode, first η^1, then η^2, followed by combined complexes with η^2 and η^3-BH_4^- ligands, and finally η^3 complexes. Depending on the system, a large or small number of frequencies are reported on the experimental papers, allowing different complexes to have a different number of frequencies represented. For

Fig. 11 B–H stretching frequencies of first-row transition metal complexes which have been characterised by X-ray diffraction. Complexes are presented from the left to the right of the periodic table, and classified according to coordination mode, first η^1, then η^2, followed by combined complexes with η^2 and η^3 BH_4^- and finally η^3 complexes. The highest frequency is represented as a *black rhomboid*, the second as a *grey square*, the third as a *white triangle* and so on ▶

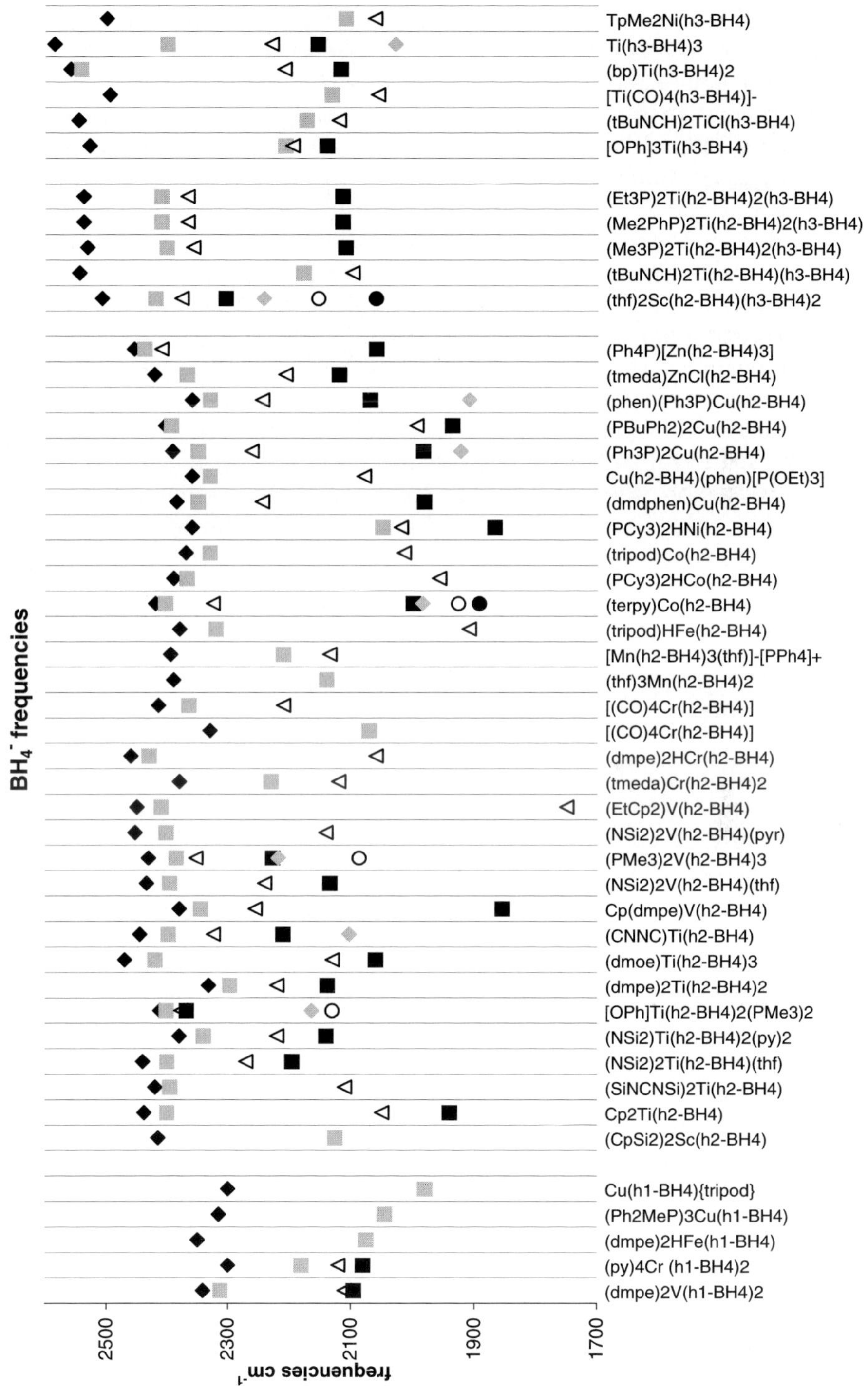
TpMe2Ni(h3-BH4)
Ti(h3-BH4)3
(bp)Ti(h3-BH4)2
[Ti(CO)4(h3-BH4)]-
(tBuNCH)2TiCl(h3-BH4)
[OPh]3Ti(h3-BH4)
(Et3P)2Ti(h2-BH4)2(h3-BH4)
(Me2PhP)2Ti(h2-BH4)2(h3-BH4)
(Me3P)2Ti(h2-BH4)2(h3-BH4)
(tBuNCH)2Ti(h2-BH4)(h3-BH4)
(thf)2Sc(h2-BH4)(h3-BH4)2
(Ph4P)[Zn(h2-BH4)3]
(tmeda)ZnCl(h2-BH4)
(phen)(Ph3P)Cu(h2-BH4)
(PBuPh2)2Cu(h2-BH4)
(Ph3P)2Cu(h2-BH4)
Cu(h2-BH4)(phen)[P(OEt)3]
(dmdphen)Cu(h2-BH4)
(PCy3)2HNi(h2-BH4)
(tripod)Co(h2-BH4)
(PCy3)2HCo(h2-BH4)
(terpy)Co(h2-BH4)
(tripod)HFe(h2-BH4)
[Mn(h2-BH4)3(thf)]-[PPh4]+
(thf)3Mn(h2-BH4)2
[(CO)4Cr(h2-BH4)]
[(CO)4Cr(h2-BH4)]
(dmpe)2HCr(h2-BH4)
(tmeda)Cr(h2-BH4)2
(EtCp2)V(h2-BH4)
(NSi2)2V(h2-BH4)(pyr)
(PMe3)2V(h2-BH4)3
(NSi2)2V(h2-BH4)(thf)
Cp(dmpe)V(h2-BH4)
(CNNC)Ti(h2-BH4)
(dmoe)Ti(h2-BH4)3
(dmpe)2Ti(h2-BH4)2
[OPh]Ti(h2-BH4)2(PMe3)2
(NSi2)Ti(h2-BH4)2(py)2
(NSi2)2Ti(h2-BH4)(thf)
(SiNCNSi)2Ti(h2-BH4)
Cp2Ti(h2-BH4)
(CpSi2)2Sc(h2-BH4)
Cu(h1-BH4){tripod}
(Ph2MeP)3Cu(h1-BH4)
(dmpe)2HFe(h1-BH4)
(py)4Cr (h1-BH4)2
(dmpe)2V(h1-BH4)2
2500
2300
2100
1900
1700
frequencies cm-1
BH4- frequencies

each complex, each frequency has a different mark. The highest frequency is represented as a black rhomboid, the second as a grey square, the third as a white triangle and so on. For all the complexes, the highest frequency corresponds to the $B-H_t$ stretching (represented by a black rhomboid). This frequency is found at higher values when the denticity of the tetrahydroborate ligand increases. Complexes with η^1-BH_4 ligands have the higher frequency, lower than 2400 cm^{-1}, while the η^2 coordination mode gives the higher frequency, around 2400 cm^{-1}, and the η^3 coordination mode has the highest frequency located between 2500–2600 cm^{-1}. Another important characteristic is that while for η^1 and η^3 coordination modes this frequency is a singlet, for the η^2 coordination mode it is a doublet.

Other particular frequencies can be observed. For the η^1 coordination mode, a frequency appears around 2000 cm^{-1} or a bit higher (grey or black squares). For η^2 complexes, frequencies in this region can also be seen, however they span a broad range, up and down, between 2150 and 1850 cm^{-1}. BH_4^- ligands coordinated in a η^3 mode present a doublet around 2100 cm^{-1}.

Analysis of the frequencies can allow one to distinguish between complexes with BH_4^- coordinated in different coordination modes. Moreover, several complexes that do not exactly follow the trends stated above can be located in Fig. 11. This is the case for complexes $(dmpe)_2V(\eta^1-BH_4)_2$, $(CpSi2)_2Sc(\eta^2-BH_4)$, $(tmeda)Cr(\eta^2-BH_4)_2$, $[(CO)_4Cr(\eta^2-BH_4)]$, $(thf)_3Mn(\eta^2-BH_4)_2$, $[Mn(\eta^2-BH_4)_3(thf)]^-[PPh_4]^+$, $(PCy_3)_2HNi(\eta^2-BH_4)$ and $(bp)Ti(\eta^3-BH_4)_2$.

The $(dmpe)_2V(\eta^1-BH_4)_2$ complex presents quite ambiguous signals. The presence of two bands, like a doublet, between 2300–2350 cm^{-1} does not seem to fit into a η^1 coordination, but at the same time, they are too low for a η^2 coordination. Looking at the $V-B$ distances (2.833 Å), the coordination mode is clearly η^1. The distances are too long for η^2 coordination and consequently the complex must have the two tetrahydroborate ligands coordinated as η^1-BH_4.

$(CpSi2)_2Sc(\eta^2-BH_4)$ does not show the doublet between 2400–2600 cm^{-1} characteristic of the η^2 mode [77] Moreover, it is a 16-electron complex and a η^3-BH_4 coordination could be expected in order to obey the 18-electron rule (see Sect. 3.1). Nevertheless, its frequencies are too low for η^3 coordination and the $Sc-B$ distance also agrees with η^2 coordination. This behaviour has been interpreted as an indication of the importance of electrostatic interaction for early transition-metal tetrahydroborate complexes [113].

$(tmeda)Cr(\eta^2-BH_4)_2$ and $[(CO)_4Cr(\eta^2-BH_4)]$ do not show the expected doublet for the $B-H_t$ stretching according to the η^2 coordination of its BH_4^- ligands. The first vibrational spectrum of $[(CO)_4Cr(\eta^2-BH_4)]$ did not fit in the η^2 coordination frequencies [54]. Frequencies were reinvestigated and the new data obtained (represented in Fig. 11 the new and the old values) is in perfect agreement with a η^2 coordination of the BH_4^- ligands. The signals of the $(tmeda)Cr(\eta^2-BH_4)_2$ complex could be interpreted as corresponding to two BH_4^- ligands coordinated in a monodentate manner [93]. However, the Cr – B distances for this complex are 2.44 and 2.42 Å too short for a monodentate coordination and typical for bidentate tetrahydroborate ligands. Two manganese complexes, $(thf)_3Mn(\eta^2-BH_4)_2$ and $[Mn(\eta^2-BH_4)_3(thf)]^-[PPh_4]^+$, do not show the doublet either. However they are complexes with $\eta^2-BH_4^-$ ligands [103, 104]. The $[Mn(\eta^2-BH_4)_3(thf)]^-[PPh_4]^+$ complex was discussed by Xu and Lin [4] as a complex with three η^1-BH_4^- ligands. The complex has been described by Makhaev [3] as a η^2 complex. The absence of a doublet could support the η^1 structure, but the high value of the higher frequency and the crystallographic data support the η^2 one. Also, the $(PCy_3)_2HNi(\eta^2-BH_4)$ complex does not show the doublet. However, this complex has a 18 electron-configuration being η^2, and its Ni – B distance corresponds to a η^2 ligand. The $(bp)Ti(\eta^3-BH_4)_2$ complex presents a doublet which could be explained by a η^2 coordination of the BH4-ligands. However, the frequencies are too high for this kind of coordination. Considering the η^3 coordination, the complex has a 16 electron configuration in agreement with a Ti d^0 metal centre.

3 Theoretical Analysis of the Coordination Modes

3.1 Qualitative Orbital Interaction Analysis

The way in which a BH_4^- ligand interacts with a transition metal depends on its coordination mode (η^1, η^2 or η^3). Basically, this ligand is isoelectronic with methane and is characterised by four occupied MOs (a_1 and t_2). The low-lying fully symmetric a_1 orbital can be neglected in the bonding scheme, as shown by Hori et al. in the ab initio study of $Be(BH_4)_2$ [136]. Therefore, the BH_4^- ligand has three t_2 orbitals susceptible of interacting with a transition metal fragment. Due to the low electronegativity of the boron atom, t_2 orbitals are mainly developed on hydrogen atoms. A significant d-t_2 overlap is therefore found only if the hydrogen atom involved in a particular t_2 orbital points toward the metal. As a consequence, the interaction scheme depends strongly on the coordination mode. A single t_2 orbital interacts with the metal in the η^1 coordination mode, while two and three orbitals are involved in the η^2 and η^3 modes, respectively. The three t_2 orbitals can be labelled as σ, $\pi_{||}\pi_{\perp}$ and are depicted in Scheme 11

The ligand orbital involved in a η^1 coordination mode is σ, those in η^2 and η^3 being $(\sigma, \pi_{||})$ and $(\sigma, \pi_{||}\pi_{\perp})$, respectively. As long as low-lying orbitals of proper symmetry are present on the metal fragment, η^1-BH_4 acts as a 2-electron donor and η^2-BH_4 and η^3-BH_4 act as 4- and 6-electron donors, respectively. From this orbital interaction scheme, the coordination mode can be easily predicted

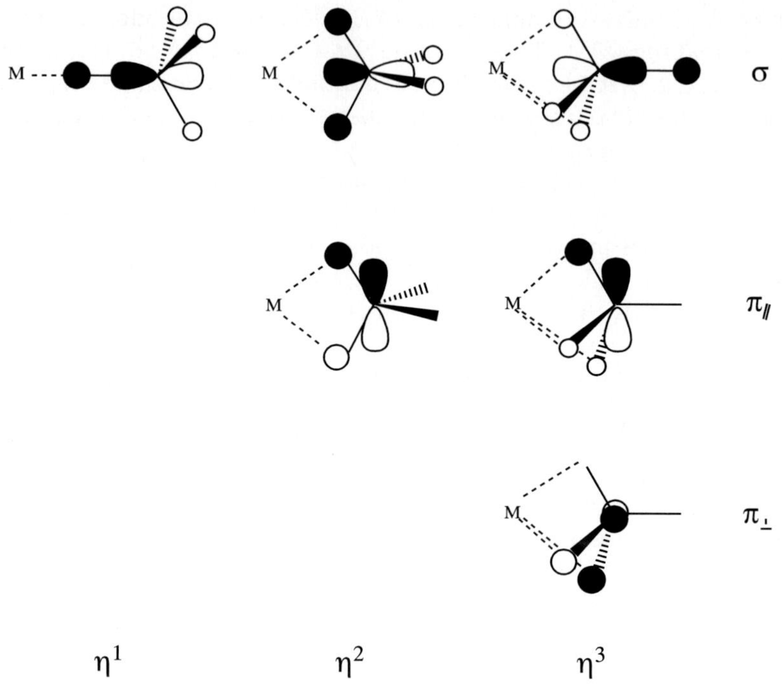

Scheme 11 BH_4^- orbitals involved in η^1, η^2 and η^3 coordination modes

from a simple electron count. For closed-shell complexes, the optimal coordination is obtained in order to get a formally 18-electron complex. The ideal electron number may be different from 18 in paramagnetic complexes: for instance, it is 16 in high-spin d^2 complexes and 17 in d^1 ones.

Numerous examples of mono(tetrahydroborate) complexes illustrate this simple relationship between the BH_4^- coordination mode and the 18-electron rule. A nice case can be found in the series of Cu-borohydride complexes. The tetrahydroborate ligand is bound in an η^1 fashion in the $Cu(BH_4)(PMePh_2)_3$ complex, in which the total number of electrons is $10(Cu) + 3 \times 2(phosphines) + 2(BH_4^-)$. The loss of a phosphine group in $Cu(BH_4)(PPh_3)_2$ entails a change in the coordination mode of $BH_4^-(\eta^1 \to \eta^2)$ in order to keep the 18-electron count.

The situation becomes more complex when several BH_4^- ligands are bound to the metal. For instance, in $Zr(\eta^3 - BH_4)_4$ or in $Hf(\eta^3 - BH_4)_4$ it has been shown that each of the four groups cannot act as a 6-electron donor (leading to a 24-electron complex!) because there are not enough acceptor orbitals on the metal. However, a 18-electron count is still obtained when symmetry restrictions are taken into account for orbital interactions.

To illustrate the way the electron counting can be achieved in complexes with more than one tetrahydroborate group bound to the metal, let us consider $Ti(BH_4)_3$. This complex with three BH_4^- ligands adopts a trigonal-planar geometry with respect to the boron atom of BH_4^- [116]. We will only examine the most symmetrical structures in which the BH_4^- ligands are all bound in the same way to the metal (mono-, bi, and tridentate, **1–4** Scheme 12), the metal and B atoms being coplanar.

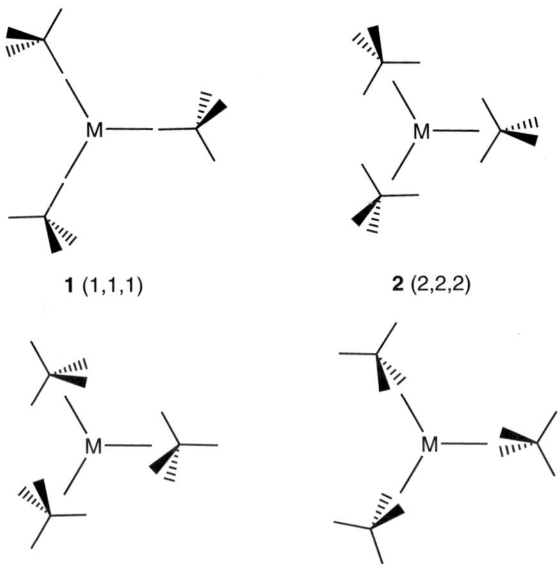

1 (1,1,1) **2** (2,2,2) **3** (3,3,3) **4** (2*,2*,2*)

Scheme 12 Most symmetrical structures in $Ti(BH_4)_3$

In the (η^1,η^1,η^1) or (1,1,1) structure (**1**), only the σ orbitals on the ligands are involved. Their proper symmetry combinations can interact with three metal orbitals (for instance, s, p_x and p_y; Scheme 13) so that the three BH_4^- ligands act as a 6-electron donor in this coordination mode. Since one electron is remaining in the d block (d^1 complex), the total number of electrons is 7 in the (1,1,1) structure.

In the (2,2,2) complex **2**, with all the bridging hydrogen atoms lying in the MB_3 plane, the $\pi_{||}$ orbitals must be considered in addition to the σ ones. The two lowest symmetry combinations can interact with the xy and x^2–y^2 d metal orbitals (Scheme 14). The highest one is of f symmetry (three nodal planes) and cannot find any symmetry-adapted orbital in a transition metal. Consequently, the three BH_4^- ligands act as a 4-electron donor through their $\pi_{||}$ orbitals. The (2,2,2) structure can be described as a $6(\sigma) + 4(\pi_{||}) + 1(\text{Ti}) = 11$-electron complex.

In the (3,3,3) structure **3**, the three $\pi_\perp$ orbitals are involved. They mix to give three symmetry-adapted orbitals which can interact with the p_z, xz

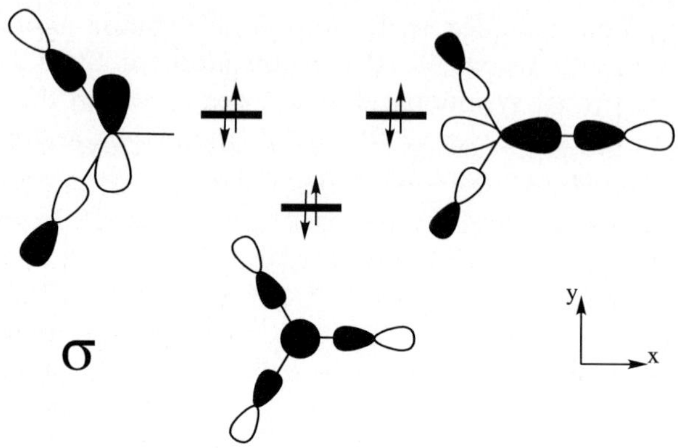

Scheme 13 Interactions between the symmetry-adapted combinations of the σ orbitals of the BH_4^- ligands and the metal orbitals. For the sake of clarity, the σ orbitals are pictured as a combination of simple boron p orbitals

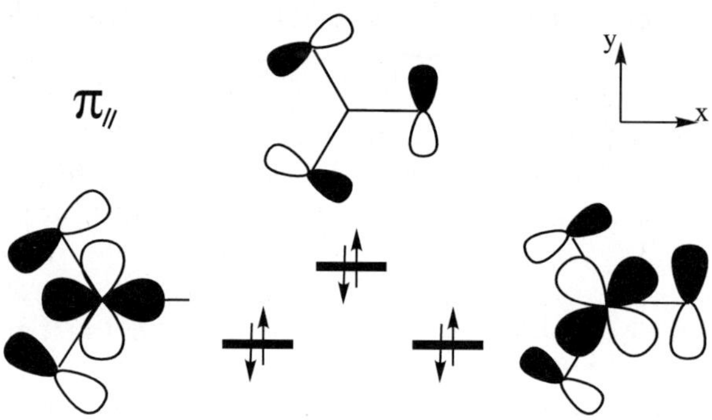

Scheme 14 Interactions between the symmetry-adapted combinations of the $\pi_{||}$ orbitals of the BH_4^- ligands and the metal orbitals. For the sake of clarity, the $\pi_{||}$ orbitals are pictured as a combination of simple boron p orbitals

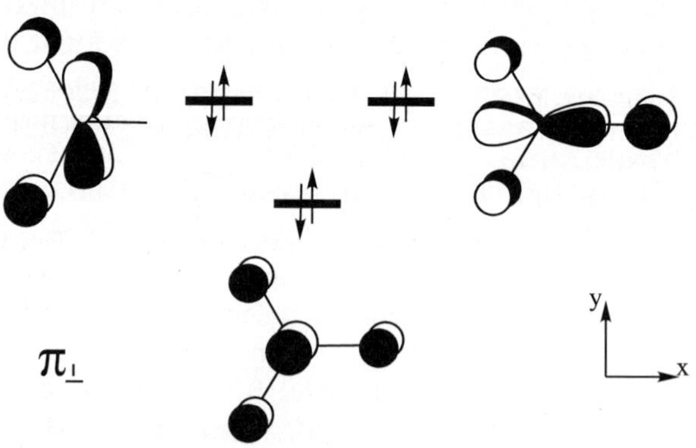

Scheme 15 Interactions between the symmetry-adapted combinations of the $\pi_{\perp}$ orbitals of the BH_4^- ligands and the metal orbitals. For the sake of clarity, the $\pi_{\perp}$ orbitals are pictured as a combination of simple boron p orbitals

and *yz* metal orbitals (Scheme 15). The three BH_4^- groups act as a 6-electron donor through their $\pi_\perp$ orbitals. The (3,3,3) structure can be described as a $6(\sigma) + 4(\pi_{||}) + 6(\pi_\perp) + 1(\mathrm{Ti}) = 17$-electron complex [118].

In structure **4**, the three tetrahydroborate ligands are coordinated in a η^2 mode, but the H_bBH_b planes are orthogonal to the MB_3 plane (this type of structure is actually found for the $Al(BH_4)_3$ complex). In this structure, noted (2*,2*,2*), σ and $\pi_\perp$ orbitals on the BH_4^- ligands are involved, so that $6(\sigma) + 6(\pi_\perp) = 12$ electrons are given to the metal and the (2*,2*,2*) complex is therefore a $12 + 1 = 13$-electron one. Note that the experimental structure (3,3,3) is a 17-electron one [116], which is the expected ideal electron count for a d^1 complex.

3.2
Energy Ordering vs. Electron Count

The qualitative molecular orbital analysis described above allows assigning an electron count to any structure. Quantitative ab initio and DFT calculations give the relative energies of the possible structures of a complex depending on the BH_4^- coordination mode. These calculations have established that in most of the transition-metal tetrahydroborate complexes, there is a clear relationship between the energy of the structure and the electron count around the metal, which depends on the coordination mode of the BH_4^- ligand. Two nice examples of systems which present different coordination modes of the tetrahydroborate, despite having a similar set and arrangement of the other ligands, serve to illustrate this argument.

In the series of zerovalent $[M(CO)_4(BH_4)]^-$ (M = Ti, Cr, Mo) transition metal borohydrides, the Cr and Mo complexes (d^6) contain a bidentate borohydride ligand [16], whereas the Ti compound (d^4) contains a tridentate borohydride ligand [90]. Calculations in these systems as well as in the d^6

Table 4 Electron count and relative energies (kcal/mol) of the η^1, η^2 and η^3 isomers in the $[M(CO)_4(BH_4)]^{n-}$ ($n = 1$, M = Ti, Cr, Mo; $n = 0$, M = Mn) complexes

	Ti[a]		Cr[b]		Mo[b]		Mn[c]	
	Electron count	Energy kcal mol^{-1}	Electron count	Energy kcal mol^{-1}	Electron count	Energy kcal mol^{-1}	Electron count	Energy kcal mol^{-1}
η^1	14	31.1	16	22.7	16	21.4	16	24.3
η^2	16	10.2	18	0.0	18	0.0	18	0.0
η^3	18	0.0	20	19.4	20	15.0	20	14.4

[a] [90]
[b] [16]
[c] [12]

neutral parent Mn compound [12] give an energy ordering of the η^1, η^2 and η^3 isomers that is in agreement with the electron count. The structures with a deficiency or excess of electrons regarding the ideal electron count are found to be notably higher in energy.

The basic relationship between the coordination mode and the number of electrons around the metal still applies in rather complicated systems such as those with three BH_4^- ligands. Complexes of general formula $(MBH_4)_3L_2$ have been characterised with scandium, titanium and vanadium. All the complexes adopt distorted trigonal-bipyramidal geometry with the BH_4^- ligands at the equatorial positions (Scheme 16).

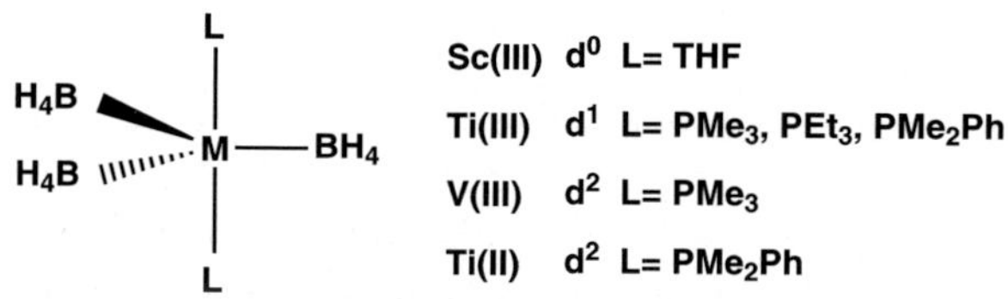

Scheme 16 Experimentally characterised $M(BH_4)_3L_2$ complexes

The coordination modes of the tetrahydroborato ligands differ in these complexes (X-ray diffraction) as shown in Scheme 17. In the scandium complex (d^0), two groups are η^3 and one η^2, giving a (2,3,3) structure [112, 130]. The structure of the titanium d^1 complex was initially described as (1,1,2) [135], while that of the high-spin complex of vanadium (d^2) is (2,2,2) [92, 105]. The titanium (II) $[Ti(BH_4)_3(PMe_2Ph)]^-$ unit also has three bidentate BH_4^- ligands [131].

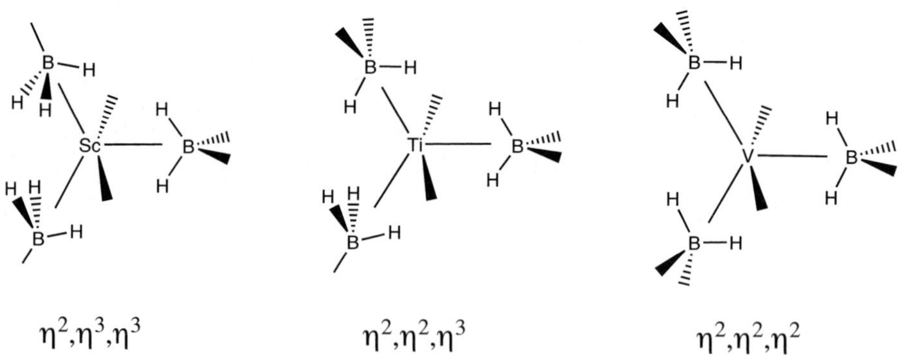

Scheme 17 Lowest energy structures in $M(BH_4)_3(PH_3)_2$ (M = Sc, Ti, V) complexes

Using the analysis developed in Sect. 3.1 and taking symmetry restrictions into account in order to obtain the proper electron count, it is possible to give an electron count for each of the ten structures (from (1,1,1) to (3,3,3)) which differ in the coordination mode of the BH_4^- ligands [106, 128, 137]. The relative energies as a function of the electron count of a selected set of structures for the three complexes are depicted in Fig. 12.

In the diamagnetic d^0 scandium complex, the lowest energy structure (2,3,3) is formally an 18-electron complex [137]. In the d^2 high-spin vanadium complex, the ideal electron count is 16 and this number is attained for the most stable (2,2,2) structure [106]. For the d^1 titanium complex, the ideal electron number is 17 and the lowest energy structure (2,2,3) belongs to the 17-electron species family [128]. There is a clear relationship between the energy of the structure and the usual electron count around the metal which depends on the coordination mode of the BH_4^- ligands. As it can be appreciated in Fig. 12, the energy increases with the deviation from the ideal electron count, and the larger the deviation from this ideal number, the higher the energy.

The most stable structures calculated for the $Sc(BH_4)_3(PH_3)_2$ and $V(BH_4)_3(PH_3)_2$ complexes agree with the experimental data for the $Sc(BH_4)_3(THF)_2$ and $V(BH_4)_3(PMe_3)_2$ complexes. However, the theoretically predicted (2,2,3) absolute minimum for the $Ti(BH_4)_3(PH_3)_2$ complex was in disagreement with the experimental data of $Ti(BH_4)_3(PMe_3)_2$, which had been initially interpreted in terms of η^1, η^1, and η^2 coordination. The energy of the (1,1,2) structure was calculated to be about 40 kcal/mol above the (2,2,3) minimum; furthermore, the (1,1,2) structure was not even a local minimum on the potential energy surface. Very recently, the nature of the BH_4 binding modes

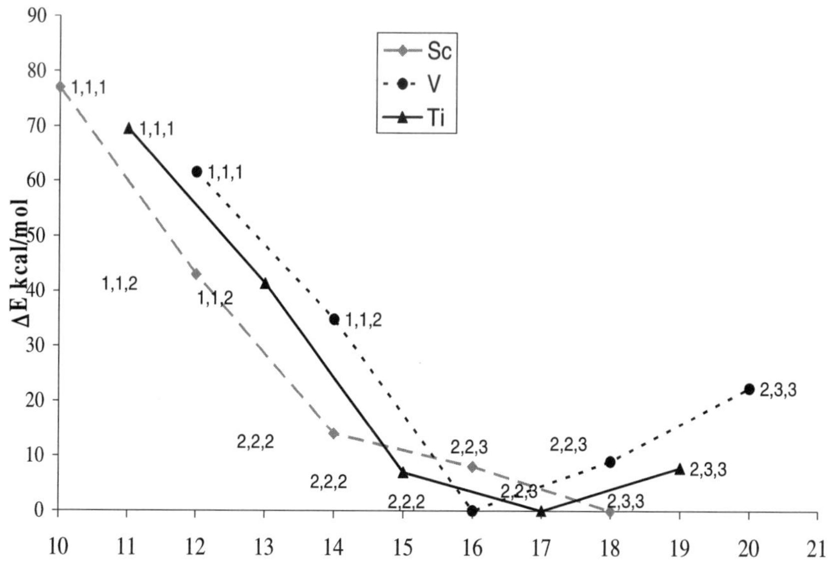

Fig. 12 Relative energies as a function of the electron count for a selected set of structures of $M(BH_4)_3(PH_3)_2$ complexes (M = Sc, Ti, V) [106, 128, 137]

in $Ti(BH_4)_3(PR_3)_2$ complexes has been reinvestigated [131]. The molecular structure of $Ti(BH_4)_3(PEt_3)_2$ and $Ti(BH_4)_3(PMe_2Ph)_2$ have been determined using new methods that permit collecting diffraction data from thermally and air-sensitive solids. These crystallographic studies clearly are consistent with the theoretical results and led to reformulate the coordination modes for the borohydrate ligands in the $Ti(BH_4)_3(PR_3)_2$ species as (2,2,3). These results support the importance of electronic factors in determining the coordination modes of tetrahydroborate ligands.

The electron count has been shown to be a powerful tool to analyze the problem of coordination mode in tetrahydroborate complexes. However, a more detailed analysis of orbital interactions is required to explain why complexes with the same electron number can have very different energies. First, a formal 2-electron transfer from the BH_4^- ligand to the metal is associated with an interaction between a ligand orbital and a low-lying orbital on the metal. This interaction can be stronger or weaker, depending on the overlap and the energy gap between the interacting orbitals. A second point arises from the orbital interaction scheme which can differ for the same electron count. Finally, electron counting does not take into account 4-electron interactions between ligand orbitals and low-lying occupied orbitals of the metal fragment.

Although the prevalence of the electronic factor in the M-BH_4 interaction has been demonstrated, electrostatic effects can't be neglected. Lin et al. have demonstrated the almost equal importance of both electronic and electrostatic effects in $[Y(thf)_4(BH_4)_2]^+$ [113]. To maximize the electronic interaction between the metal atom and the two axial BH_4^- ligands, both *xz* and *yz* orbitals should be fully utilized in the metal–BH_4^-π bonding (Scheme 18). To achieve

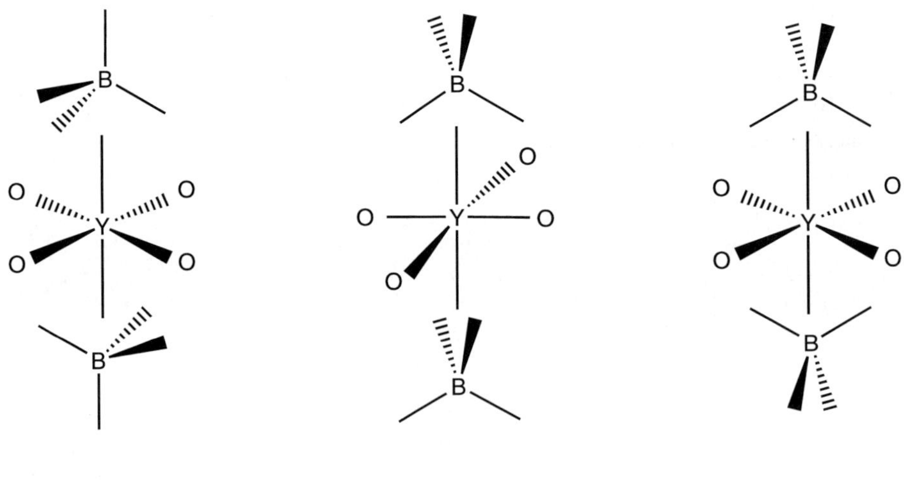

Scheme 18 Different coordination modes of the two BH_4^- ligands in $[Y(thf)(BH_4)_2]^+$: **a** two η^3; **b** two η^2 perpendicular to each other; **c** two η^2 parallel to each other

this maximization, both BH_4^- ligands are expected to coordinate to the central atom either through a η^3 mode (**a**) or through a η^2 mode by having the two bridging units perpendicular to each other (**b**) (Scheme 18). Structures **a** and **b** have the same electron count and can achieve the optimal metal–BH_4^- electronic interaction. Therefore, their energy difference (6.8 kcal/mol at the CISD level) can be approximately considered as resulting from the different electrostatic interactions between these two modes (η^2 and η^3). When the two $\eta^2-BH_4^-$ ligands are parallel to each other (**c**), the two π orbitals from both BH_4^- ligands can only interact with one of the two d orbitals of the metal atom, and thus the optimal electronic interaction is not satisfied. The energy difference between the two η^2 structures (**b** and **c**) (7.0 kcal/mol at the CISD level) can be viewed as the electronic interaction energy difference. Calculations have shown that in the yttrium complex, the electronic and electrostatic stabilizations are of almost equal importance. The importance of electrostatic effects in the metal-tetrahydroborate interactions have also been stressed in a theoretical study of $(PH_3)_nCu(BH_4)$ ($n = 1, 2, 3$) complexes [138].

4 Terminal-bridging Hydrogen Exchange

Transition-metal tetrahydroborate complexes are often fluxional, and commonly a single resonance is observed for all of the four B–H hydrogens in the ^{1}H NMR spectrum at ambient temperatures, pointing out that a fast intramolecular exchange on the NMR time scale of the terminal (t) and bridging hydrogens has taken place. The mechanisms proposed for the hydrogen exchange involve a change on the coordination mode of the tetrahydroborate ligand. For instance, for bidentate complexes, two different possible pathways for bridge-terminal hydrogen permutation are possible: via a monodentate BH_4^- ligand (dissociative mechanism) and via a tridentate BH_4^- ligand (associative mechanism) (Scheme 19).

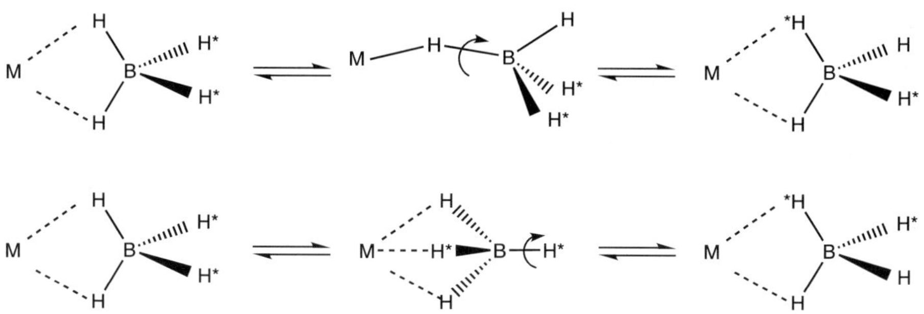

Scheme 19 Dissociative (*top*) and associative (*bottom*) mechanisms for the $H_{bridging}$/$H_{terminal}$ exchange in a (η^2-BH_4) ligand

NMR measurements do not allow discrimination between the two exchange pathways. This has been possible with theoretical studies in which the transition state for the hydrogen exchange has been located. In this section, we will first present the experimental information available on the activation barriers obtained from variable-temperature NMR measurements. Then, the computational studies of these processes which have allowed the determination of the transition states for the exchanges will be revised and the possible mechanisms will be discussed.

4.1 Experimental Data on Activation Barriers

The number of complexes for which the $\Delta G^{\neq}$ have been determined from NMR data is rather limited, making it difficult to establish clear trends (Table 5). For many compounds, the exchange is too fast on the NMR time scale to distinguish separate signals for the terminal and bridging hydrogens, even at low temperatures. Compounds of Group 3 and 4 transition metals

Table 5 $\Delta G^{\neq}$ (kcal/mol) for the $H_{bridging}/H_{terminal}$ hydrogen exchange determined from temperature-variable NMR measurements

Compound	$\Delta G^{\neq}$	Refs.
$Cp_2V(\eta^2\text{-}BH_4)$	7.6	[68]
$[V\{(C_5H_3^tBu)C_2Me_4(C_5H_3^tBu)\}(\eta^2\text{-}BH_4)]$	10.0	[72]
$Cp_2Nb(\eta^2\text{-}BH_4)$	14.6	[146]
$Cp^*_2Nb(\eta^2\text{-}BH_4)$	16.4	[146]
$Cp'_2Nb(\eta^2\text{-}BH_4)$	13.6	[78]
$[Nb\{(C_5H_4)CMe_2(C_5H_4)\}(\eta^2\text{-}BH_4)]$	8.4	[72]
$[Nb\{(C_5H_4)Cet_2(C_5H_4)\}(\eta^2\text{-}BH_4)]$	8.6	[72]
$[Nb\{(C_5H_4)C(C_5H_{10})(C_5H_4)\}(\eta^2\text{-}BH_4)]$	8.4	[72]
$[Nb\{(C_5H_4)C_2Me_4(C_5H_4)\}(\eta^2\text{-}BH_4)]$	12.0	[72]
$[Nb\{(C_5H_3^tBu)C_2Me_4(C_5H_3^tBu)\}(\eta^2\text{-}BH_4)]$	11.8	[72]
$[Nb\{(C_5H_4)SiMe_2(C_5H_4)\}(\eta^2\text{-}BH_4)]$	11.8	[72]
$Ind_2Nb(\eta^2\text{-}BH_4)$	13.2	[147]
$Cp^*CpTa(\eta^2\text{-}BH_4)$	16.4	[139]
$[Mo(CO)_4((\eta^2\text{-}BH_4))^-$	10.1	[55]
$Co(\eta^2\text{-}BH_4)terpy$	11.1	[148]
$[Ti(CO)_4((\eta^3\text{-}BH_4))^-$	8.8	[83]
$Zr(\eta^3\text{-}BH_4)_4$	7.3 [a]	[149]
$Hf(\eta^3\text{-}BH_4)_4$	8.1 [a]	[149]
$(C_5H_4Me)_2Hf((\eta^2\text{-}BH_4)_2$	< 4.9	[97]

[a] Solid state NMR

possess a clear-cut fluxionality. Only recently, individual signals from the terminal and bridging hydrogens were observed and the activation energy was determined for a titanium complex [83]. In addition, this is the only experimental activation energy available for a tridentate tetrahydroborate group. Tetrahydroborates of metal Groups 5–8 are much less fluxional than complexes formed by Group 3 and 4 transition metals. For many of them, separate signals for H_t and H_b can be detected in low-temperature NMR spectra. In several osmium and iridium complexes, the hydrogen exchange does not proceed at a noticeable rate even at room or higher temperatures. On the contrary, for compounds of Group 11 metals, separate signals for bridging and terminal hydrogens have not been identified.

The $\Delta G^{\neq}$ values also suggest that the fluxionality of the tetrahydroborate complexes decreases upon passing from first-row transition metals to second- or third row metals. This effect can be appreciated by comparing the values for vanadium, niobium and tantalum metallocenes. Indeed, most of the $\Delta G^{\neq}$ values collected in Table 5 have been determined in niobium metallocenes. Substitution of Cp for the bulkier Cp* ligand increases the barrier, pointing out that the fluxionality decreases with an increase in the size of the ligand. On the contrary, the data in a series of *ansa*-bridged tetrahydroborate niobocene compounds indicate that the free energy barrier to bridge-terminal hydrogen exchange is considerably reduced relative to the non-bridged species when the *ansa*-bridge imposes a significant change in geometry of the metallocene. The barrier decreases when the inter-ring angle increases. Exchanges in complexes with more than one tetrahydroborate ligand appear to be faster, and very few experimental data are available for such compounds.

4.2
Computational Studies of the Hydrogen Exchanges

Systems for which the terminal-bridging exchange has been theoretically studied, with location and characterization of the corresponding transition states, are presented in Table 6.

Calculated values are in very good agreement with the experimental ones in complexes for which the experimental data are available. Calculations also reproduce the trends along the periodic table discussed in the preceding subsection. The low activation barriers obtained in the series of copper complexes agree with the non- experimental detection of separate signals for bridging and terminal hydrogens of Group 11 tetrahydroborate complexes. The high barrier calculated in the osmium complex is also in accordance with the non- observance of H_t/H_b exchange, even at 360 K in $OsH_3(BH_4)(PPr^i{}_3)_2$ [15].

The comparison of the geometries of the located transition states and the analysis of the transition vectors allow one to obtain a deeper insight into the

Table 6 Computed energy barriers (kcal/mol) for the $H_{bridging}/H_{terminal}$ exchange

Compound	Minimum	Transition State	$\Delta E^{\neq}$	Refs.
$Cp_2Nb(\eta^2\text{-}BH_4)$	η^2	η^3	13.9; 15.4	[78, 140]
$[Nb\{(C_5H_4)SiH_2(C_5H_4)\}(\eta^2\text{-}BH_4)]$	η^2	η^3	12.5	[140]
$[Nb\{(C_5H_4)CH_2(C_5H_4)\}(\eta^2\text{-}BH_4)]$	η^2	η^3	8.5	[140]
$[M(CO)_4(\eta^2\text{-}BH_4)]^-$				
M = Cr	η^2	η^1	15.6	[16]
M = Mo	η^2	η^1	13.1	[16]
M = Ti	η^3	η^2	10.1	[90]
$Cu(BH_4)(PH_3)_n$				
$n = 1$	η^3	η^2	2.5	[11]
$n = 2$	η^2	η^3	11.7	[11]
$n = 3$	η^1	η^2	3.0	[11]
$OsH_3(\eta^2\text{-}BH_4)(PH_3)_2$	η^2	η^1	20.8	[15]
$Ti(BH_4)_3$	(η^3, η^3, η^3)	(η^2, η^3, η^3)	5.1	[118]

mechanism of the process. In complexes with a bidentate tetrahydroborate, both dissociative (via η^1 transition states) and associative (via η^3 transition states) mechanisms have been found. The mechanism for the hydrogen exchange in the $Cp_2Nb(\eta^2-BH_4)$ is depicted in Scheme 20. The exchange takes place with an associative mechanism via a η^3-BH_4 transition state. Despite that it has been proposed that the bridge-terminal hydrogen exchange in metallocene tetrahydroborate complexes involves a $\eta^5-\eta^3$ ring shift of one of the cyclopentadienyls rings [72, 139], analysis of the eigenvalue associated with the only imaginary frequency of the transition state structure shows that the transition vector has no components in the Cp rings [78]. The reaction coordinate is essentially the rotation around one of the $B-H_b$ bonds in the minimum. Very similar transition states have been located for the exchange in *ansa*-bridged tetrahydroborate niobocene complexes [140]. The same mechanism was found for the hydrogen exchange in $Cu(\eta^2-BH_4)(PH_3)_2$ [11, 12].

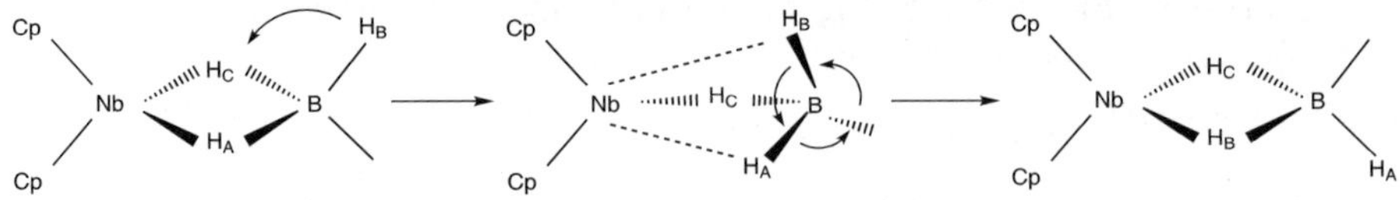

Scheme 20 Mechanism for the hydrogen exchange in $Cp_2Nb(\eta^2-BH_4)$ [78]

In the (η^2-BH_4) complexes $[M(CO)_4(\eta^2-BH_4)]^-$ (M = Cr, Mo) and $OsH_3(\eta^2-BH_4)(PH_3)_2$ the hydrogen exchange occurs via a dissociative pathway.

The transition states can be described as monodentate BH_4 complex-like structures, with only one bridging hydrogen atom [14–17]. Despite the transition states for the hydrogen exchange in this set of bidentate transition metal tetrahydroborate complexes have been labelled as η^3 or η^1, it must be stressed that the transition state structures do not fit into the idealized η^1 or η^3 coordinations with C_{3v} local symmetry for the $M-BH_4$ fragment. Indeed, all the transition states present a similar arrangement of the M-BH_4 unit, with one bridging hydrogen and two at distances intermediate between η^1 or η^3 coordinations (Scheme 21).

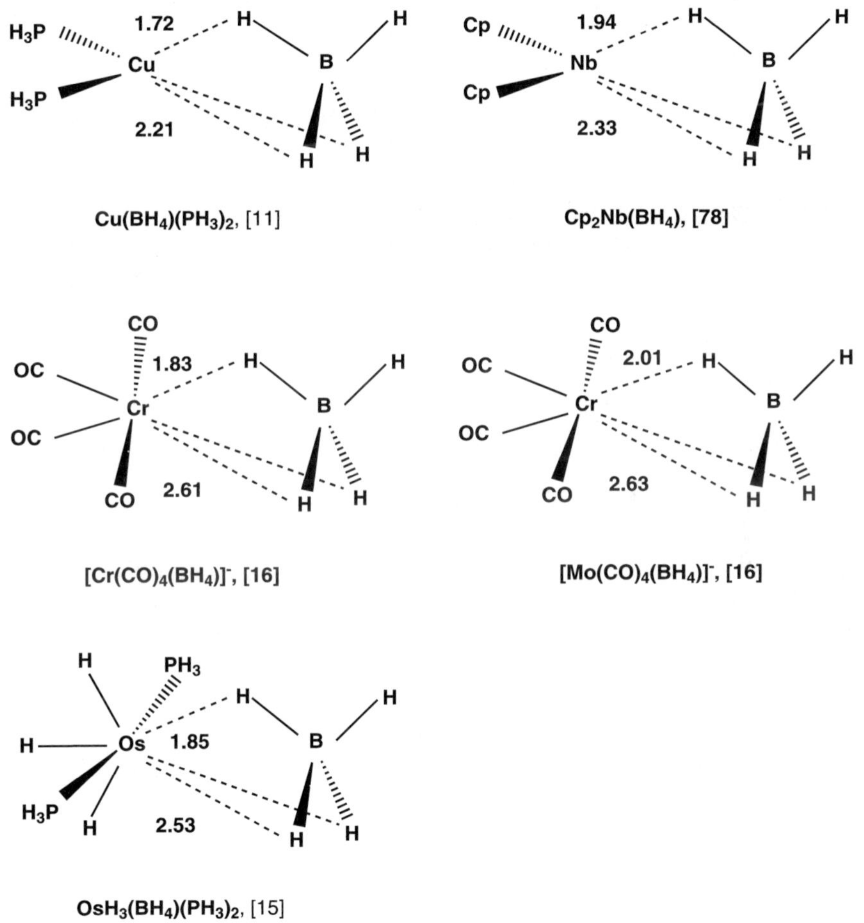

Scheme 21 Transition state geometries for the bridge-terminal hydrogen exchange in η^2-BH_4 complexes

These structures can be described either as distorted η^3 complexes with one bridging hydrogen much closer to the metal than the other two, or as strongly nonlinear η^1 complexes, depending on how far from the metal that the two weakly further H_b atoms are. Calculations performed by Ariafard et al. on the activation volumes support a dissociative mechanism for the hydrogen exchange in $[M(CO)_4(\eta^2-BH_4)]^-$ (M = Cr, Mo) [16, 17].

The mechanism found for the exchange of bridging and terminal hydrogen in complexes containing one η^3 tetrahydroborate involves a η^2 transition state (Scheme 22) [11].

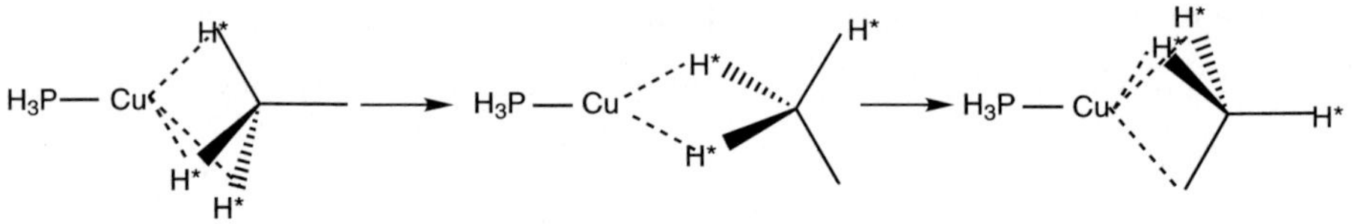

Scheme 22 Mechanism for the exchange of bridging and terminal hydrogens in $Cu(\eta^3-BH_4)(PH_3)$ [11]

The only η^1-BH_4 complex for which the hydrogen exchange has been studied also shows a η^2 transition state. In this case, the reaction coordinate can be mainly described as the rotation around the M···B axis (Scheme 23) [11].

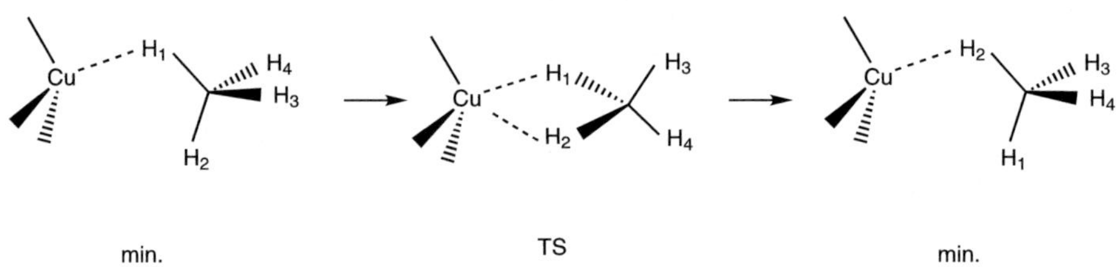

Scheme 23 Mechanism for the exchange of bridging and terminal hydrogens in $Cu(\eta^1-BH_4)(PH_3)_3$ [11]

Mechanisms for the hydrogen exchange in complexes with more than one BH_4 ligand are more complicated and have been less studied. For these systems, other structures, in addition to the absolute minimum, fulfil the optimal electron count number. These structures, usually not very high with respect to the most stable one, are providing low energy routes for the terminal-bridging exchange. This point has been demonstrated in a thorough study of the coordination modes of the tetrahydroborate ligand in $Ti(BH_4)_3$ [118]. In this d^1 complex the absolute minimum adopts a C_{3h} (η^3, η^3, η^3) structure which corresponds to an ideal 17-electron count [118]. Another structure (η^2, η^3, η^3) with the bidentate ligand orthogonal to the TiB_3 plane (structure (2*,3,3) is also a 17-electron complex. This low-energy structure (2*,3,3) allows an easy exchange mechanism between bridging and terminal hydrogen atoms with an activation energy of only 5.1 kcal/mol (Scheme 24).

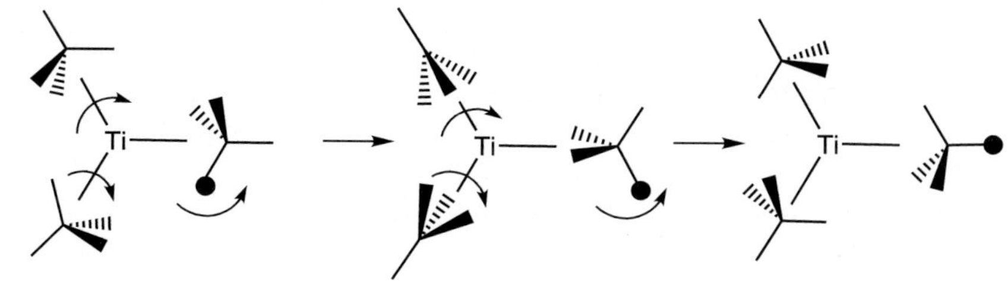

(3,3,3) (2*,3,3) (3,3,3)

Scheme 24 Mechanism for the exchange between bridging and terminal hydrogens in $Ti(\eta^3-BH_4)$ [118]

The scope of dynamic intramolecular processes involving the coordinated terahydroborate is not limited to the bridge-terminal hydrogen exchange. In addition to the H_b/H_t exchange, complexes with several η^3-BH_4 ligands exhibit a reach dynamics related to the fast rotation of the η^3 groups. For instance, in $Ti(\eta^3-BH_4)_3$ an activation energy of only 1.3 kcal/mol has been computed for the concerted rotation of the three η^3 groups (Scheme 25) [118]. The energy barriers for the concerted rotation of the four BH_4 groups into a staggered orientation of $M(\eta^3-BH_4)_4$ (M = Ti, Zr, Hf) are also very low [127].

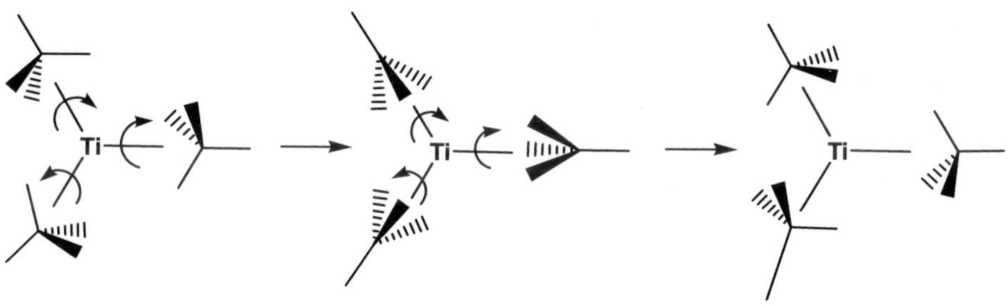

min. TS min.

Scheme 25 Concerted rotation of the three η^3-BH_4 ligands in $Ti(\eta^3-BH_4)$ [118]

It has also been shown that rapid hydrogen exchange can occur between BH_4^- and other ligands coordinated to transition metals. The existence of an intramolecular exchange process between the bridging hydrogen atoms of a BH_4^- ligand and a hydride ligand has been postulated for the $Cp_2ZrH(\eta^2-BH_4)$ [150] and $OsH_3(\eta^2-BH_4)(P(c-C_5H_9)_3)_2$ complexes [8].

The feasibility of such a process was demonstrated in $OsH_3(\eta^2-BH_4)(P^iPr_3)_2$ and the theoretical study unravelled the mechanism of the H(hydride)/H(tetrahydroborate) exchange [14, 15]. The bridging tetrahydroborate hydrogen is first transferred to the metal, leading to a $OsH_4(BH_3)(PR_3)_2$

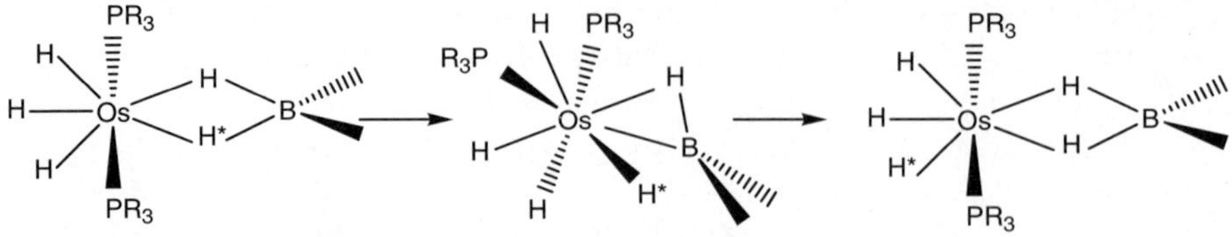

Scheme 26 Mechanism of the exchange process between the bridging hydrogens of the BH_4^- ligand and a hydride ligand in $OsH_3(\eta^2-BH_4)(PR_3)_2$ [8]

intermediate. This 7-coordinate intermediate contains a BH_3 ligand coordinated in a η^2 H – B fashion (Scheme 26)

The calculated energy barrier (19.0 kcal/mol) is in reasonable agreement with the experimentally measured $\Delta G^{\neq}$ for the osmium complex (13±2 kcal/mol) and also close to the value of 20.9 kcal/mol measured for $Cp_2ZrH(\eta^2-BH_4)$ [150]. It must be pointed out that in $OsH_3(\eta^2-BH_4)(P^iPr_3)_2$, the H(hydride)/H(tetrahydroborate) exchange takes places with a lower barrier than the $H_{bridging}/H_{terminal}$ exchange, which is not directly observed in the NMR spectra in the range of investigated temperatures [15].

5
Summary and Perspectives

This contribution has reviewed two basic aspects of the chemistry of tetrahydroborate transition metal complexes: the bonding modes and the hydrogen exchange dynamics. First, the structurally characterised mononuclear transition metal tetrahydroborate complexes have been classified according to the coordination geometry of the metal. Then, the metal–boron distances and the vibrational features of the coordinated borohydrides – which are the key experimental data usually used to determine the hapticity of tetrahydroborate binding in transition metal complexes – have been surveyed. Detailed analyses of the metal–boron distance in the X-ray characterised complexes have allowed relating distances and coordination modes and establishing trends along the periodic table. In addition, the influence of ligand effects in the M – B distances have been recognized. The difficulties and limitations in univocally assigning the coordination mode of the tetrahydroborate with the experimental information available, especially in complexes with more than one BH_4^-, have been illustrated. Several examples of wrongly assigned complexes which can be found in the literature have been reported.

Theoretical studies summarized in Section 3 have shown that electronic factors are prevalent in determining the coordination mode of the tetrahydroborate ligands. These factors have been rationalized by means of simple orbital interaction arguments further supported by quantitative calculations which have given the relative energies of the possible structures of a complex, depending on the BH_4^- coordination mode. However, the importance of electrostatic and steric effects in the metal-tetrahydroborate interactions has only begun to be recognized. Further theoretical studies separating and quantifying electronic, electrostatic and steric effects on the $M-BH_4$ bonding should help to clarify this point.

The analysis of the dynamic phenomena associated with the exchange between bridging and terminal hydrogens of the coordinated BH_4^- ligands constitutes the last section of this contribution. Initially, the exchange free energy barriers determined from temperature-variable NMR measurements have been collected. The limited number of experimental values available draws difficult extracting conclusions regarding trends in barriers and factors affecting them. Several detailed theoretical studies exist on the mechanism which has proven that the $H_{bridging}/H_{terminal}$ exchange involves a change on the coordination mode of the tetrahydroborate ligand. The process entails the sequential breaking and forming of M – H – B bridges at a usually low energy cost and could have important consequences in the reactivity of coordinated terahydroborates. Further computational studies of the processes of rupture of M – HB and H – B bonds are required in order to get a deeper insight on the dynamic properties of tetrahydroborate complexes.

Most of the experimental data discussed in this contribution were reported in the eighties and early nineties. It may seem that almost all have already been done in transition metal tetrahydroborate chemistry, but this is not the case. Borohydride chemistry is still a very active field of research. As recent examples illustrating a renewed interest in the field can be cited, such as the preparation of thermally stable tetrahydroborate complexes with nickel [89] and iron [151] transition metals, and the characterization of paramagnetic iron [15] and nickel [13, 152] borohydride complexes. In addition, some intriguing questions concerning $M\text{-}BH_4$ bonding have emerged. Short M – B distances observed together with theoretical analysis of bond orders in several η^3-BH_4 structures support a substantial bonding interaction between the metal and boron nuclei, suggesting the description of the binding as a η^4 interaction [151]. In complexes with hydride ligands, the possible existence of M – H–boron interactions entailing hypercoordinate boron centres have been pointed out and octahedral BH_6 supported through M – H – B bridges have been described [153, 154].

In recent years, emphasis has also been put on controlling the reactivity of tetrahydroborate complexes. In this way, the use of the tetrahydroborate ligand as a "gate-keeper" and protected hydride ligand in catalytic reaction sequences have been investigated [1, 155]. A M – HB bridge can be broken at a low energy cost, as indicated by the low energy barriers determined for the hydrogen exchange processes. A η^2-bonded tetrahydroborate ligand can, by switching to the η^1-bonding mode, create a vacant coordination site for the uptake of an organic substrate or other reactant. It has been demonstrated that it is possible to perform such processes with nucleophiles, avoiding the abstraction of BH_3 [156]. Quantitative calculations of the metal–tetrahydroborate bonding energies, which are still lacking, should allow the comparison of the bonding energy of a BH_4^- ligand with that of common organometallic ligands, assisting with the design of more efficient systems. Computational studies of the reactivity of BH_4^- ligands, which are also still scarce, will also contribute to a better understanding of the reaction mechanisms, allowing an improved control of the reactivity. Significant advances in tetrahydroborate chemistry can be expected in the near future.

Acknowledgements Part of the computational work cited in this review was carried out in our group at the UAB in collaboration with the Université de Paris-Sud. We express our gratitude to our co-workers Y. Jean, F. Volatron, M. Duran and A. Jarid. Financial support from the Spanish MEC through Project CTQ2005-09000-C02-01 and from Generalitat de Catalunya through grant 2005SGR00896 is gratefully acknowledged. A.L. thanks the Generalitat de Catalunya for a "Distinció per a la Promoció de la Recerca Universitària".

References

1. Marks TJ, Kolb JR (1977) Chem Rev 77:263
2. Ephritikhine M (1997) Chem Rev 97:2193
3. Makhaev VD (2000) Uspekhi Khimii 69:795
4. Xu Z, Lin Z (1996) Coord Chem Rev 156:139
5. Basuli F, Bailey BC, Watson LA, Tomaszewski J, Huffman JC, Mindiola DJ (2005) Organometallics 24:1886
6. Atwood JL, Hunter WE, Carmona-Guzman E, Wilkinson G (1980) J Chem Soc, Dalton Trans 3:467
7. Barron AR, Salt JE, Wilkinson G, Motevalli M, Hursthouse MB (1986) Polyhedron 5:1833
8. Frost PW, Howard JAK, Spencer JL (1984) J Chem Soc, Chem Commun, p 1362
9. Musaev DG, Morokuma K (1995) Organometallics, p 3327
10. Jandl S, Dufour P, Richard P, Nekvasil V, Zhigunov DI, Barilo SN, Shiryaev SV, Aldridge S, Downs AJ, Hofmann M, Pulham CR, Schleyer PV (1998) J Mol Struct 444:29
11. Jarid A, Lledós A, Jean Y, Volatron F (1995) Chem Eur J 1:436
12. Oishi Y, Albright TA, Fujimoto H (1995) Polyhedron 14:2603
13. Kandiah M, McGrady GS, Decken A, Sirsch P (2005) Inorg Chem 44:8650
14. Esteruelas MA, Jean Y, Lledos A, Oro LA, Ruiz N, Volatron F (1994) Inorg Chem 33:3609
15. Demachy I, Esteruelas MA, Jean Y, Lledos A, Maseras F, Oro LA, Valero C, Volatron F (1996) J Am Chem Soc 118:8388
16. Ariafard A, Amini MM (2005) J Organomet Chem 690:84
17. Ariafard A, Amini MM, Azadmehr A (2005) J Organomet Chem 690:1147
18. Makhaev VD, Borisov AP, Lobkovskii EB, Semenenko KN (1980) Izv Akad Nauk SSSR, Ser Khim, p 2614 (Russ Chem Bull)
19. Lobkovskii EB, Antipin MY, Borisov AP, Makhaev VD, Semenenko KN, Struchkov YT (1981) Koord Khim, p 307 (Russ J Coord Chem)
20. Takusagawa F, Fumagalli A, Koetzle TF, Shore SG, Schmitkons T, Fratini AV, Morse KW, Wei CY, Bau R (1981) J Am Chem Soc 103:5165

21. Bommer JC, Morse KW (1980) Inorg Chem 19:587
22. Dapporto P, Midollini S, Orlandini A, Sacconi L (1976) Inorg Chem 15:2768
23. Ghilardi CA, Midollini S, Orlandini A (1982) Inorg Chem 21:4096
24. Baker MV, Field LD (1984) J Chem Soc, Chem Commun, p 996
25. Bau R, Yuan HSH, Baker MV, Field LD (1986) Inorg Chim Acta 114:L27
26. Baker MV, Field LD (1990) Appl Organomet Chem 4:543
27. Ohkuma T, Koizumi M, Muñiz K, Hilt G, Kabuto C, Noyori R (2002) J Am Chem Soc 124:6508
28. Guo R, Morris RH, Song D (2005) J Am Chem Soc 127:516
29. Yoshida T, Adachi T, Ueda T, Akao H, Tanaka T, Goto F (1995) Inorg Chim Acta 231:95
30. Liang F, Schmalle HW, Fox T, Berke H (2003) Organometallics 22:3382
31. Liang F, Jacobsen H, Schmalle HW, Fox T, Berke H (2000) Organometallics 19:1950
32. Höck J, Jacobsen H, Schmalle HW, Artus GRJ, Fox T, Amor JI, Bäth F, Berke H (2001) Organometallics 20:1533
33. Lippard SJ, Melmed KM (1967) Inorg Chem 6:2223
34. Green BE, Kennard CHL, Hawkins CJ, Smith G, James BD, White AH (1980) Acta Crystallogr B 36:2407
35. Green BE, Kennard CHL, Smith G, Elcombe MM, Moore FH, James BD, White AH (1984) Inorg Chim Acta 83:177
36. Lobkovskii EB, Makhaev VD, Borisov AP (1984) Zh Strukt Khim (Russ J Struct Chem) 25:496
37. Makhaev VD, Borisov AP, Lobkovskii EB, Polyakova VB, Semenenko KN (1985) Izv Akad Nauk SSSR, Ser Khim (Russ Chem Bull) 34:1731
38. Moncol J, Gembicky M, Coppens P (2005) Acta Crystallogr E 61:m242
39. Li CY, Zhang YG, Zhou Y, Zhang LF (2000) Chinese J Struct Chem 19:91
40. Dick DG, Duchateau R, Edema JJH, Gambarotta S (1993) Inorg Chem 32:1959
41. Prust J, Hohmeister H, Stasch A, Roesky HW, Magull J, Alexopoulos E, Usón I, Schmidt H-G, Noltemeyer M (2002) Eur J lnorg Chem, p 2156
42. Berno P, Moore M, Minhas R, Gambarotta S (1996) Can J Chem 74:1930
43. Makhaev VD, Borisov AP, Antsyshkina AS, Sadikov GG, Poraikoshits MA, Kedrova NS, Maltseva NN, Istomin SY (1993) Koord Khim (Russ J Coord Chem) 19:858
44. Reger DL, Mason SS, Rheingold AL (1993) J Am Chem Soc 115:10406
45. Koutsantonis GA, Lee FC, Raston CL (1994) J Chem Soc, Chem Commun, p 1975
46. Gerlach CP, Arnold J (1997) J Chem Soc, Dalton Trans, p 4795
47. Corey EJ, Cooper NJ, Canning WM, Lipscomb WN, Koetzle TF (1982) Inorg Chem 21:192
48. Saito T, Nakajima M, Kobayashi A, Sasaki Y (1978) J Chem Soc, Dalton Trans, p 482
49. Green MLH, Munakata H, Tato S (1971) J Chem Soc(A), Inorg Phys Theor, p 469
50. Nakajima M, Moriyama H, Kobayashi A, Saito T, Sasaki Y (1975) J Chem Soc, Chem Commun, p 80
51. Nakajima M, Saito T, Kobayashi A, Sasaki Y (1977) J Chem Soc, Dalton Trans, p 385
52. Scoles L, Gambarotta S (1995) Inorg Chim Acta 235:375
53. Nückel S, Burger P (2001) Organometallics 20:4345
54. Darensbourg MY, Bau R, Marks MW, Burch RR, Deaton JC, Slater S (1982) J Am Chem Soc 104:6961
55. Kirtley SW, Andrews MA, Bau R, Grynkewich GW, Marks TJ, Tipton DL, Whittlesey BR (1977) J Am Chem Soc 99:7154
56. Thomas SA (1989) J Cryst Spectrosc 19:1017
57. Ghilardi CA, Innocenti P, Midollini S, Orlandini A (1982) J Organomet Chem 231:C78

58. Ghilardi CA, Innocenti P, Midollini S, Orlandini A (1985) J Chem Soc, Dalton Trans, p 605
59. Statler JA, Wilkinson G, Thornton-Pett M, Hursthouse MB (1984) J Chem Soc, Dalton Trans, p 1731
60. van der Zeijden AAH, Shklover V, Berke H (1991) Inorg Chem 30:4393
61. Liu X-Y, Bouherour S, Jacobsen H, Schmalle HW, Berke H (2002) Inorg Chim Acta 330:250
62. Tamm M, Drebel B, Lügger T, Fröhlich R, Grimme S (2003) Eur J lnorg Chem, p 1088
63. Tamm M, Drebel B, Urban V, Lügger T (2003) Inorg Chem Commun 5:837
64. Hessen B, Teuben JH, Lemmen TH, Huffman JC, Caulton KG (1985) Organometallics 4:946
65. Hessen B, Lemmen TH, Luttikhedde HJG, Teuben JH, Petersen JL, Huffman JC, Jagner S, Caulton KG (1987) Organometallics 6:2354
66. You YJ, Wilson SR, Girolami GS (1994) Organometallics 13:4655
67. Fryzuk MD, Mylvaganam M, Zaworotko MJ, McGillivray LR (1993) J Am Chem Soc 115:10360
68. Marks TJ, Kennelly WJ (1975) J Am Chem Soc 97:1439
69. Liu F-C, Plecnik CE, Liu S, Liu J, Meyers EA, Shore SG (2001) J Organomet Chem 627:109
70. Kirillova NI, Gusev AI, Struchkov YT (1974) Zh Strukt Khim (Russ J Struct Chem) 15:718
71. Bailey NJ, Green MLH, Leech MA, Saunders JF, Tidswell HM (1997) J Organomet Chem 538:111
72. Conway SLJ, Doerrer LH, Green MLH, Leech MA (2000) Organometallics 19:630
73. Lantero DR, Ward DL, Smith MR III (1997) J Am Chem Soc 119:9699
74. Melmed KM, Coucouva D, Lippard SJ (1973) Inorg Chem 12:232
75. Mamaeva GI, Hargittai I, Spiridonov VP (1977) Inorg Chim Acta 25:L123
76. Hartwig JF, De Gala SR (1994) J Am Chem Soc 116:3661
77. Lappert MF, Singh A, Atwood JL, Hunter WE (1983) J Chem Soc, Chem Commun, p 206
78. Antinolo A, Carrillo-Hermosilla F, Fernandez-Baeza J, Garcia-Yuste S, Otero A, Rodriguez AM, Sanchez-Prada J, Villasenor E, Gelabert R, Moreno M, Lluch JM, Lledos A (2000) Organometallics 19:3654
79. Schumann H, Keitsch MR, Demtschuk J, Muhle S (1998) Z Anorg Allg Chem 624:1811
80. Choukroun R, Douziech B, Donnadieu B (1997) Organometallics 16:5517
81. Laske DA, Duchateau R, Teuben JH, Spek AL (1993) J Organomet Chem 462:149
82. Qian CT, Zou G, Nie W-L, Sun J, Lemenovskii DA, Borzov MV (2000) Polyhedron 19:1955
83. Fischer PJ, Young VG, Ellis JE (2000) Angew Chem Int Ed 39:189
84. Garcia A, Mercero JM, Ugalde JM (1997) J Phys Chem A 101:5953
85. Garcia A, Mercero JM, Fowler JE, Ugalde JM (1998) J Phys Chem A 102:2055
86. Herrmann WA, Denk M, Scherer W, Klingan FR (1993) J Organomet Chem 444: C21
87. Noth H, Schmidt M (1995) Organometallics 14:4601
88. Kazani A, Gambarotta S, Bensimon C (1997) Can J Chem 75:1494
89. Desrochers PJ, LeLievre S, Johnson RJ, Lamb BT, Phelps AL, Cordes AW, Gu W, Cramer SP (2003) Inorg Chem 42:7945
90. Ariafard A, Fazaeli R, Aghabozorg HR, Monajjemi M (2003) J Mol Struct (Theochem) 625:305

91. Jensen JA, Girolami GS (1989) Inorg Chem 28:2107
92. Jensen JA, Girolami GS (1988) J Am Chem Soc 110:4450
93. Dionne M, Hao S, Gambarotta S (1995) Can J Chem 73:1126
94. Marsh RE (2004) Acta Crystallogr B60:252
95. Makhaev VD, Borisov AP, Gnilomedova TP, Lobkovskii EB, Chekhlov AN (1987) Izv Akad Nauk SSSR, Ser Khim (Russ Chem Bull) 36:1518
96. Jensen JA, Wilson SR, Schultz AJ, Girolami GS (1987) J Am Chem Soc 109:8094
97. Johnson PL, Cohen SA, Marks TJ, Williams JM (1978) J Am Chem Soc 100:2709
98. Smith BE, James BD, Dilts JA (1976) J Inorg Nuc Chem 38:1973
99. Sztaray B, Rosta E, Bocskey Z, Szepes L (1999) J Organomet Chem 582:267
100. Schaper F, Brintzinger H-H (2000) Acta Crystallogr C56:e78
101. Saidov BI, Borisov AP, Makhaev VD, Boiko GN, Antsyshkina AS, Kedrova NS, Mal'tseva NN (1990) Zh Neorg Khim (Russ J Inorg Chem) 35:626
102. Antsyshkina AS, Poraikoshits MA, Saidov BI, Makhaev VD, Borisov AP, Maltseva NN, Kedrova NS (1991) Koord Khim (Russ J Coord Chem) 17:405
103. Lobkovskii EB, Chekhlov AN, Makhaev VD, Borisov AP (1989) Koord Khim (Russ J Coord Chem) 15:377
104. Makhaev VD, Borisov AP, Karpova TP, Lobkovskii EB, Tarasov BP, Chekhlov AN (1989) Izv Akad Nauk SSSR, Ser Khim (Russ Chem Bull) 38:377
105. Jensen JA, Girolami GS (1989) Inorg Chem 28:2114
106. Lledos A, Duran M, Jean Y, Volatron F (1991) Inorg Chem 30:4440
107. Borisov AP, Makhaev VD, Lobkovskii EB, Semenenko KN (1986) Zh Neorg Khim (Russ J Inorg Chem) 31:86
108. Lobkovskii EB, Makhaev VD, Borisov AP, Semenenko KN (1986) Zh Strukt Khim (Russ J Struct Chem) 27:161
109. Corazza F, Floriani C, Chiesivilla A, Guastini C (1991) Inorg Chem 30:145
110. Cloke FGN, Hitchcock PB, Love JB (1995) J Chem Soc, Dalton Trans, p 25
111. Lobkovskii EB, Kravchenko SE, Kravchenko OV (1982) Zh Strukt Khim (Russ J Struct Chem) 23:582
112. Kravchenko OV, Kravchenko SE, Polyakova VB, Semenenko KN (1980) Koord Khim (Russ J Coord Chem) 6:76
113. Xu Z, Lin Z (1996) Inorg Chem 35:3964
114. Belskii VK, Sobolev AN, Bulychev BM, Alikhanova TK, Kurbonbekov A, Mirsaidov U (1990) Koord Khim (Russ J Coord Chem) 16:1693
115. Wang H, Wang H, Li H-W, Xie Z (2004) Organometallics 23:875
116. Dain CJ, Downs AJ, Goode MJ, Evans DG, Nicholls KT, Rankin DWH, Robertson HE (1991) J Chem Soc, Dalton Trans, p 967
117. Borisenko KB, Downs AJ, Robertson HE, Rankin DWH, Tang CY (2004) J Chem Soc, Dalton Trans, p 967
118. Jarid A, Lledos A, Jean Y, Volatron F (1993) Inorg Chem 32:4695
119. Broach RW, Chuang IS, Marks TJ, Williams JM (1983) Inorg Chem 22:1081
120. Keiderling TA, Wozniak WT, Gay RS, Jurkowitz D, Bernstein ER, Lippard SJ, Spiro TG (1975) Inorg Chem 14:576
121. Davies N, Wallbridge MGH, Smith BE, James BD (1973) J Chem Soc, Dalton Trans 2:162
122. Smith BE, Shurwell HF, James BD (1977) J Chem Soc, Dalton Trans, p 710
123. Bird PH, Churchill MR (1967) J Chem Soc, Chem Commun, p 403
124. Green JC, de Simone M, Coreno M, Jones A, Pritchard HMI, McGrady GS (2005) Inorg Chem 44:7781
125. Jensen JO (2003) Spectrochim Acta A 59:379

126. Jensen JO (2003) Vib Spectrosc 31:227
127. Haaland A, Shorokhov DJ, Tutukin AV, Volden HV, Swang O, McGrady GS, Kaltsoyannis N, Downs AJ, Tang CY, Turner JFC (2002) Inorg Chem 41:6646
128. Volatron F, Duran M, Lledos A, Jean Y (1993) Inorg Chem 32:951
129. Joseph SCP, Cloke FGN, Cardin CJ, Hitchcock PB (1995) Organometallics 14:3566
130. Lobkovskii EB, Kravchenko SE, Semenenko KN (1977) Zh Strukt Khim (Russ J Struct Chem) 18:312
131. Goedde DM, Girolami GS (2006) Inorg Chem 45:1380
132. Segal BG, Lippard SJ (1978) Inorg Chem 17:844
133. Edelstein N (1981) Inorg Chem 20:297
134. Allen FH (2002) Acta Crystallogr B58:380
135. Jensen JA, Girolami GS (1986) Chem Comm 15:1160
136. Hori K, Saito G, Teramae H (1988) J Phys Chem 92:3796
137. Lledos A, Duran M, Jean Y, Volatron F (1992) Bull Soc Chim Fr 129:216
138. Nissen-Sobocinska B, Jezowska-Trzebiatowska B (1993) J Organomet Chem 452:277
139. Green MLH, Wong L-L (1989) J Chem Soc, Dalton Tans, p 2133
140. Ashworth NJ, Conway SLJ, Green JC, Green MLH (2000) J Organomet Chem 609:83
141. Jensen JA, Wilson SR, Girolami GS (1988) J Am Chem Soc 110:4977
142. Engelhardt LM, Pakawatchai C, White AH, Healy PC (1985) J Chem Soc, Dalton Trans, p 125
143. Green BE, Kennard CHL, Smith G, White AH (1981) Cryst Struct Commun 10:1245
144. Kutal C, Grutsch P, Atwood JL, Rogers RD (1978) Inorg Chem 17:3558
145. Antsyshkina AS, Poraikoshits MA, Makhaev VD, Borisov AP, Kedrova NS, Maltseva NN (1992) Koord Khim (Russ J Coord Chem) 18:474
146. Bell RA, Cohen SA, Doherty NM, Threlkel RS, Bercaw JE (1986) Organometallics 5:972
147. Green MLH, Hughes AK (1992) J Chem Soc, Dalton Trans, p 527
148. Wink DJ, Cooper NJ (1984) J Chem Soc, Dalton Tans, p 1257
149. Chuang I-S, Marks TJ, Kennelly WJ, Kolb JR (1977) J Am Chem Soc 99:7536
150. Marks TJ, Kolb JR (1975) J Am Chem Soc 97:3397
151. Mehn MP, Brown SD, Paine TK, Brennessel WW, Cramer CJ, Peters JC, Que L (2006) Dalton Trans 1347
152. Journaux Y, Lozan V, Klingele J, Kersting B (2006) Chem Commun 83
153. Hillier AC, Jacobsen H, Gusev D, Schmalle HW, Berke H (2001) Inorg Chem 40:6334
154. Guilera G, McGrady GS, Steed JW, Kaltsoyannis N (2004) New J Chem 28:444
155. Chamberlain B, Duckett SB, Lowe JP, Mawby RJ, Stott JC (2003) Dalton Trans 2603
156. Duckett SB, Lowe JP, Mawby RJ (2006) Dalton Trans 2661

Author Index Volumes 101–130

Author Index Vols. 1–100 see Vol. 100

The volume numbers are printed in italics

Subject Index

Printing: Krips bv, Meppel, The Netherlands
Binding: Stürtz, Würzburg, Germany